11
Topics in Heterocyclic Chemistry

Series Editor: R. R. Gupta

Topics in Heterocyclic Chemistry
Series Editor: R. R. Gupta

Recently Published and Forthcoming Volumes

Bioactive Heterocycles V

Volume Editor: Mahmud Tareq Hassan Khan

With contributions by

M. J. Carlucci · H. Cerecetto · E. B. Damonte · O. Demirkiran
M. González · N. Hamdi · J. H. Jung · M. T. H. Khan · Y. Liu
M. C. Matulewicz · J. C. Menéndez · A. Monge · I. Orhan · B. Özcelik
C. A. Pujol · A. Romerosa · M. Saoud · N. Saracoglu · B. Şener
S. Süzen · G. Topcu · T. Xu · S. Zhang

The series *Topics in Heterocyclic Chemistry* presents critical reviews on "Heterocyclic Compounds" within topic-related volumes dealing with all aspects such as synthesis, reaction mechanisms, structure complexity, properties, reactivity, stability, fundamental and theoretical studies, biology, biomedical studies, pharmacological aspects, applications in material sciences, etc. Metabolism will be also included which will provide information useful in designing pharmacologically active agents. Pathways involving destruction of heterocyclic rings will also be dealt with so that synthesis of specifically functionalized non-heterocyclic molecules can be designed.

The overall scope is to cover topics dealing with most of the areas of current trends in heterocyclic chemistry which will suit to a larger heterocyclic community.

As a rule contributions are specially commissioned. The editors and publishers will, however, always be pleased to receive suggestions and supplementary information. Papers are accepted for *Topics in Heterocyclic Chemistry* in English.

In references *Topics in Heterocyclic Chemistry* is abbreviated *Top Heterocycl Chem* and is cited as a journal.

Springer WWW home page: springer.com
Visit the THC content at springerlink.com

ISSN 1861-9282
ISBN 978-3-540-73405-5 Springer Berlin Heidelberg New York
DOI 10.1007/978-3-540-73406-2

Springer is a part of Springer Science+Business Media

springer.com

© Springer-Verlag Berlin Heidelberg 2007

Cover design: WMX Design GmbH, Heidelberg
Typesetting and Production: LE-TEX Jelonek, Schmidt & Vöckler GbR, Leipzig

Printed on acid-free paper   02/3100 YL – 5 4 3 2 1 0

Topics in Heterocyclic Chemistry
Also Available Electronically

For all customers who have a standing order to Topics in Heterocyclic Chemistry, we offer the electronic version via SpringerLink free of charge. Please contact your librarian who can receive a password or free access to the full articles by registering at:

springerlink.com

If you do not have a subscription, you can still view the tables of contents of the volumes and the abstract of each article by going to the SpringerLink Homepage, clicking on "Browse by Online Libraries", then "Chemical Sciences", and finally choose Topics in Heterocyclic Chemistry.

You will find information about the

– Editorial Board
– Aims and Scope
– Instructions for Authors
– Sample Contribution

at springer.com using the search function.

Dedicated to my family – my wife Arjumand, my son Araz and my daughter Aurora, who gave me the love, free time and inspiration, to make these volumes successful.

Preface

This volume contains 10 chapters. The contributions are from researchers famous in their respective fields and the chapters contain high quality reviews on topics related to the chemo-biological studies of several different heterocyclic groups.

The first chapter from Saracoglu reviews the functionalization of indoles and the pyrroles via Michael additions, as these compounds have potential for their biological activities.

In second chapter Menéndez reviews the chemistry of the welwitindolinones.

Topcu and Demirkiran, in the third chapter, describe the chemistry and biological studies of lignans from *Taxus* species, including their biosynthesis and recent strategies for the synthesis of lignans. Lignans are a very important class of molecules, as they have a very diverse spectrum of biological activities, such as antitumour, antiviral, hepatoprotective, antioxidant, antiulcer, anti-allergen, anti-platelet, and anti-osteoporotic activities.

In next chapter Süzen describes the antioxidant activities of synthetic indole derivatives and their possible mechanisms of action.

In chapter five González et al. presents a comprehensive review on the chemistry and biology of the quinoxaline 1,4-dioxide and phenazine 5,10-dioxide type molecules. They also discuss the mode of action, structure-activity studies and other relevant chemical and biological properties for such molecules.

In the chapter six, Khan briefly discusses the anti-angiogenic and telomerase inhibitory activities of quinoline and its analogues.

Liu et al., in chapter seven, briefly reviews some aspects of studies of bioactive marine sponge furanosesterterpenoids from the last 10 years, including their total syntheses.

Pujol et al. in chapter eight, reviews the chemistry, origin and antiviral activities of naturally occurring sulfated polysaccharides for the prevention and control of viral infections such as HIV-1 and -2, human cytomegalovirus (HCMV), dengue virus (DENV), respiratory syncytial virus (RSV), and influenza A virus.

In chapter nine Hamdi et al. describes the synthesis and biological activities of the heterocyclic and vanillin ether coumarins.

In the last chapter, Orhan et al. reviews their recent findings on antiviral and antimicrobial heterocyclic compounds from Turkish plants.

Tromsø, Norway 2007   Mahmud Tareq Hassan Khan

Contents

Contents of Volume 9

Bioactive Heterocycles III

Volume Editor: Khan, M. T. H.
ISBN: 978-3-540-73401-7

Contents of Volume 10

Bioactive Heterocycles IV

Volume Editor: Khan, M. T. H.
ISBN: 978-3-540-73403-1

Top Heterocycl Chem (2007) 11: 1–61
DOI 10.1007/7081_2007_073
© Springer-Verlag Berlin Heidelberg
Published online: 4 July 2007

Functionalization of Indole and Pyrrole Cores via Michael-Type Additions

Nurullah Saracoglu

Department of Chemistry, Atatürk University, 25240 Erzurum, Turkey
nsarac@atauni.edu.tr

Abstract The Michael reaction has been the subject of numerous reviews. Furthermore, the first review on anti-Michael addition has been published. The present review focuses only on the functionalization of indoles and the pyrroles via Michael additions because of the potential biological activity exhibited by these compounds.

Keywords Catalyst · Conjugate addition · Indole · Michael addition · Pyrrole

Abbreviations

Ac	Acetyl
bmim	1-Butyl-3-methylimidazolium
Bn	Benzyl
Boc	*tert*-Butoxycarbonyl
BOPCl	(Bis(2-oxo-3-oxazolidinyl)phosphinic chloride)
BOX	Bisoxazoline
Bu	Butyl
^{t}Bu	*tert*-Butyl

^{t}BuOK	Potassium *tert*-butoxide
Bz	Benzoyl
BzR	Benzodiazepine receptor
CAN	Ceric ammonium nitrate
CNS	Central nervous system
DAIB	(Diacetoxyiodo)benzene
DCM	Dichloromethane
DDQ	2,3-Dichloro-5,6-dicyano-1,4-benzoquinone
DEAD	Diethyl azodicarboxylate
DIBAL-H	Diisobutylaluminum hydride
DMAc	*N,N*-Dimethyl acetamide
DMAD	Dimethyl acetylenedicarboxylate
DMAP	4-(Dimethylamino)pyridine
DME	1,2-Dimethoxyethane
DMF	Dimethylformamide
DMSO	Dimethyl sulfoxide
DPMU	*N,N'*-Dimethylproplenurea
ee	Enantiomeric excess
Et	Ethyl
EWG	Electron-withdrawing group
HOMO	Highest occupied molecular orbital
Me	Methyl
MVK	Methylvinylketone
MW	Microwave
NBS	*N*-Bromosuccinimide
NMR	Nuclear magnetic resonance
NSCLC	Non-small-cell lung carcinoma
Nu	Nucleophile
Ph	Phenyl
PPSE	Polyphosphoric acid trimethylsilyl ester
i-Pr	Isopropyl
SET	Single electron transfer
Tf	Trifluoromethanesulfonyl
TFA	Trifluoroacetic acid
THF	Tetrahydrofuran
TLC	Thin layer chromatography
TMSCl	Chlorotrimethylsilane
TMSCN	Trimethylsilylcyanide
Tol-BINAP	2,2'-Bis(di-4-tolylphosphino)-1,1'-binaphthyl
p-TsOH	*para*-Toluenesulfonic acid

1
Introduction

Pyrrole (**1**) and indole (**2**) moieties occur widely in synthetic and natural products, either as a simple structural unit or as part of more complex annulated systems [1–9]. The pyrrole derivatives **3** and **4** display antibacterial activity [1–3]. Marine alkaloids ($\pm$)-B-norrhazinal (**5**) and (–)-rhazinilam

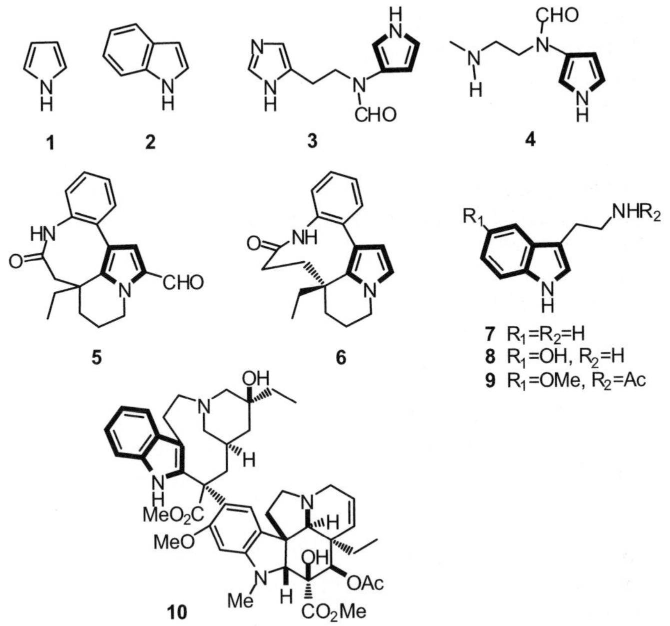

Scheme 1

(6) possess intriguing antimitotic properties [10–14]. Tryptamine (7) and its derivatives are present in many naturally and synthetically derived molecules with interesting biological activities [15]. Serotonin (5-hydroxytryptamine) (8) is a monoamine neurotransmitter synthesized in serotonergic neurons in the central nervous system (CNS) and enterochromaffin cells in the gastrointestinal tract [16]. Melatonin (9) is an important hormone [16]. Vinblastine (10), isolated from *Catharanthus roseus*, has been widely used as an agent for cancer chemotherapy [17–19] (Scheme 1).

1.1
Indole and Pyrrole Rings: Structure and Reactivity

Pyrrole (1) is an aromatic heterocycle with a five-membered, electron-rich ring [2]. There are three possible points of fusion to the pyrrole ring, a fact that has important ramifications for the stability of the systems resulting from fusion with aromatic nuclei. Fusion of a benzene ring at the b-bond, as in indole (2), does not perturb the benzene nucleus and thus gives rise to stable compounds [15]. In contrast, fusion at the c-bond, as in isoindole (11), interrupts the benzene π-sextet, reducing the aromaticity and, consequently, the stability of the system. Isoindole (benzo[c]pyrrole, 11) is highly reactive [15, 20–24]. Although it belongs to a ten-electron aromatic system, 11 is very reactive towards cycloaddition reactions. Due to its instability, the isoindole usually has to be generated in situ and used immediately. Indolizine (**12**) is an isomer of indole (**2**) obtained by the transposition of adjacent carbon and nitrogen atoms at the a-bond [15] (Scheme 2).

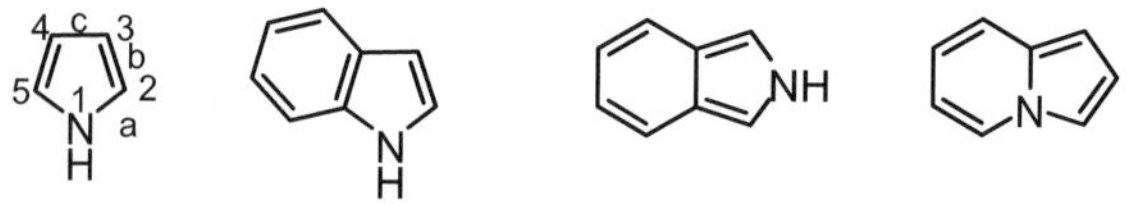

Scheme 2

Indole (**2**) undergoes electrophilic substitution preferentially at the β(C3)-position whereas pyrrole (**1**) reacts predominantly at the α(C2)-position [15]. The positional selectivity in these five-membered ring systems is well explained by the stability of the Wheland intermediates for electrophilic substitution. The intermediate cations from β (for indole, **2**) and α (for pyrrole, **1**) are the more stabilized. Pyrrole compounds can also participate in cycloaddition (Diels–Alder) reactions under certain conditions, such as Lewis acid catalysis, heating, or high pressure [15]. However, calculations of the frontier electron population for indole and pyrrole show that the HOMO of indole exhibits high electron density at the C3 while the HOMO of pyrrole is high at the C2 position [25–28] (Scheme 3).

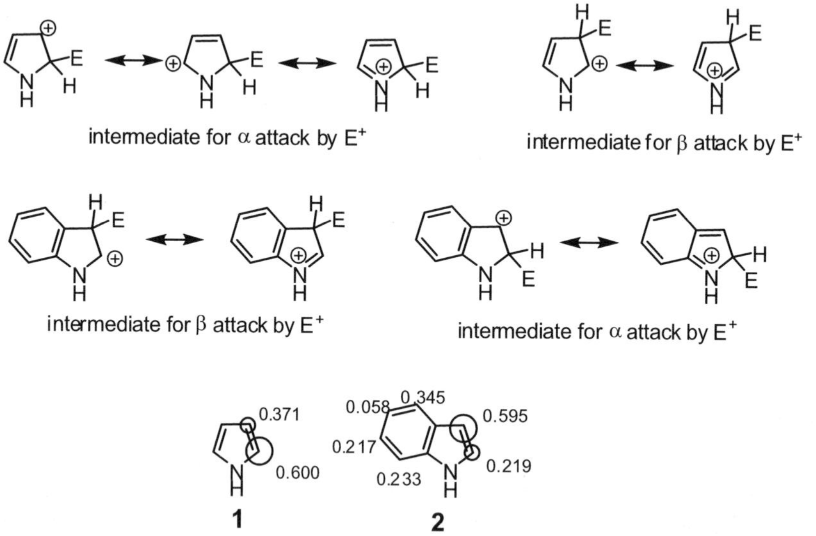

intermediate for α attack by E⁺

intermediate for β attack by E⁺

intermediate for β attack by E⁺

intermediate for α attack by E⁺

Scheme 3

1.2
Michael Addition and its Mechanism

Michael addition, also termed 1,4-, or conjugate-addition, or Friedel–Crafts alkylation, is the conjugate addition of nucleophilic species (Michael donors) to α,β-unsaturated systems (Michael acceptor; α,β-unsaturated carbonyl compounds, nitriles, esters, phosphates, sulfones, nitroalkenes and alkynoates among others) creating a new bond at the β-position [29–36]. The Michael reaction is one of the most important carbon–carbon and carbon–heteroatom bond-forming reactions in organic synthesis. In some circumstances, addition at the carbonyl atom occurs (i.e., 1,2-addition). The reactivity of Michael acceptors can be altered so that 1,4-addition would be circumvented in favor of the α-position of an α,β-unsaturated system. This is known as anti-Michael addition, contra-Michael addition, abnormal Michael reaction or substitution at carbon-α [37]. The regioselectivity of the Michael reaction can be inverted by groups with strongly electron-withdrawing properties at the β-position and the reaction gives the α-addition product (Scheme 4).

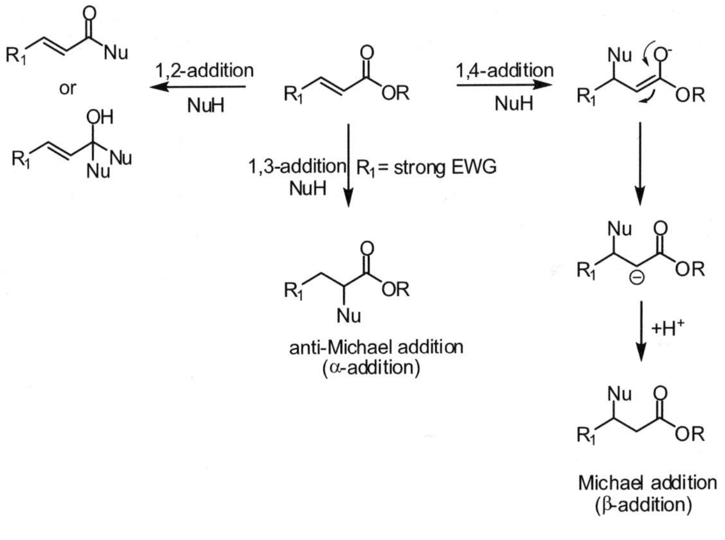

Scheme 4

Normally, the nucleophile or the Michael acceptor needs to be activated in the Michael additions. To achieve this activation, either the nucleophile is deprotonated with strong bases or the acceptor is activated in the presence of Lewis acid catalysts under much milder conditions. Recently, important advances have been made with Lewis acid catalysts and these developments continue. Four possible mechanisms are suggested for the catalyst action in conjugate additions to enones under nonbasic conditions [38]. First is the

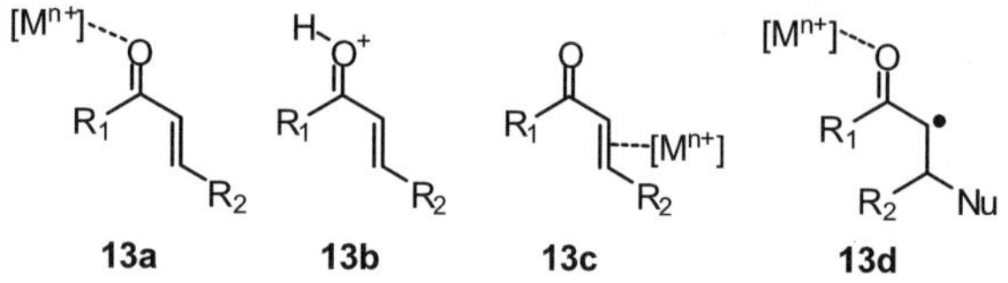

13a　　　**13b**　　　**13c**　　　**13d**

Scheme 5

formation of carbonyl of enone–metal ion complexes, such as **13a**. In the second mechanism, the carbonyl oxygen can be protonated by acid to give **13b**. Another is the direct interaction between double bond and transition-metal catalysts to yield the activated complex **13c**. Finally, free radicals, such as intermediate **13d**, form with the interactions between metal ion, enone and nucleophile (Scheme 5).

2
Michael Addition Applications for Indole

2.1
Michael Additions of Indoles to Conjugate Systems by Various Acid Catalysts

The lanthanide salts are used increasingly as Lewis acids in organic synthesis. For the first time, the addition of indole and methyl indole to a series of Michael acceptors in the presence of $Yb(OTf)_3 \cdot 3H_2O$ has been achieved at both high and ambient pressure (Table 1) [39, 40] (Scheme 6). While the more reactive and less sterically hindered electrophiles gave the 3-alkylated indoles in good to excellent yields under ambient pressure, a significant improvement in yields and reaction time was observed at high pressure.

Table 1 Reaction conditions for Michael products **17** and **18**

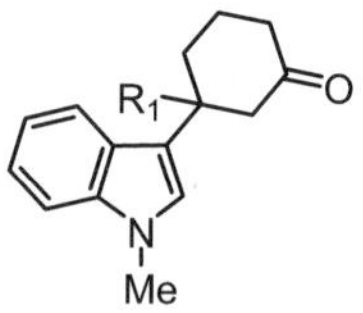

Product	Time (days)	Yield (%)	Pressure (kbar)	Catalyst
17 R_1 = H	7	37	Ambient	$Yb(OTf)_3 \cdot 3H_2O$
18 R_1 = Me	7	3	Ambient	$Yb(OTf)_3 \cdot 3H_2O$
17 R_1 = H	7	56	13	$Yb(OTf)_3 \cdot 3H_2O$
18 R_1 = Me	7	11	13	$Yb(OTf)_3 \cdot 3H_2O$

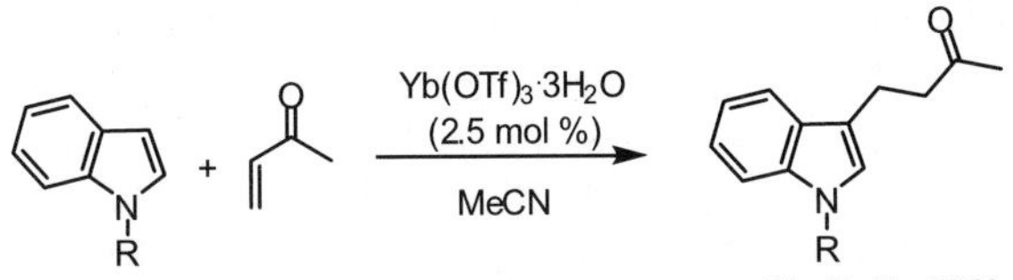

2 R=H
14 R=Me

15 R=H 73%
16 R=Me 72%

Scheme 6

Nevertheless, Michael acceptors such as phenyl vinylsulfone, ethyl cinnamate, methyl acrylate, acrylonitrile and α,β-unsaturated aldehydes failed to react in the reaction catalyzed by $Yb(OTf)_3 \cdot 3H_2O$.

Indium halides have emerged as potential Lewis acids imparting high regio- and chemoselectivity in various chemical transformations [41–43]. The reactions can be carried out under mild conditions either in aqueous or in non-aqueous media. Yadav et al. demonstrated a superior catalytic Lewis acid activity of $InCl_3$ in the conjugative addition of indole (**2**) and 2-methylindole (**19**) (Scheme 7) [44].

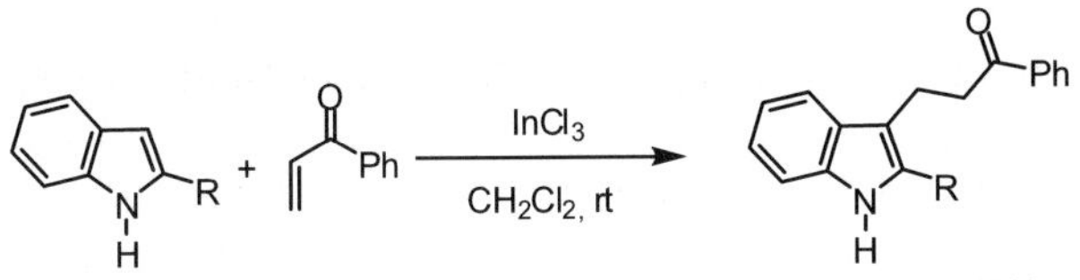

2 R=H
19 R=Me

20 R=H 87%
21 R=Me 90%

Scheme 7

2-Indolyl-1-nitroalkanes **22** are highly versatile intermediates for the preparation of several biologically active compounds such as melatonin analogs **23**, 1,2,3,4-tetrahydro-β-carbolines **24** and triptans **25** (Scheme 8) [45].

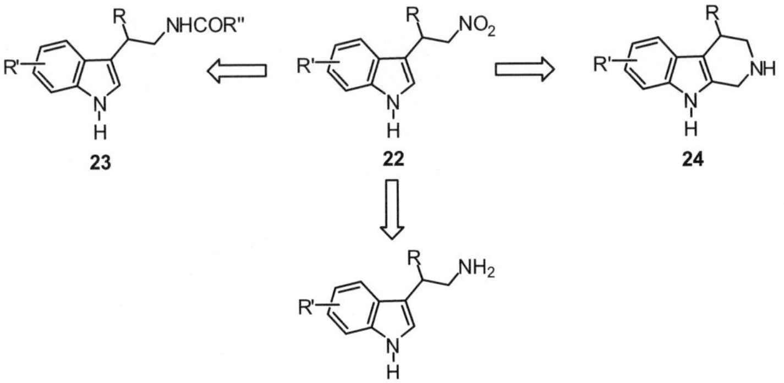

23 22 24 25

Scheme 8

A general and mild InBr$_3$-catalyzed protocol for the conjugate addition of indoles to nitroalkenes to give 2-indolyl-1-nitroalkanes was described by Bandini, Umani-Ronchi and et al. [45]. The process performed in aqueous media provides the functionalized indoles in excellent yields (99–65%) and allows catalyst to be reused several times without loss of effectiveness (Scheme 9).

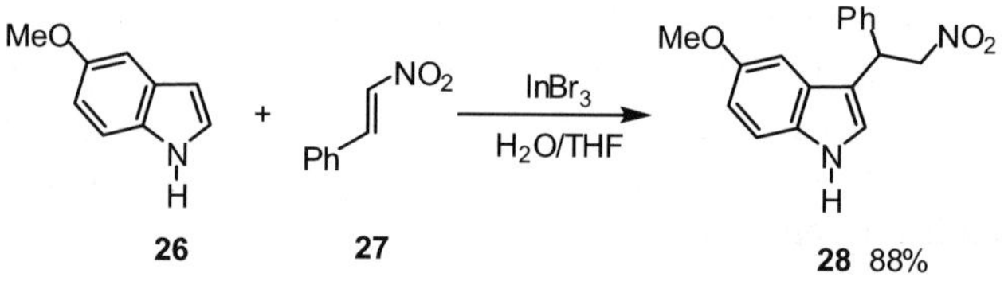

Scheme 9

The β-carboline skeleton with its 9H-pyrido[3,4-b]indole (29) is frequently encountered in pharmacology due to its activity in the central nervous system at serotonin receptors. It also shows prominent biological properties at the benzodiazepine receptor (BzR) [45]. ZK 93423 (30) remarkably amplifies the agonist activity of such compounds towards BzR. 1,2,3,9-Tetrahydro-β-carbolines are common precursors of β-carbolines [46]. 1,3,4,9-Tetrahydro-

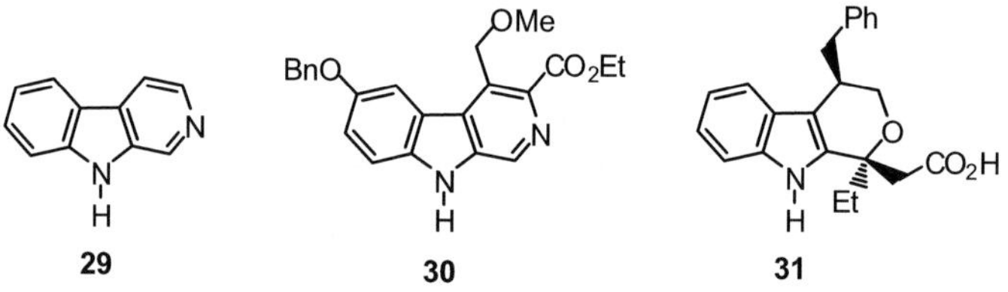

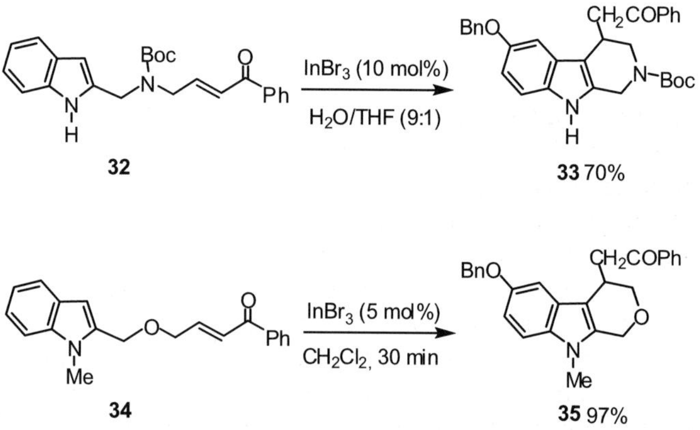

Scheme 10

pyrano[3,4-b]indoles, such as pemedolac (31) tested as an anti-inflammatory agent, are well-known potent analgesic agents [47–49]. The synthesis of tetrahydro-β-carbolines 33 and tetrahydro-pyrano[3,4-b]indoles 35 were realized via an intramolecular Michael addition catalyzed by InBr₃ in excellent yields, both in anhydrous organic and aqueous media (Scheme 10) [50].

Secondary metabolites and marine sponge alkaloids include bis(indolyl) motifs [51–54]. A general procedure for the synthesis of 1,3-bis(indol-3-yl)butane-1-ones was developed by Michael additions catalyzed by InBr₃/TMSCl (chlorotrimethylsilane) (Scheme 11) [55]. The use of 10 mol % TMSCl increased the rate of the process, as shown in Table 2. It was proposed that TMSCl promotes the reaction by dissolving the highly insoluble complex derived from the acceptor and InBr₃.

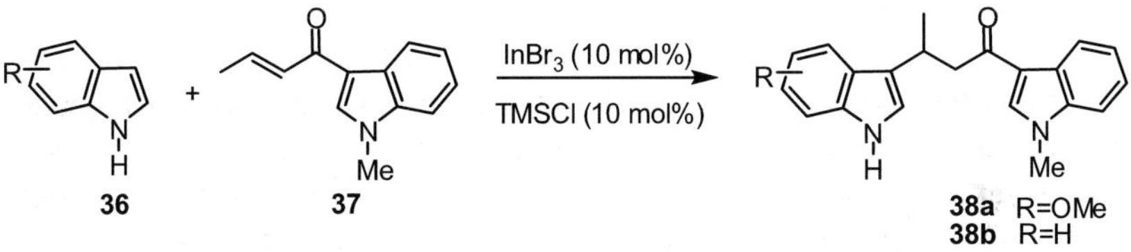

Scheme 11

Table 2 Optimization for the formation of **38a** by Michael additions

Lewis Acid (10%)	TMSCl (%)	Time (h)	Yield (%)
–	–	72	0
AlCl₃	–	48	Traces
InCl₃	–	48	75
InBr₃	–	48	85
InBr₃	10	24	96

Lewis acidity of InBr₃ toward coordination and acid nucleophiles was not affected. Cozzi et al. described a sequential, one-pot InBr₃-catalyzed 1,4-then 1,2-addition to enones, indicating that this is a versatile catalyst for the Michael additions [56]. When InBr₃ (10 mol %) catalyst is used together with trimethylsilylcyanide (TMSCN), the reactions start by the 1,4-conjugate addition of indoles to α,β-ketones and then finish by the 1,2-addition of TMSCN to the β-substituted ketones in one pot (Scheme 12).

The natural product asterriquinone A1 (41) and asterriquinone derivatives containing the 3-indolylbenzoquinone structure exhibit a wide spectrum of biological activities, including antitumor properties, inhibition of HIV reverse transcriptase and as an orally active nonpeptidyl mimetic of insulin

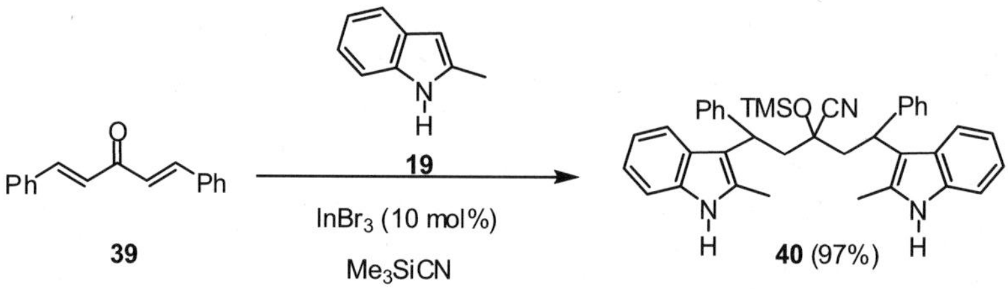

Scheme 12

with antidiabetic activity [57–63]. At the same time, InBr$_3$ catalyzes efficiently the Michael addition of indoles to *p*-benzo- and naphthoquinones under mild conditions to give 3-indolylquinones. The reactions proceed rapidly at room temperature and in dichloromethane. A probable suggested mechanism is shown in Scheme 13 [64]. Hydroquinone produced from the first addition to quinone is oxidized with another equivalent of *p*-quinone in the formation of indol-3yl-quinones. However, bis(indolyl)hydroquinone (**46**) was obtained by the reaction of the parent benzoquinone with indole under similar conditions in 80% yield (Scheme 13).

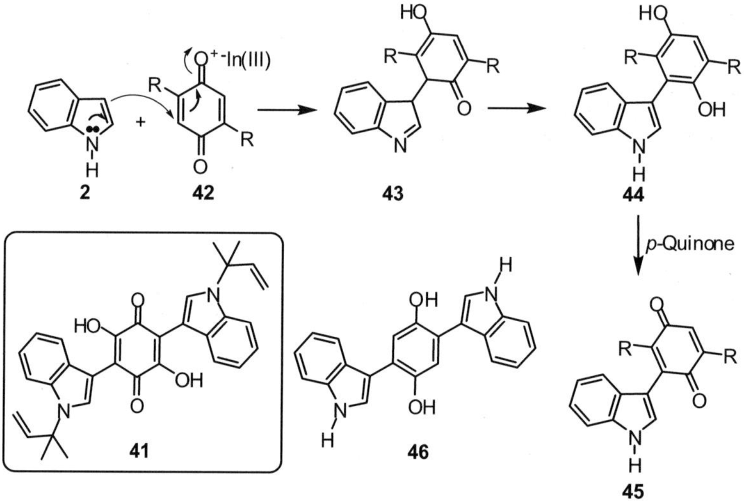

Scheme 13

Treatment of indoles with 2,5,8-quinolinetriones in the presence of a catalytic amount HCl provided the 3-vinylindole derivatives **48a** and **48b**, which could be transformed to polyheterocyclic quinone systems through Diels–Alder reactions (Scheme 14) [65].

With the exception of Yb(OTf)$_3$·3H$_2$O, indium salts and Bronsted acids, there are several metal-based Lewis acid catalysts available for these Michael reactions, such as a CeCl$_3$·7H$_2$O–NaI combination supported on

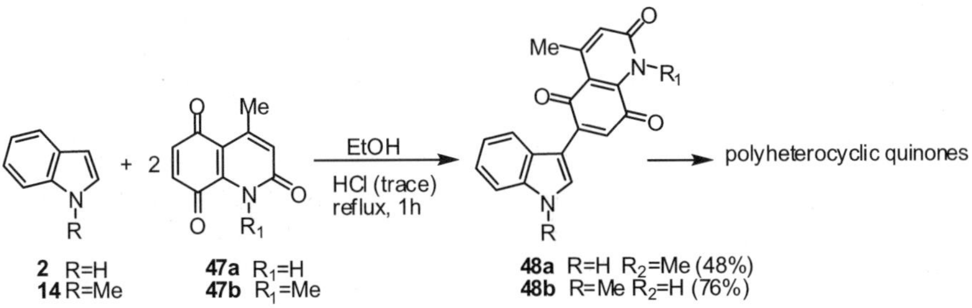

Scheme 14

silica gel [66, 67], solid acid (dry Amberlist-15 as a heterogeneous cata-lyst) [68], bismuth triflate $(Bi(OTf)_3)$ [69], aluminum dodecyl sulfate tri-hydrate $[Al(DS)_3] \cdot 3H_2O$ in water [70], $PdCl_2(CH_3CN)_2$ in ionic 1-butyl-3-methylimidazolium tetrafluoroborate ([bmim][BF$_4$]) [71], SmI$_3$ [72], het-eropoly acid $(H_4[Si(W_3O_{10})_3])$ [73], LiClO$_4$ without solvent [74], aluminum-dodecatungstophosphate (AlPW$_{12}$O$_{40}$) [75], ZnBr$_2$ supported on hydoxya-patite (Zn-HAP) [76], Fe-exchanged montmorillonitrite (K10-FeO) [77], CuBr$_2$ [78], ZrOCl$_2 \cdot 8H_2O$ as a moisture-tolerant [79], ZrCl$_4$ [80], SnCl$_2$ [81] (Scheme 15, Table 3).

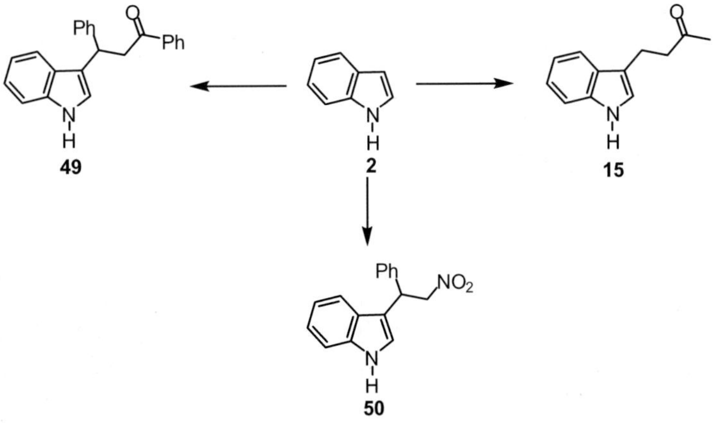

Scheme 15

The high diastereoselective synthesis of multifunctionalized 3,4-dihydro-coumarins bearing a quaternary stereocenter was developed through tan-dem Michael additions of indole and its derivatives (1-methyl, 2-methyl, 4-methoxy, 5-methoxy, 5-bromo, 6-benzyloxy) to 3-nitrocoumarines (3-nitro-chromen-2-one, 6- and 7-methyl-3-nitro-chromen-2-one) followed by methyl vinyl ketone in a one-pot step. For the tandem Michael additions, after the first Michael reaction of indole (2) with 3-nitrocoumarine (51) catalyzed

Table 3 Some catalysts for Michael addition and reaction conditions

Product	Catalyst	Solvent	Time (h)	Yield (%)	Refs.
15	$CeCl_3 \cdot 7H_2O$, NaI, SiO_2	MeCN	2	96	[66, 67]
15	$Bi(OTf)_3$	MeCN	1.5	90	[69]
15	$Al(DS)_3 \cdot 3H_2O$	Water	24	20	[70]
15	SmI_3	MeCN	1	95	[72]
15	$H_4[Si(W_3O_{10})_3]$	MeCN	15 min	85	[73]
15	$LiClO_4$	No solvent	1.5	90	[74]
15	$AlPW_{12}O_{40}$	MeCN	10 min	96	[75]
15	Zn-HAP	MeCN	4	89	[76]
15	K10-FeO	MeCN	2	83	[77]
15	$CuBr_2$	MeCN	0.25	45	[78]
15	$ZrOCl_2 \cdot 8H_2O$	No solvent	2	77	[79]
15	$ZrCl_4$	CH_2Cl_2	8 min	92	[80]
49	Amberlist-15 dry	CH_2Cl_2	24	95	[68]
49	$PdCl_2(MECN)_2$	[bmim][BF_4]	2	94	[71]
49	SmI_3	MeCN	6	85	[72]
49	Zn-HAP	MeCN	24	70	[76]
49	K10-FeO	MeCN	6	73	[77]
49	$CuBr_2$	MeCN	0.25	83	[78]
49	$ZrCl_4$	CH_2Cl_2	5 min	96	[80]
49	$SnCl_2$	No solvent	2 min	90	[81]
50	$CeCl_3 \cdot 7H_2O$, NaI, SiO_2	MeCN	8	96	[66, 67]
50	$Bi(OTf)_3$	MeCN	2.5	80	[69]
50	$Al(DS)_3 \cdot 3H_2O$	Water	12	88	[70]
50	SmI_3	MeCN	1	95	[72]
50	$H_4[Si(W_3O_{10})_3]$	MeCN	25 min	90	[73]

by $Mg(OTf)_2$ in *i*-PrOH at 20 °C was completed, methyl vinyl ketone was added directly for the second Michael addition in the presence of a base (Scheme 16) [82]. The use of Ph_3P as base provided a higher yield. In an analogous manner, pyrrole (1) was used as a good Michael donor for 3-nitrocoumarine (51).

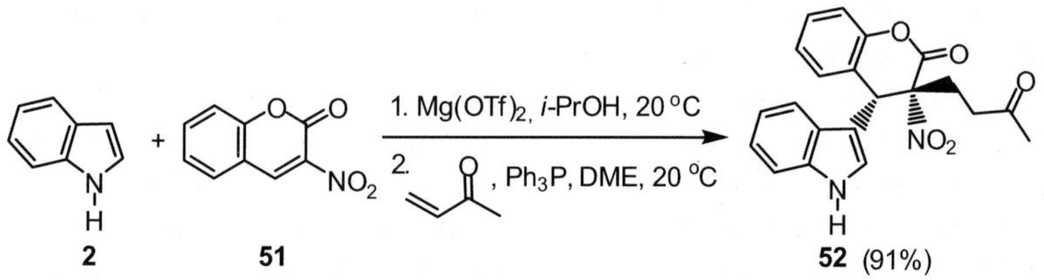

Scheme 16

The bismuth nitrate-catalyzed Michael reaction of enones with amino compounds is a very simple and efficient method. This method is totally independent of solvent or external proton source. The reaction of many structurally diverse indoles with enones under $Bi(NO_3)_3$-catalyzed conditions afforded products at the 3-position of the indole ring in high yield when this position was unoccupied. When the 3-position was occupied by a substituent, the reaction took place at the 2-position of the indole ring (Scheme 17) [83, 84].

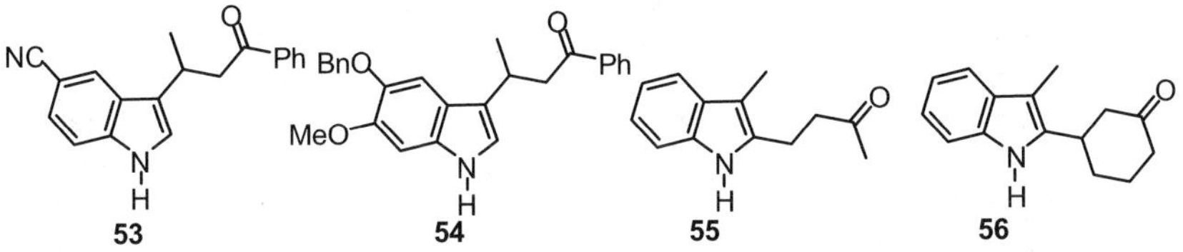

Scheme 17

2-Substituted indoles are potential intermediates for many alkaloids and pharmacologically important substances. There is a need for yet easier access to the methods for the preparation of 2-substituted indoles. Generally, restricted methods have been reported for the preparation of 2-substituted indoles. Recently, a new approach for synthesis of 2-substituted indole derivatives with Michael addition was disclosed [84]. This synthesis strategy is based on a dipole change by transforming the indole ring into a pyrrole derivative. First, indole reduces the benzene ring but not the pyrrole ring to form 4,7-dihydroindole (57) and 4,5,6,7-tetrahydroindole. The $Bi(NO_3)_3$-catalyzed Michael reaction of 4,7-dihydroindole (57), which is now a pyrrole derivative, with α,β-unsaturated carbonyl compounds, followed by oxidation gave the 2-substituted indole derivatives (60–63) (Scheme 18).

Several iodine-catalyzed organic transformations have been reported. Iodine-catalyzed reactions are acid-induced processes. Molecular iodine has received considerable attention because it is an inexpensive, nontoxic and readily available catalyst for various organic transformations under mild and convenient conditions. Michael additions of indoles with unsaturated ketones were achieved in the presence of catalytic amounts of iodine under both solvent-free conditions and in anhydrous EtOH (Scheme 19) [85, 86]. I_2-catalyzed Michael addition of indole and pyrrole to nitroolefins was also reported (Scheme 20) [87].

Recently, many organic reactions have been accelerated by ultrasonic irradiation. The advantageous use of ultrasonic irradiation for activating various reactions proceeding via single electron transfer or radical mechanisms is well reported. The procedure provides higher yields, shorter reaction times or milder conditions for a large number of organic reactions. Ultrasound-accelerated Michael additions of indoles to unsaturated ketones catalyzed by

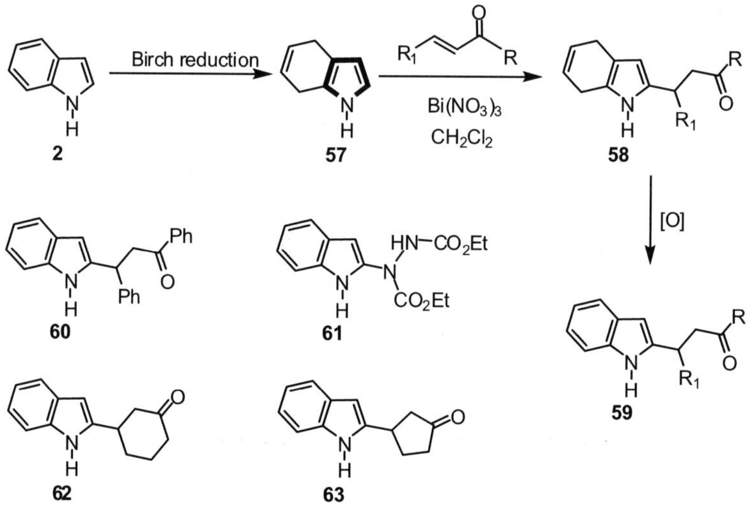

Scheme 18

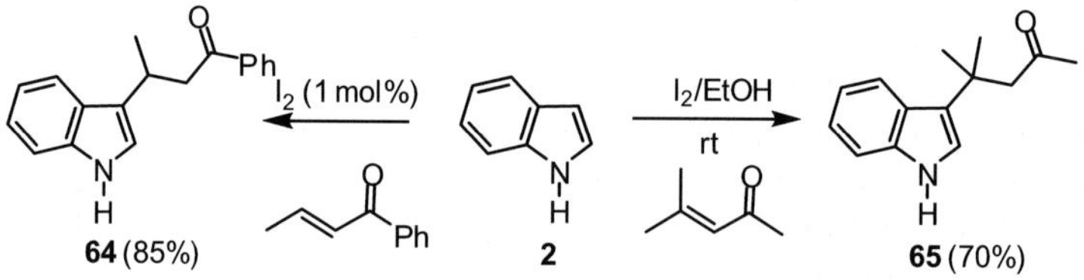

Scheme 19

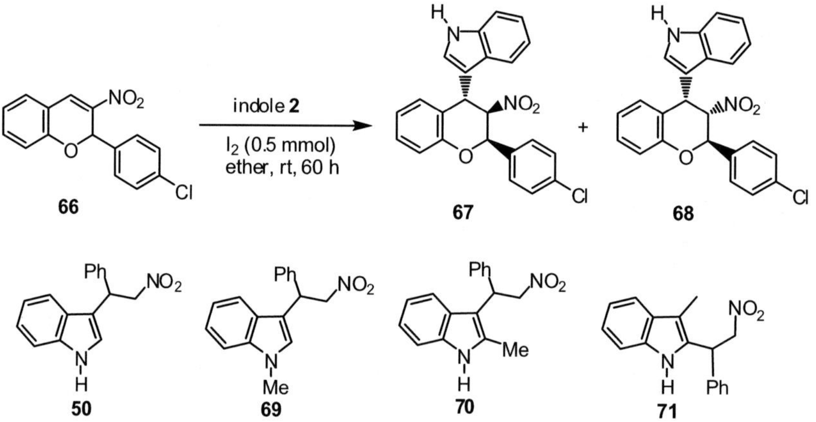

Scheme 20

ceric ammonium nitrate (CAN) [88], *para*-toluenesulfonic acid (*p*-TsOH) [89] and silica sulfuric acid [90] have been published (Scheme 21).

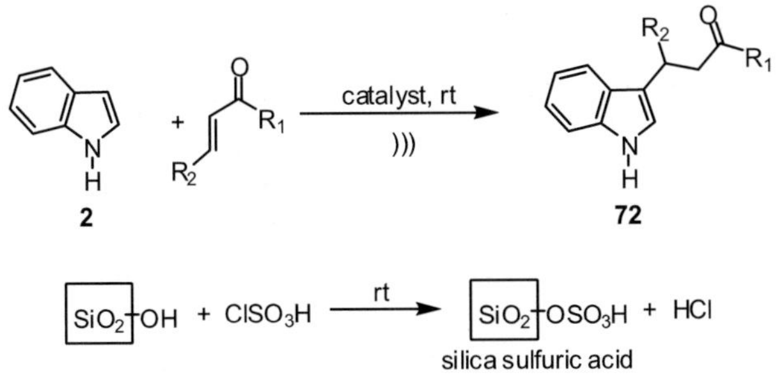

Scheme 21

2.2
Michael Additions of Indoles Under Basic and Neutral Conditions

Organophosphorus compounds are used in organic synthesis as both direct catalyst and as ligands for a number of transition metal catalysts [91, 92]. Conjugated acetylenes are used as Michael acceptors, which undergo Michael reaction with nucleophiles to yield the expected β-adduct. The regioselectivity for Michael additions using triphenylphosphine as a catalyst reoriented the reaction from the β-addition to the α-addition. Michael additions of indole and pyrrole to ethyl 2-propiolate in the presence of Ph₃P are unusual examples of α-addition of an N-nucleophile in the chemistry of indole and pyrrole. Initially, the treatment of conjugated acetylene with Ph₃P resulted in the formation of the vinylphosphonium intermediate. This intermediate serves as an α-Michael acceptor followed by proton shift and elimination of triphenylphosphine to give the α-Michael addition product (Scheme 22) [93]. It was shown that triphenylphosphine could serve as a general base catalyst for Michael additions.

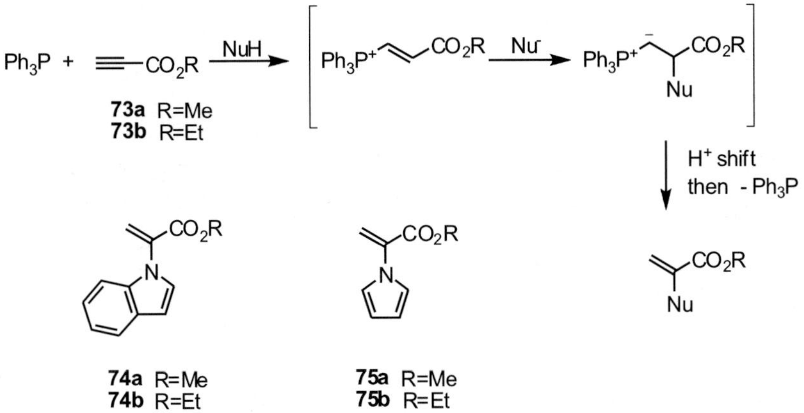

Scheme 22

The formation of 3-substituted indoles is usually achieved in the acid catalyzed conditions as mentioned above. The studies concerning the additions of indoles to electron-deficient alkenes in basic and neutral media is less known. The Michael additions of dicyclohexylammonium 2-(diethoxyphosphoryl)acrylate (76) to indole and electron-rich indoles produced regioselectively the 3-substituted indoles (78a–c) in high yield without any external catalyst. The addition of electron-deficient indoles to acrylate 76 gave N-substituted indoles (80d–e). The formation of 3- and N-substituted indoles can be explained by the fact that the regioselectivity of the addition is strongly controlled by electronic effects (Scheme 23) [94]. The conjugate additions of unsaturated phosphorous molecules 82a and 82b as Michael acceptors toward indoles have been investigated in both acidic (glacial acetic acid) and basic (NaH) conditions (Scheme 24 and Table 4) [95]. The results show that the electron-donating character of the substituent at the C-5 favors the localization of negative charge at the C-3 atom. On the other hand, an electron-withdrawing substituent (e.g., nitro) orients the reaction to aza-Michael addition.

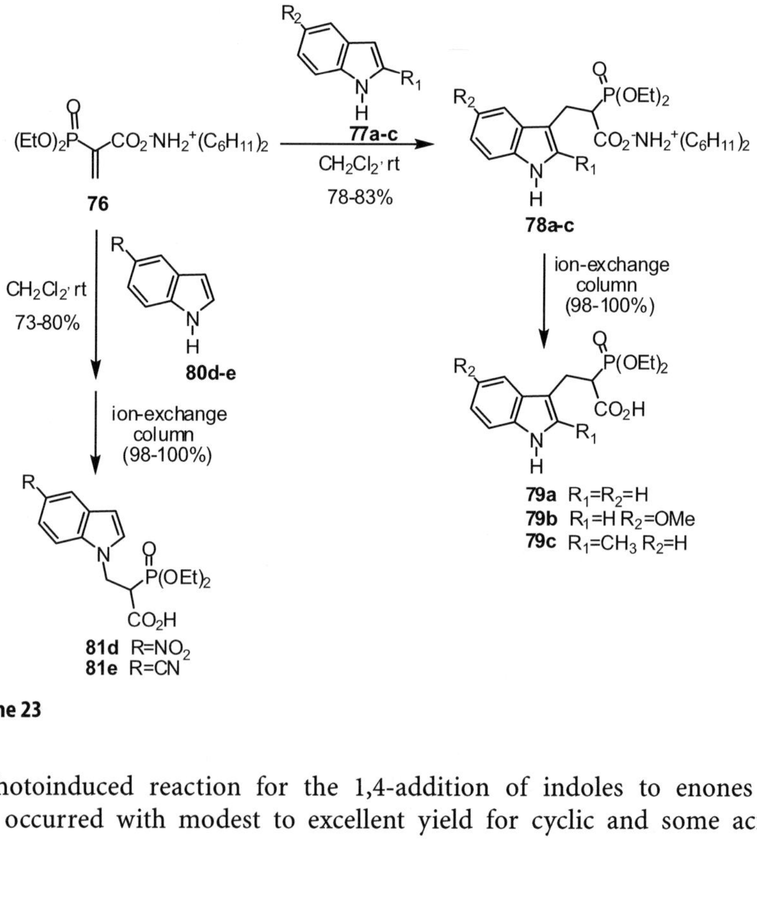

Scheme 23

Photoinduced reaction for the 1,4-addition of indoles to enones has also occurred with modest to excellent yield for cyclic and some acrylic

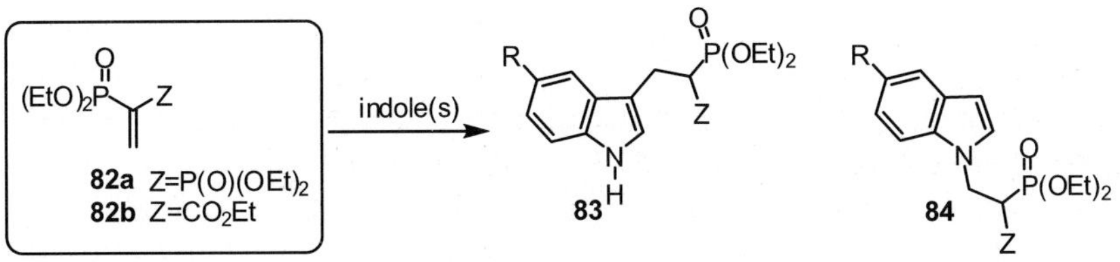

Scheme 24

Table 4 Reaction conditions for Michael adducts **83** and **84**

Reaction condition	R	Z	Product	Ratio of crude products (%)
AcOH, reflux	H	$P(O)(OEt)_2$	**83a + 84a**	56 + 44
	F	$P(O)(OEt)_2$	**83b + 84b**	63 + 27
	OMe	$P(O)(OEt)_2$	**83c + 84c**	80 + 20
	H	CO_2Et	**83d**	ca. 100
	F	CO_2Et	**83e**	ca. 100
	OMe	CO_2Et	**83f**	ca. 100
NaH, THF, rt	H	$P(O)(OEt)_2$	**84a**	ca. 100
	F	$P(O)(OEt)_2$	**84b**	ca. 100
	OMe	$P(O)(OEt)_2$	**84c**	ca. 100
	H	CO_2Et	**83d + 84d**	82 + 18
	F	CO_2Et	**83e + 84e**	65 + 35
	OMe	CO_2Et	**83f + 84f**	40 + 60
	NO_2	$P(O)(OEt)_2$	**84g**	ca. 100

enones [96]. The reaction is performed by irradiation (UVA lamps, ca. 350 nm) of the reactants in CH_2Cl_2 at room temperature under neutral conditions and avoids the necessity to use a Lewis acid or any base. CH_2Cl_2 and $CHCl_3$ are the optimal solvents and electron-withdrawing groups on the indole moiety can suppress this reaction. The photoinduced electron transfer mechanism for the reaction as shown in Scheme 25 has been proposed. According to this mechanism, the indoles can react through pathways involving single electron transfer (SET). SET between the triplet exited state of enone and indole results in the formation of the radical ions, which combine to give the 1,4-adduct followed by tautomerization.

Conjugate addition of indole to nitroolefins was carried out by thermal heating in a sealed tube or by addition of indolyl magnesium iodide or microwave irradiation. The results indicated that the microwave technique is most efficient with respect to time and yield (Scheme 26) [97].

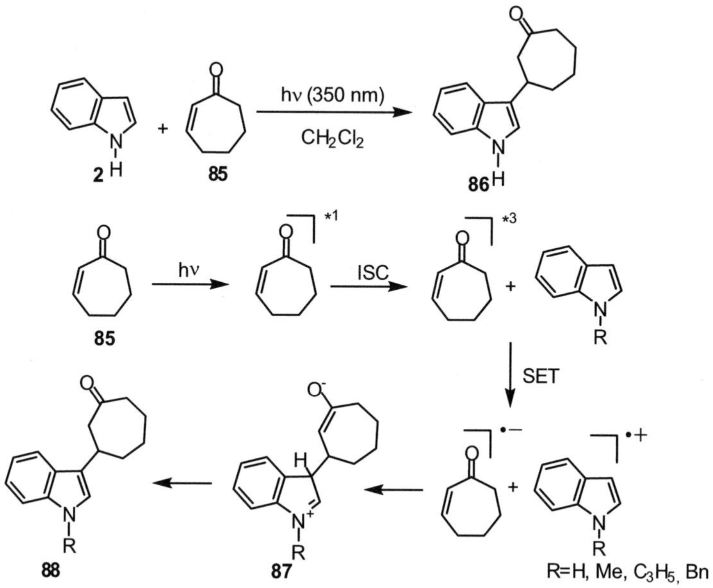
hν (350 nm)
CH2Cl2
2
85
86
hν
ISC
SET
88
87
R=H, Me, C3H5, Bn

Scheme 25

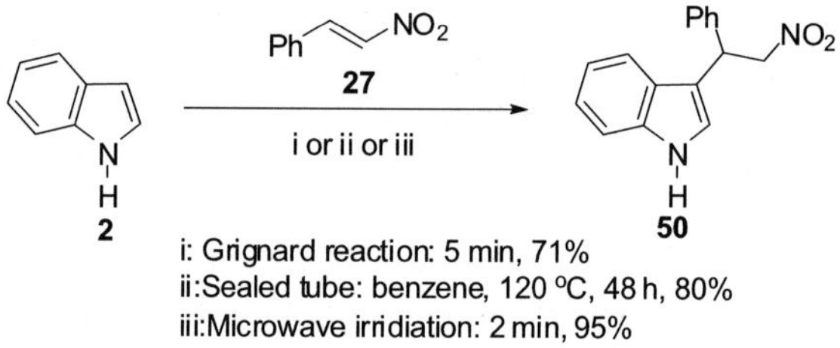
27
i or ii or iii
2
50
i: Grignard reaction: 5 min, 71%
ii:Sealed tube: benzene, 120 °C, 48 h, 80%
iii:Microwave irridiation: 2 min, 95%

Scheme 26

2.3 Enantioselective Michael Additions

The catalytic enantioselective addition of aromatic C – H bonds to alkenes would provide a simple and attractive method for the formation of optically active aryl substituted compounds from easily available starting materials. The first catalytic, highly enantioselective Michael addition of indoles was reported by Jorgensen and coworkers. The reactions used α,β-unsaturated α-ketoesters and alkylidene malonates as Michael acceptors catalyzed by the chiral bisoxazoline (BOX)-metal(II) complexes as described in Scheme 27 [98, 99].

Catalyst (*S*)-**90c**, which is more stable toward air than the copper catalyst (*S*)-**90a**, is an effective catalyst for the formation of **91**, with up to 87% ee.

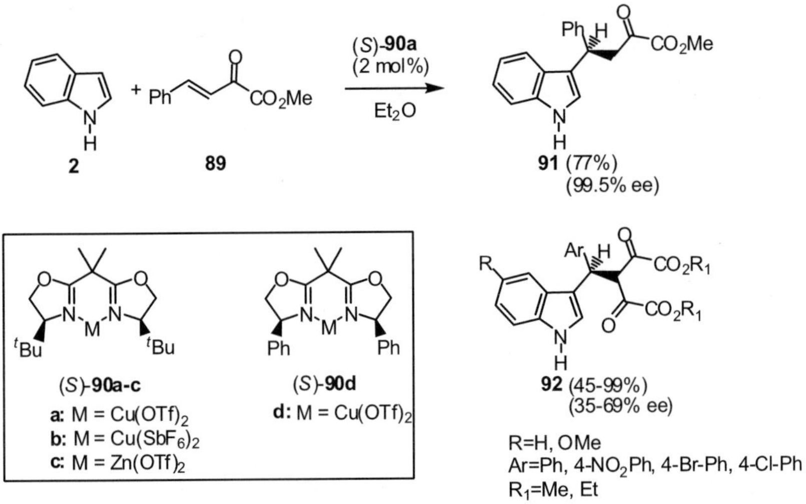

Scheme 27

The enantioselective reactions of indoles using alkylidene malonates gave the alkylation products **92** in excellent yield and with moderate ee.

The enantioselective alkylation of indoles catalyzed by C2-symmetric chiral bisoxazoline-metal complexes **90** encouraged many groups to develop superior asymmetric catalysts which are cheap, accessible, air-stable and water-tolerant. Other analogs of the bisoxazoline–metal complex **90** as chiral catalysts and new Michael acceptors have also been studied. The enantioselective alkylations of indole derivatives with α'-hydroxy enones using Cu(II)-bis(oxazoline) catalysts **93** and **94** provided the adducts in good yields

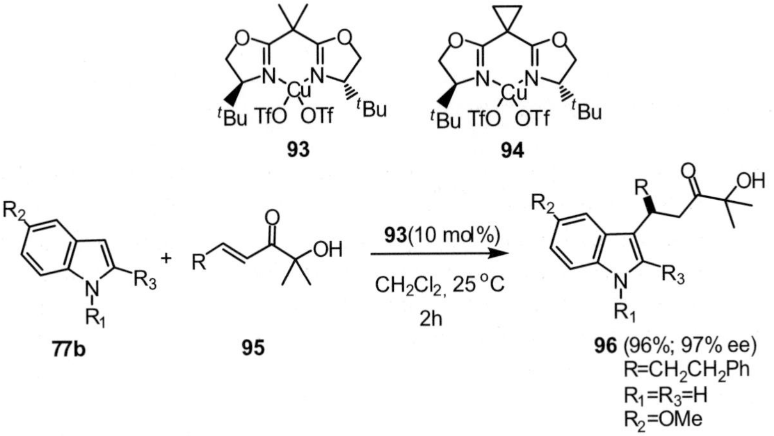

Scheme 28

and enantiomeric excess. The catalytic system showed remarkable performance at 0 or 25 °C, or even at refluxing conditions (40 °C) (Scheme 28) [100].
Enones bearing a branched β-position showed lower selectivity using catalyst
93, but high enantiomeric excesses using catalyst **94**. A distorted square planar geometry around copper is believed to elicit the selectivity as shown in
Fig. 1.

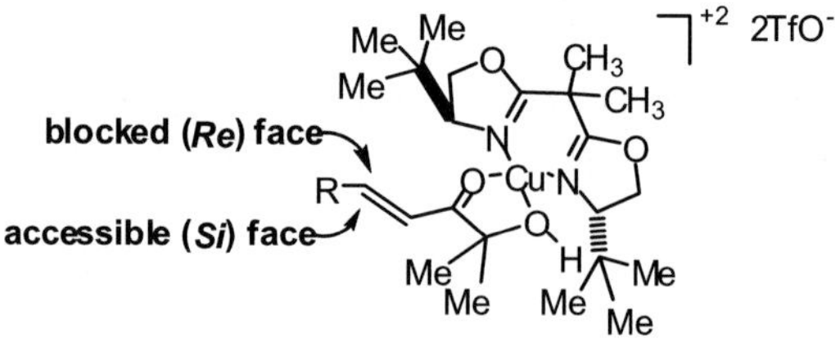

Fig. 1 Stereochemical model for acceptor–catalyst complex

The additions of indoles to ethenetricarboxylates as Michael acceptors in
the presence of copper(II) complexes (10%) of chiral bisoxazolines (**97–100**)
under mild conditions gave the alkylated products in high yield and up to
96% ee [101]. The observed enantioselectivity could be explained by secondary orbital interaction on approach of indole to the less hindered side of
the **102**-Cu(II)-ligand complex. The chiral ligands **97–99** of the catalyst gave
similar ee%. The phenyl derivative **100** produced inferior results compared
to **97–99**, while (S,S)-2,6-bis(4-isopropyl-2-oxazoline-2-yl)pyridine (**101**) gave
no reaction (Scheme 29) [56]. The enantioselective alkylation of indoles with
arylidene malonates catalyzed by i-Pr-bisoxazoline-Cu(OTf)$_2$ was also reported [102].

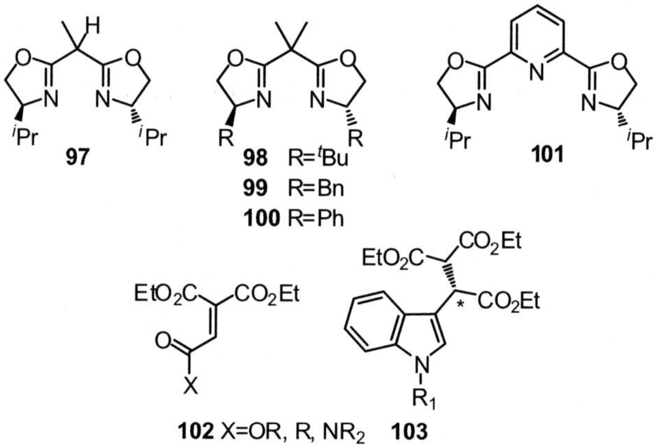

Scheme 29

A variety of chiral Lewis acid catalysts generated in situ from (S)-Ph-bisoxazoline **100** and metal salts, such as $Fe(ClO_4)_2 \cdot xH_2O$, CuOTf, $Mg(OTf)_2$, AgOTf, $Pd(OAc)_2$, $Cu(OTf)_2$, $Ni(OTf)_2$ and $Zn(OTf)_2$, were used to produce nitroalkylated indoles. The results show that bisoxazoline-Zn(II) complex **109** is the best catalyst and gave fast reaction in excellent yield and high enantioselectivities. In order to test ligand effect, asymmetric Friedel–Crafts alkylations of indoles with nitroalkenes were subsequently investigated by different chiral bisoxazoline **98–101, 104–107** and oxazoline **108** ligands, and $Zn(OTf)_2$ as Lewis acid. The ligand **100** was found to be the best ligand for providing the highest yields and enantioselectivity, whereas the tridentate pyridine-oxazoline ligand **101** was completely inactive (Scheme 30) [103].

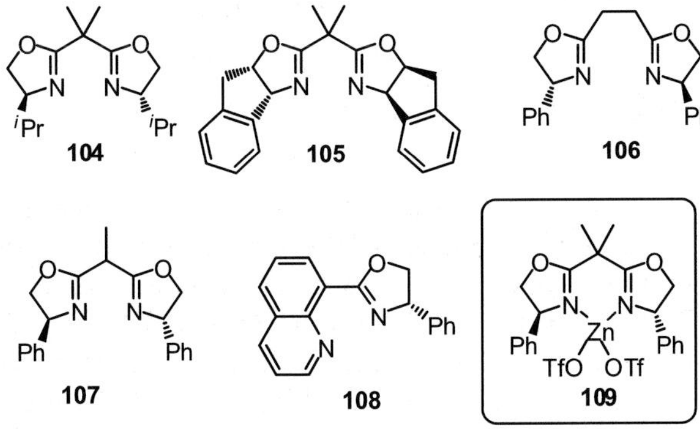

Scheme 30

A series of chiral C2-symmetric tridentate bis(oxazoline) **110, 112** and bis(thiazoline) **111**, in which a diphenylamine unit links the two chiral oxazolines, was used for a more practical and efficient catalytic asymmetric alkylation of indoles with nitroalkenes (Scheme 31) [104]. The alkylation of indole with trans-β-nitrostyrene in toluene at room temperature has been accomplished with **110d**-$Zn(OTf)_2$ or **113**-$Zn(OTf)_2$ as the catalyst. The obtained results show that the NH group in ligands **110–112** is crucial. Indole

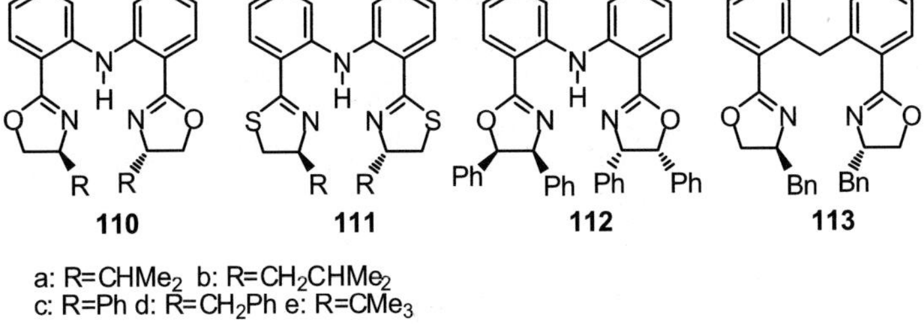

a: R=CHMe₂ b: R=CH₂CHMe₂
c: R=Ph d: R=CH₂Ph e: R=CMe₃

Scheme 31

π-interaction with the NH group between two phenyl groups would direct the indole to attack the nitroalkenes preferentially.

Enantioselective additions of α,β-unsaturated 2-acyl imidazoles, catalyzed by bis(oxazolinyl)pyridine–scandium(III)triflate complex, were used for the asymmetric synthesis of 3-substituted indoles. The complex **114** was one of the most promising catalysts. The choice of acetonitrile as the solvent and the use of 4 Å molecular sieves were also found to be advantageous. The 2-acyl imidazole residue in the alkylation products of α,β-unsaturated 2-acyl imidazoles could be transformed into synthetically useful amides, esters, carboxylic acid, ketones, and aldehydes (Scheme 32) [105]. Moreover, the catalyst **114** produced both the intramolecular indole alkylation and the 2-substituted indoles in good yield and enantioselectivity (Scheme 33) [106]. The complex

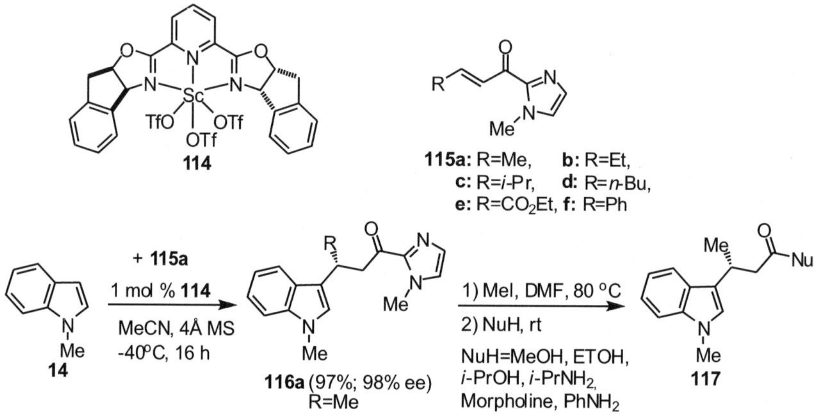
TfO
OTf
OTf
Sc
114
R
Me
115a: R=Me, b: R=Et,
c: R=i-Pr, d: R=n-Bu,
e: R=CO2Et, f: R=Ph
+ 115a
1 mol % 114
MeCN, 4Å MS
-40°C, 16 h
14
116a (97%; 98% ee)
R=Me
1) MeI, DMF, 80 °C
2) NuH, rt
NuH=MeOH, ETOH,
i-PrOH, i-PrNH2,
Morpholine, PhNH2
Nu
117

Scheme 32

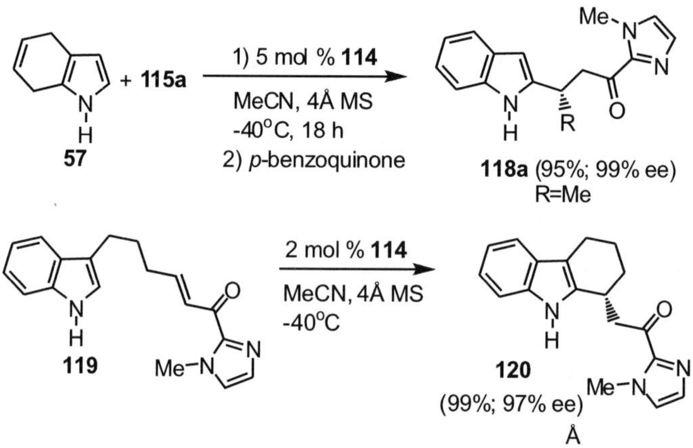
+ 115a
1) 5 mol % 114
MeCN, 4Å MS
-40°C, 18 h
2) p-benzoquinone
57
118a (95%; 99% ee)
R=Me
2 mol % 114
MeCN, 4Å MS
-40°C
119
120
(99%; 97% ee)
Å

Scheme 33

114 was also an efficient catalyst for the Michael addition of electron-rich indoles to α,β-unsaturated phosphonates [107].

Several pseudo-C3-symmetric homo- and hetero-trisoxazolines have been developed and applied successfully in the asymmetric Michael-type alkylation of indoles with alkylidene malonates. The C3-symmetric trisoxazoline **121a**-copper(ClO$_4$)$_2$ · 6H$_2$O complex, as a catalyst, is an air- and water-stable compound (Scheme 34) [108–110].

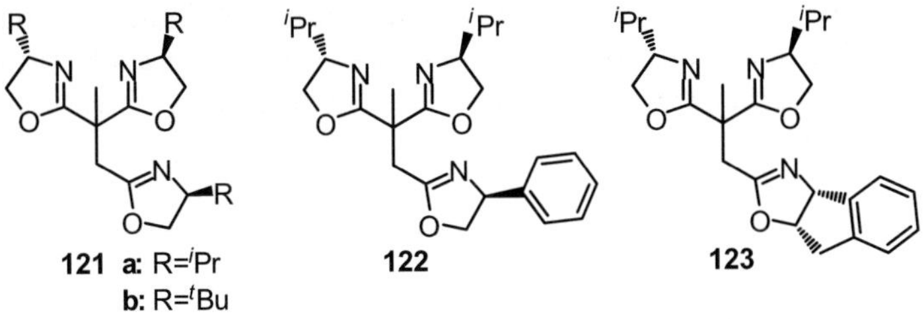

121 a: R=^{i}Pr
  b: R=^{t}Bu

122

123

Scheme 34

As a parallel to the rapid growth of asymmetric catalysis, chiral imidazolidinon-HX salts **124a–c** were used as catalysts for Michael-type alkylations between indoles and α,β-unsaturated aldehydes with high levels of enantioselectivity and reaction efficiency. This chiral catalyst system is the first reported nonchelating catalyst for indole alkylation. It was assumed that the catalyst reacts with the unsaturated aldehydes to yield the chiral and highly reactive iminium intermediate, which influences both the LUMO-lowering activation of aldehydes and stereoselectivity in the alkylation of indoles (Scheme 35) [111, 112].

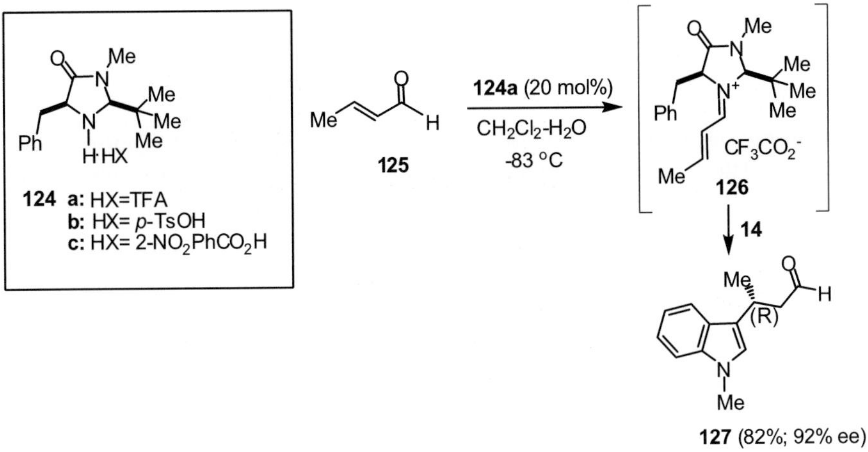

124 a: HX=TFA
  b: HX= *p*-TsOH
  c: HX= 2-NO$_2$PhCO$_2$H

125

124a (20 mol%)
CH$_2$Cl$_2$-H$_2$O
-83 °C

126

127 (82%; 92% ee)

Scheme 35

Aluminum salen complexes have been identified as effective catalysts for asymmetric conjugate addition reactions of indoles [113–115]. The chiral Al(salen)Cl complex **128**, which is commercially available, in the presence of additives such as aniline, pyridine and 2,6-lutidine, effectively catalyzed the enantioselective Michael-type addition of indoles to (*E*)-arylcrotyl ketones [115]. Interestingly, this catalyst system was used for the stereoselective Michael addition of indoles to aromatic nitroolefins in moderate enantioselectivity (Scheme 36). The Michael addition product **130** was easily reduced to the optically active tryptamine **131** with lithium aluminum hydride and without racemization during the process. This process provides a valuable protocol for the production of potential biologically active, enantiomerically enriched tryptamine precursors [116].

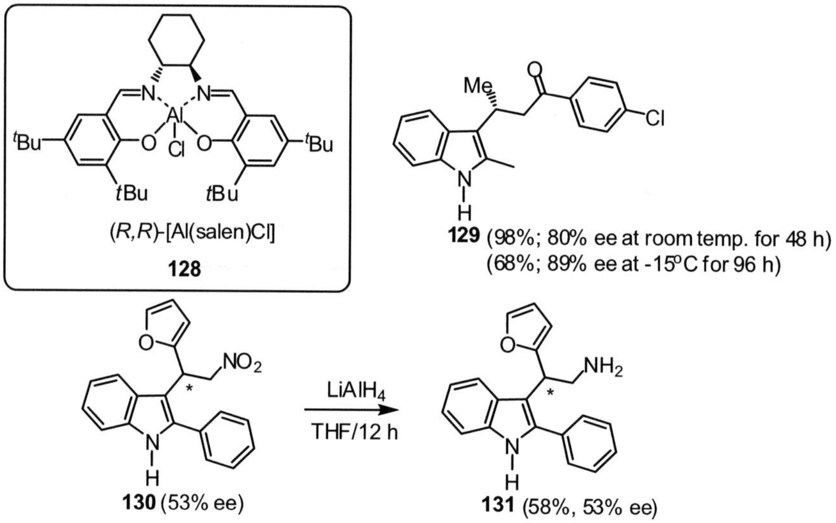

(*R*,*R*)-[Al(salen)Cl]
128

129 (98%; 80% ee at room temp. for 48 h)
(68%; 89% ee at -15°C for 96 h)

130 (53% ee)   LiAlH$_4$ / THF/12 h   **131** (58%, 53% ee)

Scheme 36

The supposed mechanism for Al(salen)Cl/amine catalyzed 1,4-addition involves a crucial stereocontrolled formation of an intermediate octahedral Schiff base–aluminum enolate **132** as depicted in Fig. 2.

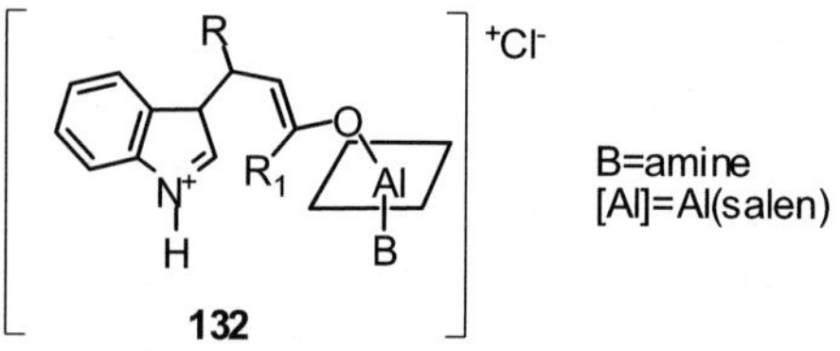

B=amine
[Al]=Al(salen)

132

Fig. 2 Intermediate for Al(salen)Cl/amine-catalyzed Michael addition

In an analogous manner, Al-Schiff base complexes possessing different chiral backbones for the 1,4-addition of both enones, nitroalkenes and intramolecular were investigated (Scheme 37) [114–116]. For example, while the base-free protocol of the Al(salen)Cl 128 gave 133 in 80% and 55% ee, Al-Schiff base complexes 134–137 showed different yields (90, 54, 54 and 38%, respectively) and enantioselectivity (39, 0, 5 and 23%, respectively). Moreover, the stereocontrolled synthesis of polycyclic indole compounds, such as tetrahydro-carbolines and their oxygen analogs, via intramolecular processes, using Al(salen)Cl and the unprecedented bimetallic Al(salen)Cl-InBr$_3$ system as promising chiral Lewis acids, has been presented [117].

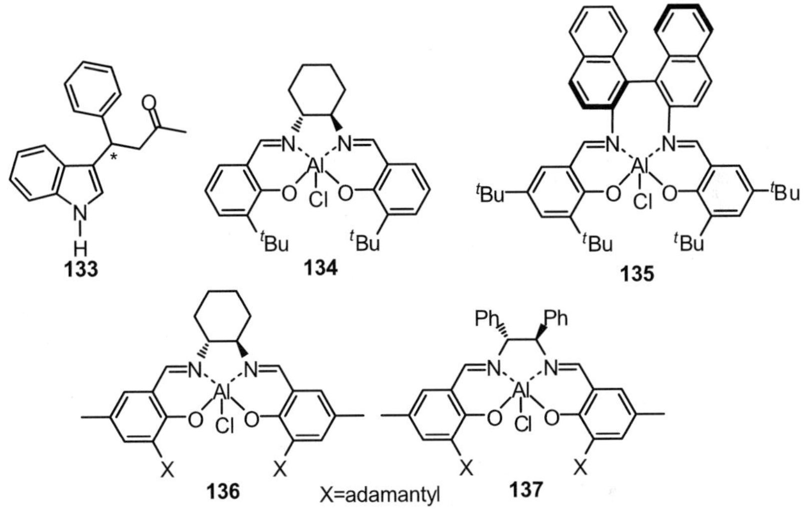

Scheme 37

α,β-Unsaturated S-(1,3-benzooxazol-2-yl) thioesters 139 have been synthesized and effectively employed as electrophiles in the stereoselective alkylations of indoles in the presence of a catalytic amount of chiral cationic Tol-BINAP-Pd(II) complex 138. The catalyst 138 was prepared in situ from PdCl$_2$(MeCN)$_2$ (1 equiv.), (S)-Tol-BINAP (1 equiv.) in toluene and AgSbF$_6$ (2 equiv.). The 1,4-adducts could be used for the synthesis of key building blocks bearing asymmetric centers as (R)-141 (Scheme 38) [118].

The simply obtainable thiourea compounds 142–145 were the first organocatalysts for the catalytic conjugate addition of indoles with nitroalkenes to yield optically active 2-indolyl-1-nitro derivative as 2R-50 in fairly good yields and enantioselectivity. The simple thiourea-based organocatalyst 145 could be easily accessed in both enantiomeric forms from the commercially available materials. At the same time, the extremely simple methodology has proved the new approach useful for the synthesis of optically active target

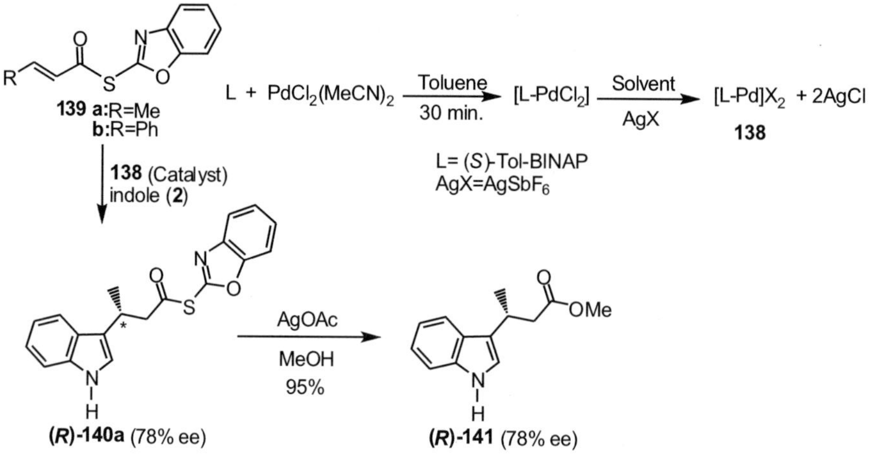

Scheme 38

compounds such as the tryptamine derivative **146** and the 1,2,3,4-tetrahydro-β-carboline core **147** (Scheme 39) [119]. The catalyst **145a** would act in a bifunctional fashion and while the nitroalkene is activated by two thiourea hydrogen atoms, the free alcoholic function of organocatalyst will interact with the indolic proton through a weak hydrogen bond. Thus, the indole as nucleophile will attack on the *Si* face of the nitroolefin, as depicted in Fig. 3.

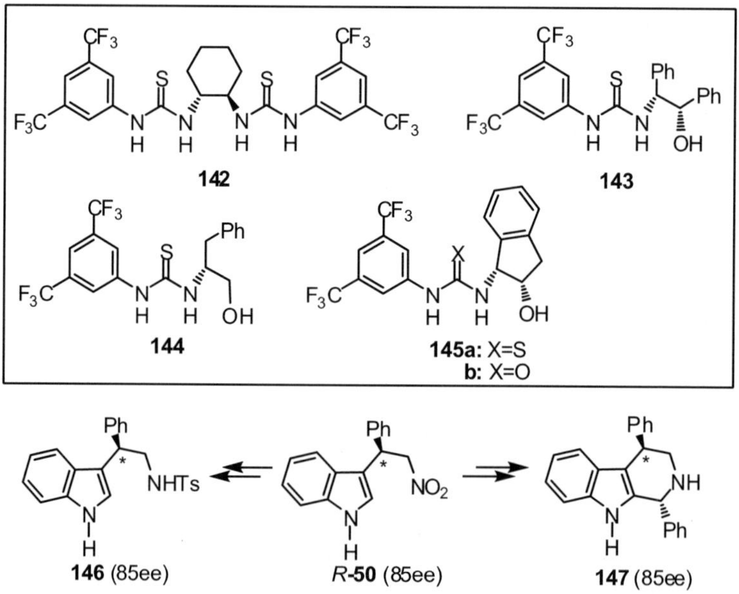

Scheme 39

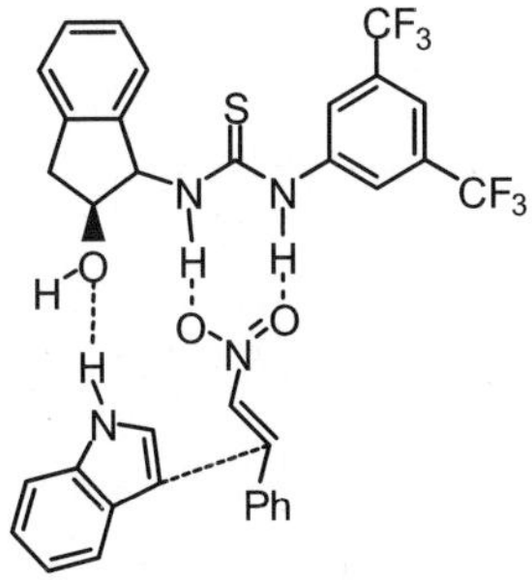

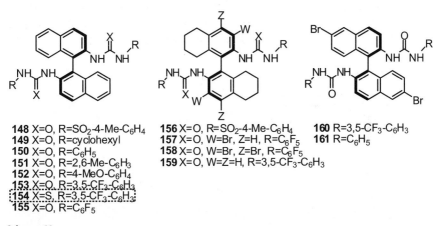

Fig. 3 Possible bifunctional approach to Michael components of the organocatalyst **145a**

Recently, a small library of novel axially chiral bis-arylthioureas **148–161** as chiral organocatalyst has been prepared and evaluated for the asymmetric addition of *N*-methylindole to nitroolefins. Initial studies have shown that the relatively simple and readily prepared (*S*)-**154** is the optimal structure (Scheme 40) [120].

Chiral bis-sulfonamides **162–163** are a new group of organocatalysts for the enantioselective Friedel–Crafts alkylation of indoles to nitroolefins. The hy-

148 X=O, R=SO$_2$-4-Me-C$_6$H$_4$
149 X=O, R=cyclohexyl
150 X=O, R=C$_6$H$_5$
151 X=O, R=2,6-Me-C$_6$H$_3$
152 X=O, R=4-MeO-C$_6$H$_4$
153 X=O, R=3,5-CF$_3$-C$_6$H$_3$
154 X=S, R=3,5-CF$_3$-C$_6$H$_3$
155 X=O, R=C$_6$F$_5$

156 X=O, R=SO$_2$-4-Me-C$_6$H$_4$
157 X=O, W=Br, Z=H, R=C$_6$F$_5$
158 X=O, W=Br, Z=Br, R=C$_6$F$_5$
159 X=O, W=Z=H, R=3,5-CF$_3$-C$_6$H$_3$

160 R=3,5-CF$_3$-C$_6$H$_3$
161 R=C$_6$H$_5$

Scheme 40

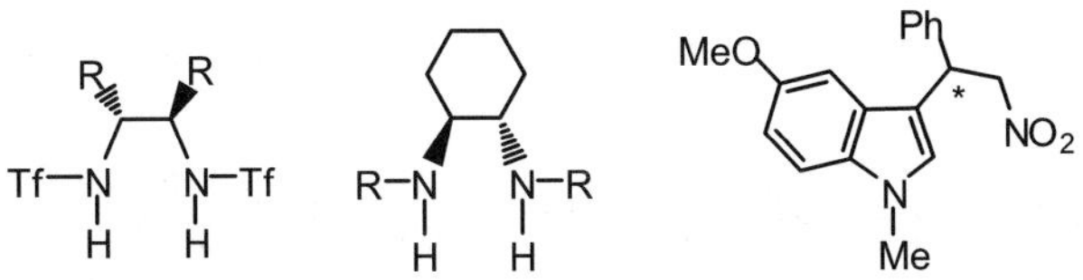

162a: R=Ph
b: R=*t*-Bu
c: R=cyclohexyl

163a: R=Tf
b: R=Tos

164

Scheme 41

drogen bond between the catalysts prepared from the chiral diamines and the oxygen atoms of the nitroolefin plays a very important role both in activating the olefin and determining the enantioselectivity of the reaction. The catalyst is used at only 2% and the reactions proceed in high yields from 64% ee to > 98% ee after crystallization (Scheme 41) [121].

2.4
Designs of Natural Products or Possible Biologically Active Molecules

The indole[2,3-a]carbazole is a fundamental nucleus of biologically active natural products such as K-252a (165), staurosporine 166 and aglycone 167. K-252a (165) is implicated in the regulation of various cellular processes including growth, differentiation and antitumor activity. Staurosporine 166

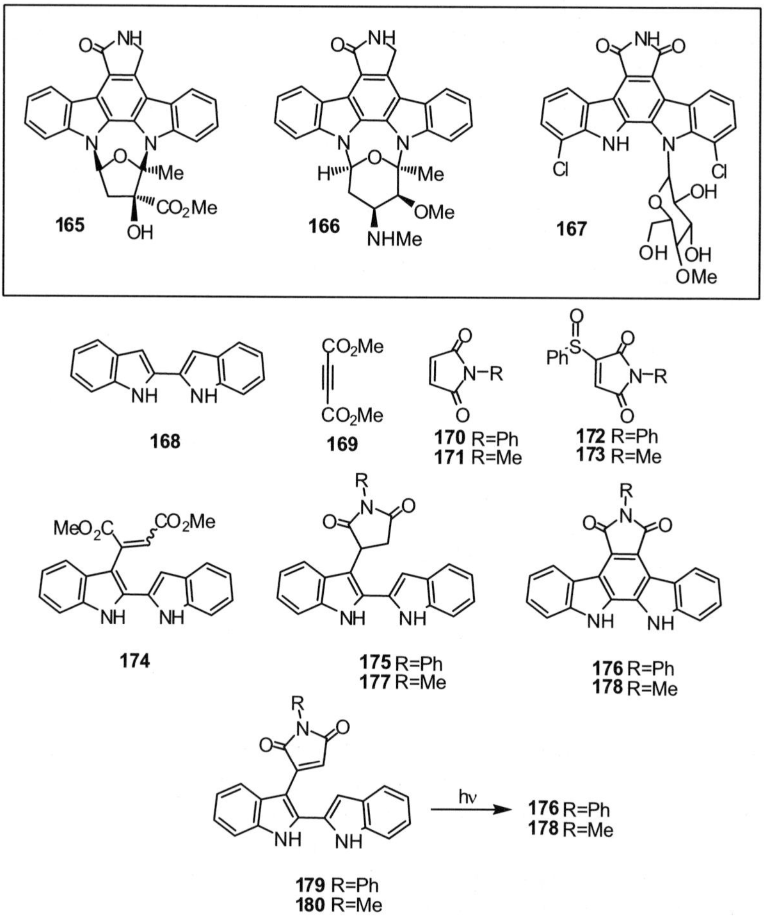

Scheme 42

exhibits hypotensive and antimicrobial activity and aglycone **167** shows interesting antitumor activity (Scheme 42) [122–132]. While 2,2′-bis-indolyl **168** reacted by dimethyl acetylenedicarboxylate (DMAD, **169**) in acetonitrile at 110 °C to give Michael addition product **174** as a single isomer in 17% yield, the reaction of **168** with N-phenylmaleimide (**170**) gave a fully aromatized product **176** and the Michael addition product **175** in low yields. N-Methylmaleimide **171** treated with bis-indolyl **168** gave a low yield of the AT2433-B aglycone **178** and the Michael product **177** (45%) as the major product. The phenylsulfinylmaleimides **172** and **173** reacted rapidly with **168** to give the Michael addition–elimination products **179** and **180**. These products were efficiently photocyclized into indolecarbazoles **176** (80%) and **178** (90%) (Scheme 42) [133, 134].

The indole[2,3-a]carbazole **185** possessing the rebeccamycin aglycone unit [122–132, 135, 136] was synthesized from the reaction of N-methylmaleimide and 2,2′-bisindolyl **181** via Michael type addition. In the presence of aluminum trichloride, the bisindole reacts to form the Michael adducts **182** and **183**. The mono-Michael adduct **182** was dehydrogenated in the presence of Pd/C in a one pot procedure to produce **184** and **185** (Scheme 43) [137].

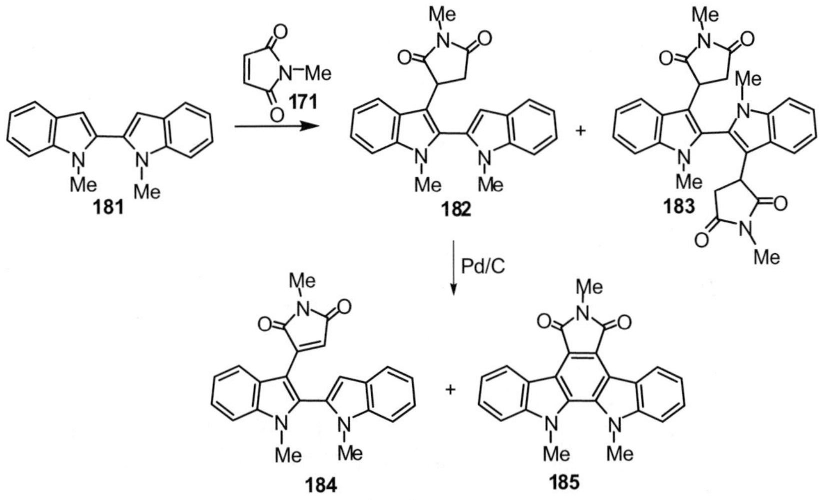

Scheme 43

Due to the widespread use of structurally diverse amino acid derivatives in practically all areas of the physical and life sciences, the synthesis and applications of these compounds are of fundamental importance. Heterocyclic β-substituted-α-alanines are non-proteinogenic amino acids that are widely found in nature [138–147]. Naturally occurring β-amino acids are also compounds with interesting pharmacological aspects. N,N-Bis(tert-butyloxycarbonyl)-dehydroalanine methyl ester (**186**) was reacted at room temperature with one equivalent of indole (2) in acetonitrile and six equivalents of potassium carbonate to give the tryptophan isoester **187**. The products of the reaction with 3-formylindole, 3-methylindole, 7-azaindole, carbazole and 3-carboxy-ethycarbazole are isoesters of the corresponding β-substituted alanine amino acids **188–192**. The products are synthesized in high yields by an aza-Michael addition of nucleophiles to the acceptor **186** using mild reaction conditions and simple work-up procedures [148]. New β-amino esters **194–195** were synthesized via conjugate additions (Scheme 44) [149].

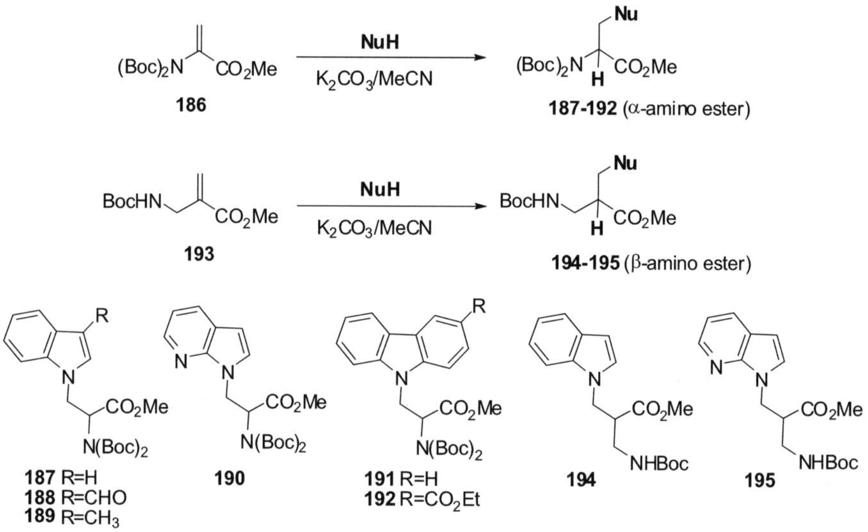

Scheme 44

Silica-supported Lewis acids are useful catalysts with microwave irradiation for conjugate additions. The silica-supported catalysts are obtained by treatment of silica with $ZnCl_2$ [Si(Zn)], Et_2AlCl [Si(Al)] or $TiCl_4$ [Si(Ti)] [150–152]. The Michael addition of methyl α-acetamidoacrylate (**196**) with indole (**2**) under Si(M) heterogeneous catalysis assisted by microwave irradiation afforded the alanine derivative **197** within 15 min and/or bis-indolyl **198**, depending on the reaction conditions (Scheme 45) [153]. While the bisindolyl product **198** is only formed when Si(Zn) was used as catalyst, the alanine derivative **197**, as a single product is formed under thermal heating in a yield of 12%. The best yields were observed with Si(Al) (Table 5). The product **198** was obtained by elimination of acetamide followed by α-Michael addition between intermediate **199** with a second mole of indole.

Tricyclic compounds containing an indole skeleton fused to another cyclic or heterocyclic moiety such as **200–202** are well known and display various pharmacological properties (Scheme 46) [154–160]. 2,3-Dihydro-1,4-benzooxazine derivatives **203–204** are promising scaffolds for pharmacolog-

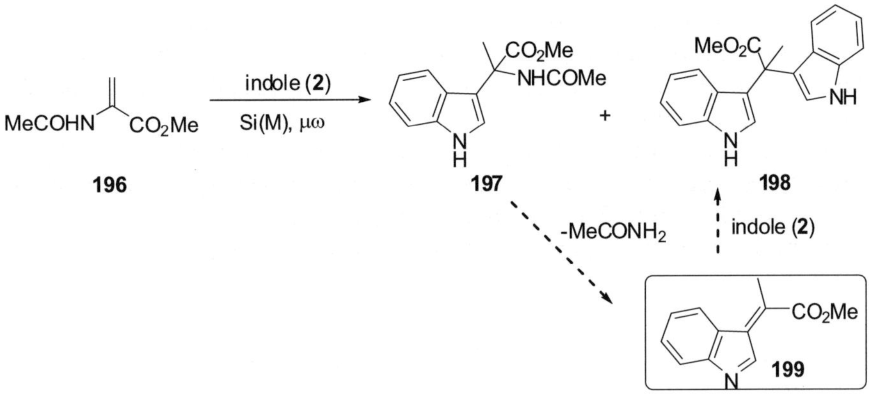

Scheme 45

Table 5 Reactions of indole 2 with **196** using silica-supported Lewis acids

Molar ratio (indole: methyl α-acetamidoacrylate)	Catalyst	Reaction conditions	Product (%)	Yield
1 : 1	Si(Zn)	15 min, 270 W, 80 °C	198	37
1 : 1	Si(Al)	15 min, 150 W, 70 °C	197 + 198	39 + 54
1 : 1	Si(Ti)	15 min, 150 W, 70 °C	197 + 198	18 + 22
1 : 2	Si(Al)	15 min, 90 W, 50 °C	197 + 198	50 + 23
1 : 2	Si(Al)	15 min, 50 °C	198	12

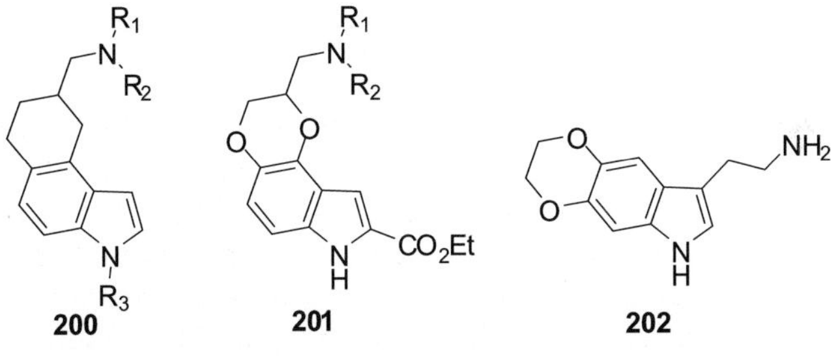

Scheme 46

ically active compounds such as anticancer or central nervous system (CNS) agents. New 1,4-oxazinoindole structures **205–208** were synthesized by the Hemetsberger reaction as depicted in Scheme 47 and the reactivity of linear derivatives was investigated. In order to generate new tetracyclic indole nuclei, aza-Michael addition of *tert*-butyl acrylate to compound **205** in the presence of a catalytic amount of Triton B in *N,N*-dimethylformamide provided the N-alkylated derivative. After acidic hydrolysis of the ester with trifluoroacetic acid in dichloromethane, electrophilic cyclization of acid **209**

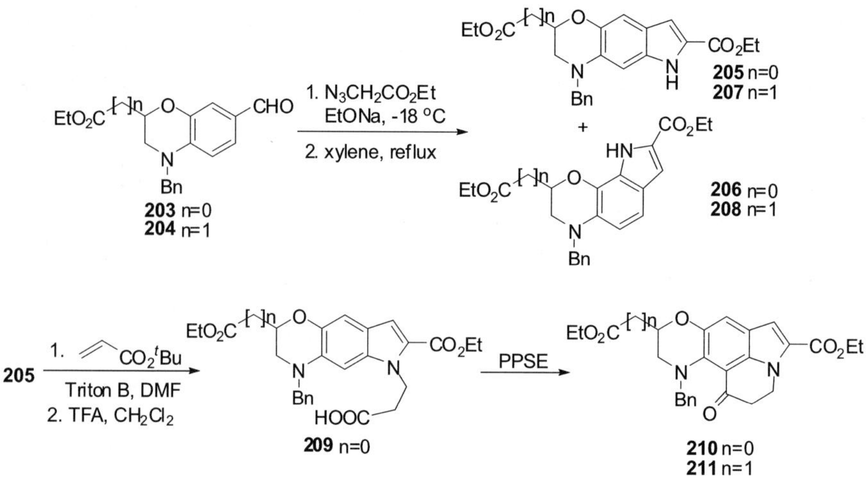

Scheme 47

was carried out in polyphosphoric acid trimethylsilyl ester (PPSE) in refluxing dichloroethane to give tetracyclic ketone **210**. According to the procedure described below, the ketone **211** was synthesized from compound **207**
(Scheme 47) [161].

Dopamine receptors can be classified into two main categories as D_1 and
D_2 receptors [162–164]. SCH 23390 (**212a**) and SKF 38393 (**212b**) are synthetic compounds that act as dopamine D_1 receptor antagonists and are used
as research tools (Scheme 48) [162–165]. As an extension of the peri-fused
congener of phenylazepinoindole derivative **213**, studies on the synthesis
and dopamine receptor binding properties of regioisomeric analogs **214–217**
have been described (Scheme 49) [165–168]. While the key reaction step for
the synthesis of **214** and **215** is Pictet–Spengler cyclization, the crucial step

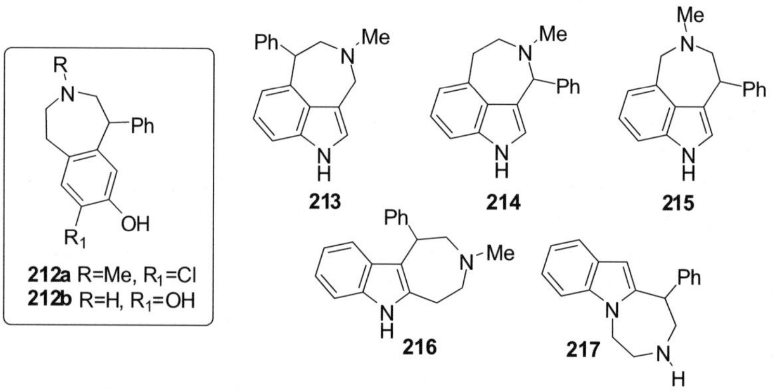

Scheme 48

for compounds **216–217** is the Michael addition reaction [165]. The synthesis of the azepino[4,5-*b*]indole **216** was achieved by starting from the indole-2-acetic acid (**218**). Michael addition product **213** obtained from **218** and β-nitrostyrene was subjected to catalytic hydrogenation using Pd/C in methanol to reduce the nitro group. Under these conditions, ring closure of the intermediate primary amine gave the lactam **220**. Borane reduction of **220** followed by the reaction of formaldehyde in the presence of NaCNBH$_3$ resulted in the formation of the *N*-methyl derivatives **216** and **221**. The starting point for the synthesis of the target compound **217** is the directed metallation of the acetal **222** and Michael addition of the corresponding indolyl lithium intermediate with β-nitrostyrene to give **223**. The compound **217** was obtained by a similar strategy (Scheme 49) [165].

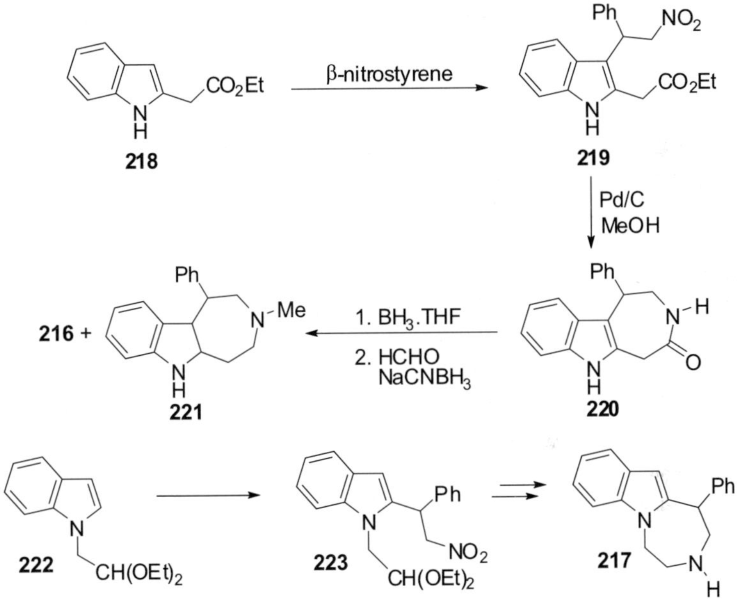

Scheme 49

Bis(indolyl)alkanes and their derivatives constitute an important part of bioactive metabolites both synthetic and natural terrestrial and marine origins [169–174]. Bioactive bis(indolyl)ethylamines **224–226** have recently been reported from marine sources, such as a tunicate and a sponge (Scheme 50) [169–173]. There are a few synthetic methods for synthesis of bis(indolyl)ethylamines [175, 176]. But methods are materializing with either strong reaction conditions or in very low or wide ranging yields.

Recently, the use of 3-(2′-nitrovinyl)indole (**227**) as a Michael acceptor with indoles as nucleophile have been reported [174, 177]. Michael reaction of **227** with eight 3-unsubstituted indoles (**2, 14, 19, 228–235**) on TLC-grade

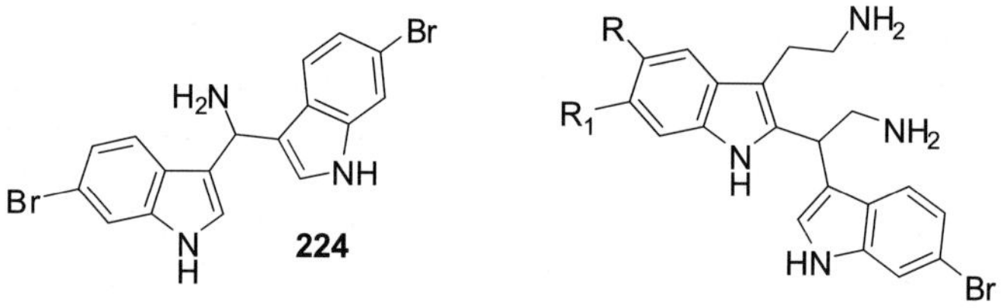

Scheme 50

silica gel, which acts as a mild acidic catalyst, under microwave irradiation or room temperature gave unsymmetrical bis(indolyl)nitroethanes (**236a–h**) in excellent yields (69–86%) (Scheme 51 and Table 6). But, the conjugate addition of the indole derivatives **233–235** with **227** failed. This situation was attributed to the fact that silica gel is not sufficiently acidic to catalyze the desired Michael addition. In order to overcome this difficulty, the reactions of **2**, **14**, **19** and **228–235** with **220** were performed in acetonitrile under reflux using *p*-TsOH (25% wt) as the catalyst. Although all indoles gave the expected unsymmetrical bis(indolyl)nitroethanes, 1-acetylindole (**235**)

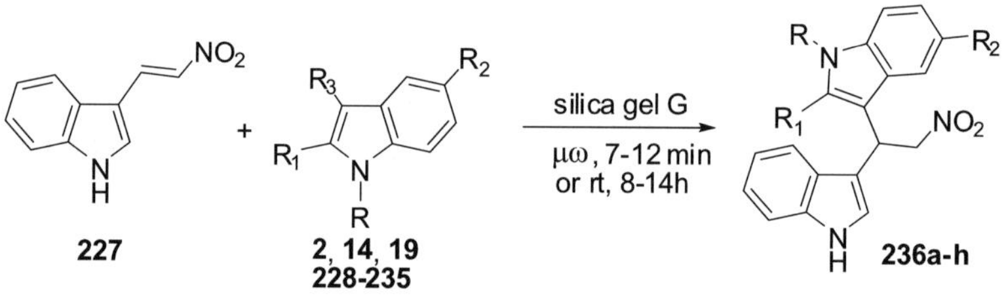

Scheme 51

Table 6 Addition products of Michael acceptor **227**

Indole	R	R_1	R_2	R_3	Product
2	H	H	H	H	236a
14	Me	H	H	H	236b
19	H	Me	H	H	236c
228	Me	H	Br	H	236d
229	Et	H	H	H	236e
230	^{i}Pr	H	H	H	236f
231	H	H	Br	H	236g
232	Me	Me	H	H	236h
233	H	H	H	Me	–
234	Me	H	H	Me	–
235	Ac	H	H	H	–

did not provide the expected product. Interestingly, 1-alkylindoles **14** and **228–230** furnished both the expected products (52–63%) and the symmetrical bis(indolyl)nitroethanes as minor products (20–27%) (Scheme 52) [174, 177]. The formation mechanism for the unexpected symmetrical product **237** is shown in Scheme 51. The corresponding unsymmetrical products for all the reactions are the initial reaction products, which underwent acid-catalyzed elimination of an indole molecule. Michael addition of the formed relevant intermediate cation with 1-alkylindole produced the symmetrical bis(indolyl)nitroethanes as secondary product. The driving force for the formation of these products is the increased electron density arising from 1-alkylation of the indole.

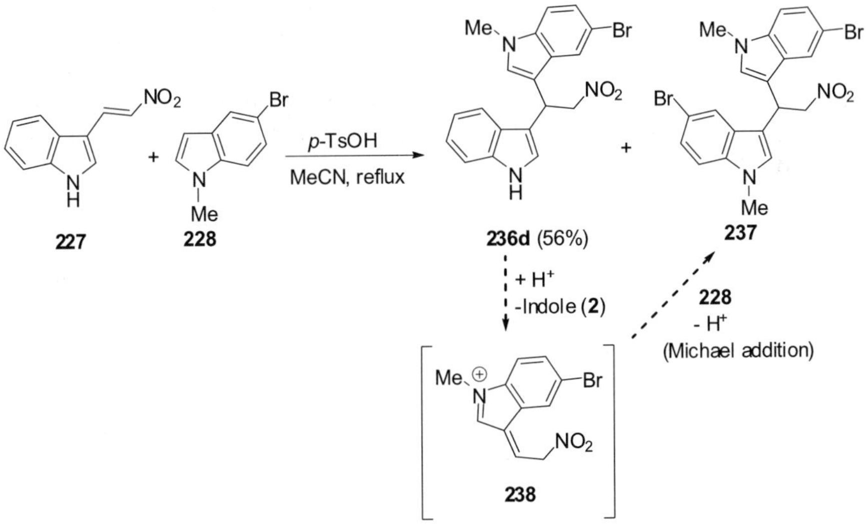

Scheme 52

The methodology described above could be successfully applied for the synthesis of compound **240**, containing the core structure of the marine metabolites **225**, and **226** starting from N-acetyltryptamine (**239**). Furthermore, the symmetrical bis(indolyl)nitroethane **236a**, as a model structure, could be reduced to the corresponding amine, isolated as its N-acetyl derivative-**241** (Scheme 53) [174, 177].

Although β-carbolines, which were mentioned earlier, could be synthesized via tyrptamines, 4-substituted β-carbolines lacking substitution at the 3-position are well represented in the literature. The conjugate additions of indole (**2**) and 5-hydroxyindole (**242**) to methyl β-nitroacrylate (**243**) under the $CeCl_3 \cdot 7H_2O – NaI – SiO_2$ catalyst system are rare examples of the α-Michael reaction for indoles. The five-step synthesis of 4-substituted β-carboline **248** was reported using Michael addition product **244** (Scheme 54) [66, 67].

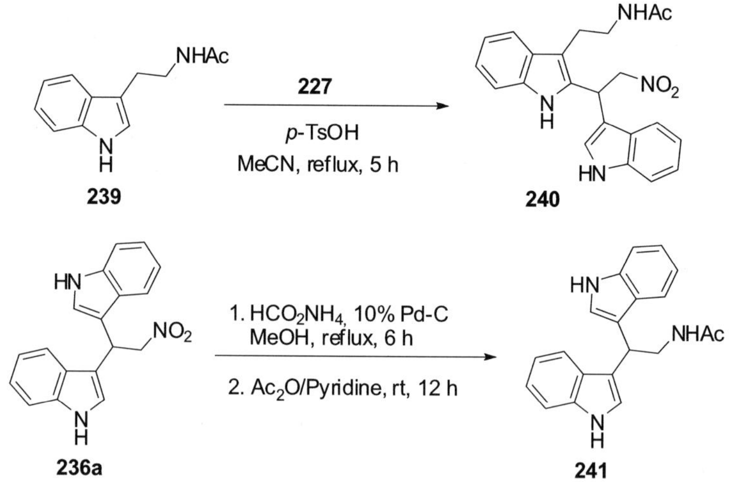

Scheme 53

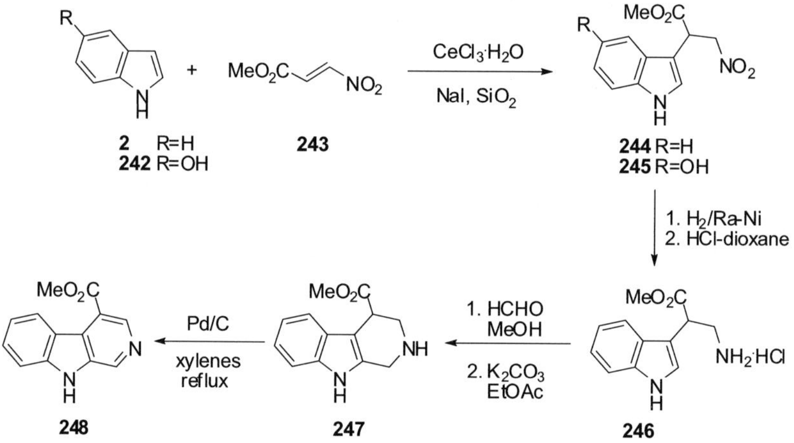

Scheme 54

The ethyl aluminum dichloride-catalyzed Michael alkylations of some indoles with *N*-(diphenylmethylene)-α,β-didehydroamino acid esters allowed successful short synthesis of the tryptophan derivative and the 1,1-diphenyl-β-carboline derivatives, as well as compounds **253** and **252** (Scheme 55) [178].

Due to their interesting complex structure and biological activity, indole alkaloids such as strychnine (**254**), vindorosine (**255**) and vindoline (**256**) have attracted considerable attention. Vindoline is a key component in the preparation of the antitumor drugs vincristine and vinblastine (**10**) [17–19, 179–183]. These alkaloids include the same tetracyclic core **257** known

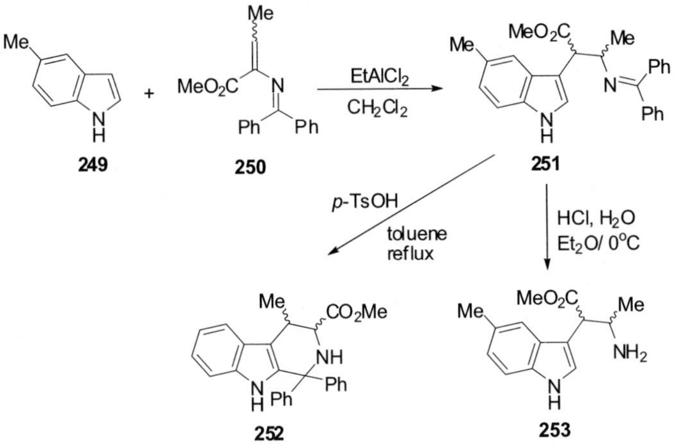

Scheme 55

as Büchi ketone (Scheme 56). As a first step for the synthesis of Büchi ketone **257** [184, 185], tryptamine derivative **259** was efficiently prepared by
coupling *N*-benzyltryptamine (**258**) with (*E*)-4-oxo-2-pentenoic acid [186].

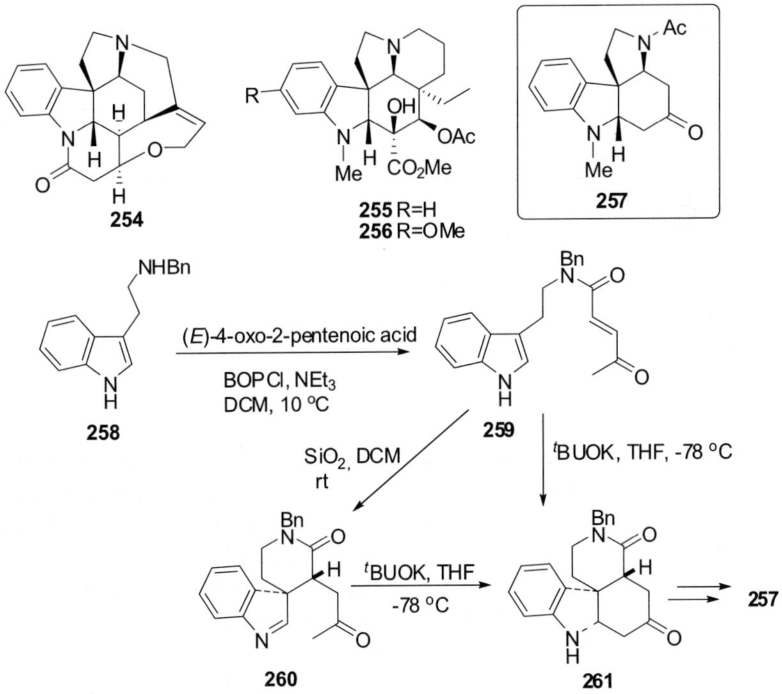

Scheme 56

Treating **259** with ^{t}BuOK in THF at $-78\,^{\circ}$C gave the polycyclic adduct **261** in low yield (14%). Unexpectedly, when **259** was treated overnight with silica gel in CH$_2$Cl$_2$, imino-spiro compound **260** was provided in good yield (79%). The base-catalyzed imino-aldol cyclization of **260** at $-78\,^{\circ}$C afforded tetracyclic compound **261** in excellent yield (85%) and as a single diastereoisomer. The intramolecular Michael addition is the key step for both of these reactions. The synthesis of the Büchi ketone **257** from compound **260** was achieved using an eight-step reaction (Scheme 56) [186].

2.5
Miscellaneous Michael Additions for Indoles

Masked *o*-benzoquinones, as the most accessible type of cyclohexa-2,4-dienones, are of immense synthetic potential [187–193]. At both room temperature and reflux, the dienones **263** generated in situ from commercially available 2-methoxyphenols **262** by adding (diacetoxyiodo)benzene (DAIB, 1 equiv.) at $0\,^{\circ}$C in MeOH underwent unusual Michael addition to indoles (**2**, **19**, **264**) followed by aromatization of the adducts to give highly functionalized 2-arylindoles **265a–c** in excellent yields (Scheme 57) [193].

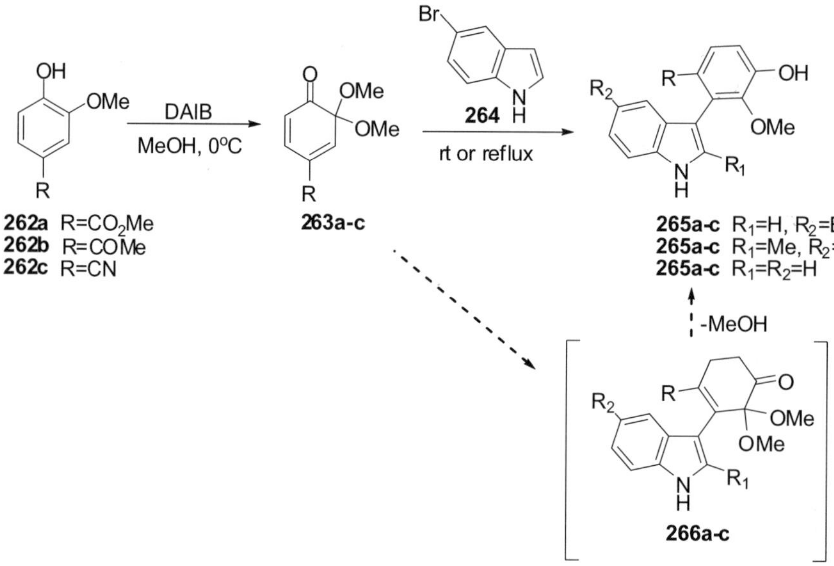

262a R=CO$_2$Me
262b R=COMe
262c R=CN

263a-c

265a-c R$_1$=H, R$_2$=Br
265a-c R$_1$=Me, R$_2$=H
265a-c R$_1$=R$_2$=H

266a-c

Scheme 57

Due to high efficiency and in some cases even enantioselectivity, solid state reactions have recently attracted considerable attention to various types of organic reactions [194–197]. The solid-state Michael reactions between 4-arylidene-3-methyl-1-phenyl-5-pyrazolones **267a–f** and indole gave rea-

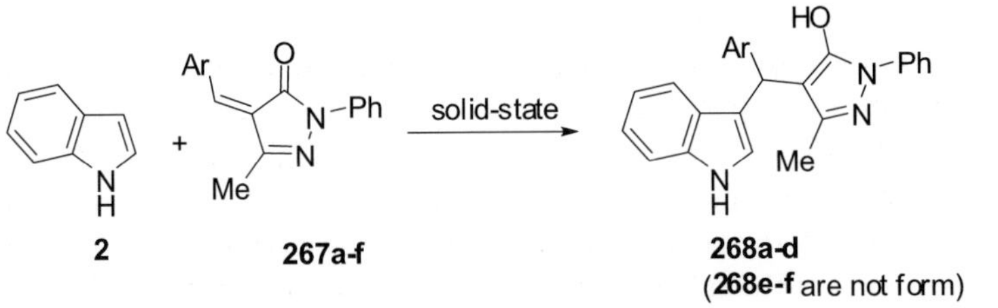

Scheme 58

a: Ar=Ph, b: Ar=4-X-C$_6$H$_4$ (X=F, Cl, Br, I, BzO, Me, NO$_2$)
c: Ar=3-O$_2$NC$_6$H$_4$, d: Ar=4-HO-3-MeOC$_6$H$_4$
e: Ar=4-Y-Ph (Y=Et$_2$N, HO), f: Ar=2-O$_2$NC$_6$H$_4$

sonable yields (40–67%) (Scheme 58). The solid state additions occurred more efficiently than in solution. While the acceptors with the electron-withdrawing substituents on the *para*-position made the addition easier, electron-donating substituents decreased the reactivity due to the increased electron density on the methylidene carbon [197].

Synthesis, structure and conformational behavior of cyclophanes are of interest. Synthesis of a novel chiral cyclophane consisting of indole as one of the core units was achieved [198]. The first bridge of the cyclophane **273** is formed by a conjugate addition of indole (**2**) to the unsaturated ketone **269**. An intramolecular N-alkylation reaction of **271** resulted in the formation of

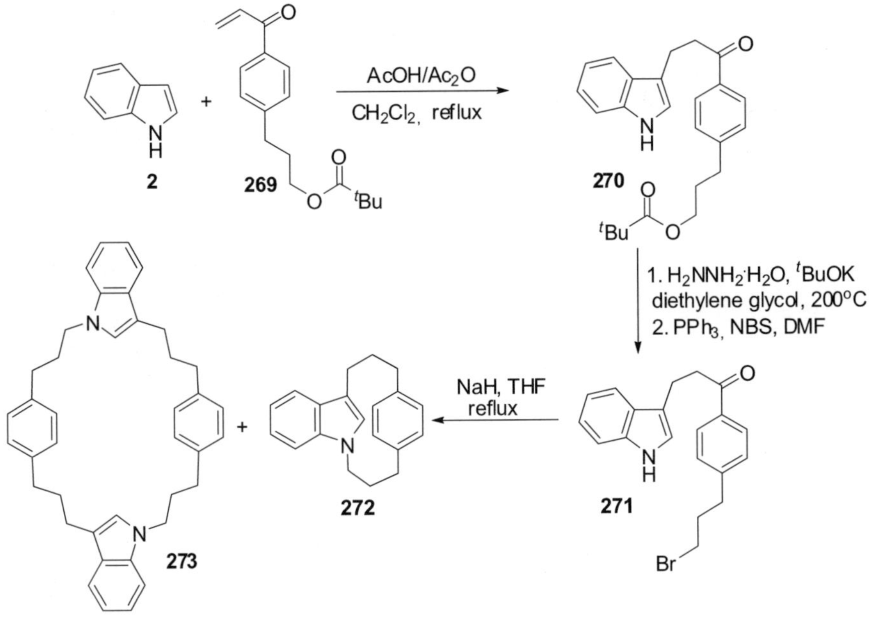

Scheme 59

the second bridge to give the targeted cyclophane **272** (44%). Under the reaction conditions, the cyclic dimer **273** is obtained as a second product in yield of 19% (Scheme 59) [198].

A dynamic NMR study of **272** revealed a ring flipping process that interconverts two enantiomeric sets of equilibrating bridge conformers. However, the conformer search of **272** was carried out at the AM1 level using Spartan. The energy differences among for the five conformers was identified as 5 kcal/mol. While the conformer (C_{chair}, N_{chair}) possessing both bridges in the *pseudo*-chair conformation is the lowest energy. The fifth conformer (tilt-C_{boat}-N_{boat}) is found as the highest energy (Fig. 4). Also, an X-ray analysis of **272** showed that the two aromatic cores are very close [198].

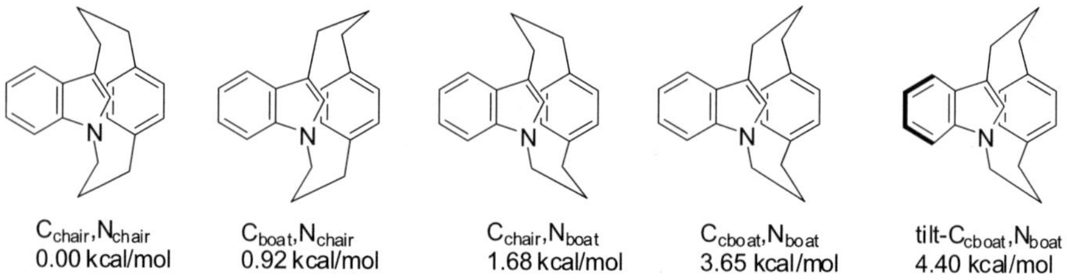

C_{chair}, N_{chair}	C_{boat}, N_{chair}	C_{chair}, N_{boat}	C_{cboat}, N_{boat}	tilt-C_{cboat}, N_{boat}
0.00 kcal/mol	0.92 kcal/mol	1.68 kcal/mol	3.65 kcal/mol	4.40 kcal/mol

Fig. 4 Conformers of **272**

Serotonin (**8**) mimics have been used effectively as treatments for a variety of psychiatric illnesses [199–201]. Homotryptamines could be constructed by multistep routes. One-pot synthesis of homotryptamines **277a–k** was achieved by the conjugate addition of indoles with acrolein in the presence of the MacMillan organocatalysts (**118a, 274**) followed by reductive amination (Scheme 60) [201].

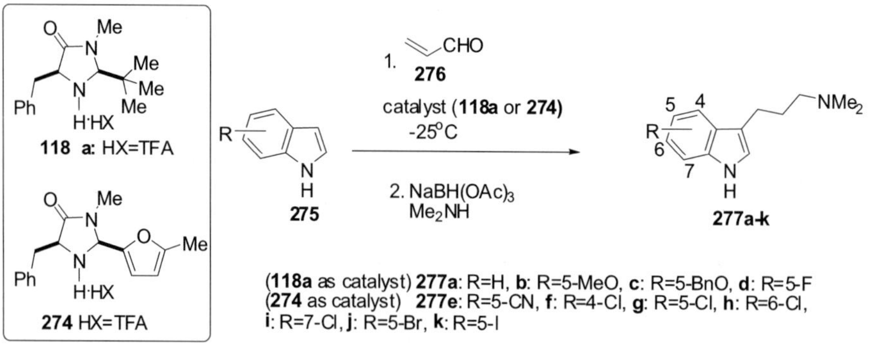

(**118a** as catalyst) **277a**: R=H, **b**: R=5-MeO, **c**: R=5-BnO, **d**: R=5-F
(**274** as catalyst) **277e**: R=5-CN, **f**: R=4-Cl, **g**: R=5-Cl, **h**: R=6-Cl,
i: R=7-Cl, **j**: R=5-Br, **k**: R=5-I

Scheme 60

N-Michael addition of indole to achiral Ni(II) complex **278** led to the synthesis of racemic α-amino acid **280** [202]. The first of the two-step reactions is realized in the presence of NaH in acetonitrile. The last step is the hydrolysis of complex **279** to obtain the desired amino acid (Scheme 61).

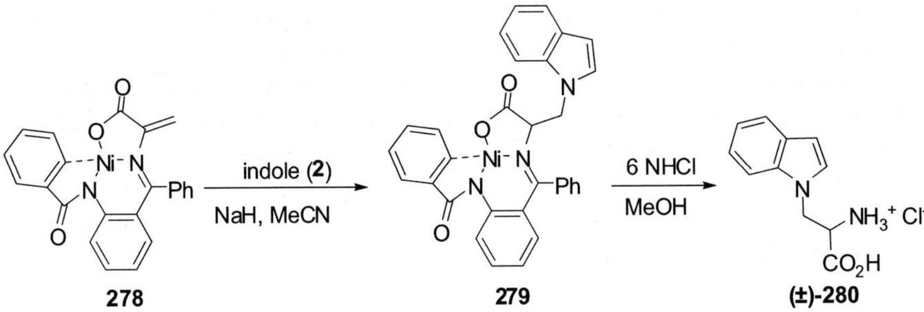
indole (2)
NaH, MeCN
6 NHCl
MeOH
Ph
$NH_3^+ Cl^-$
CO_2H
278
279
(±)-280

Scheme 61

The three-component reaction of indole (**2**) with sugar hydroxyaldehyde **281** and Meldrum's acid **282**, with a catalytic amount of D,L-proline, afforded the 3-substitution product **283** as a single isomer [203]. The substituent possesses the *cis*-fused furo[3,2-*b*]pyranone skeleton. The proline catalyzes the Knoevenagel condensation of the sugar aldehyde **281** and Meldrum's acid **282** to provide the alkylidene derivative **284** of Meldrum's acid. Then a diastereoselective Michael addition of indole and an intramolecular cyclization of this adduct **285** with evolution of carbon dioxide and elimination of acetone furnish the furopyranone in one-pot (Scheme 62).

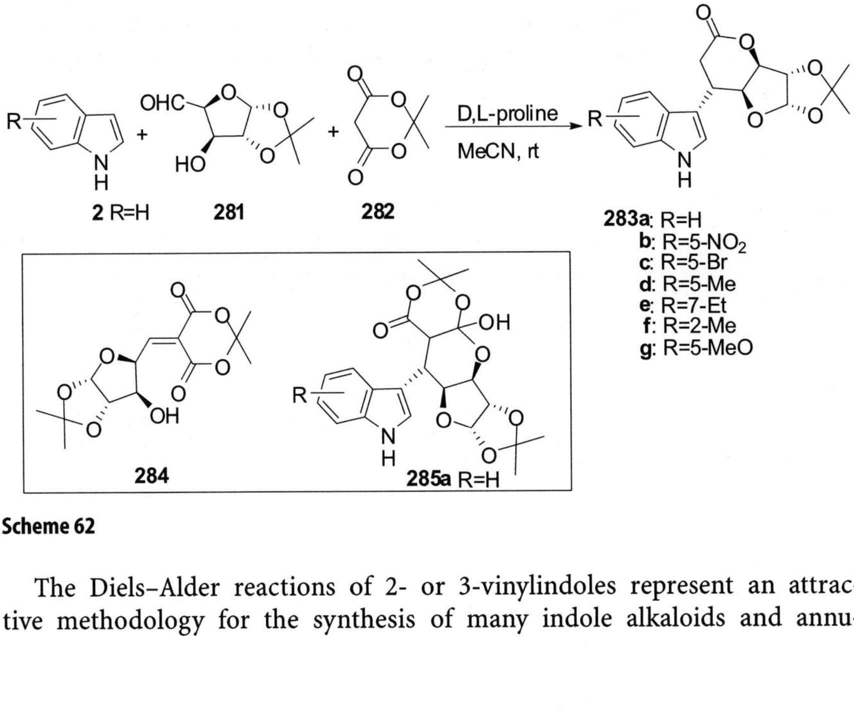
D,L-proline
MeCN, rt
2 R=H
281
282
283a: R=H
b: R=5-NO2
c: R=5-Br
d: R=5-Me
e: R=7-Et
f: R=2-Me
g: R=5-MeO
284
285a R=H

Scheme 62

The Diels–Alder reactions of 2- or 3-vinylindoles represent an attractive methodology for the synthesis of many indole alkaloids and annulated carbazole derivatives [204–217]. Compared with the corresponding 3-substituted compounds, 2-vinylindoles are not easily accessible. Recently, the elegant synthesis of new 2-vinylindole derivatives was described (Scheme 63) [217]. The synthetic approach is based on the dipole change of indole toward electrophilic substitution. Michael reaction of 4,5,6,7-tetrahydroindole (57) with DMAD (169) gave the mixture of addition products 286 and 287 in a 1 : 2.5 ratio. The next step for the 2-vinylindole synthesis is the aromatization of the cyclohexadiene ring in 286 (or 287) by 2,3-dichloro-5,6-dicyano-1,4-benzoquinone (DDQ) in dry benzene for 1 h at room temperature in a high yield. Also, Diels–Alder reactivity of these 2-vinylindoles has been investigated to provide [c]annelated 1,2-dihydro, 1,2,3,4-tetrahydro and fully aromatized carbazoles. At the same time, it is noticed that the Michael addition products (286 and 287), when heated rear-

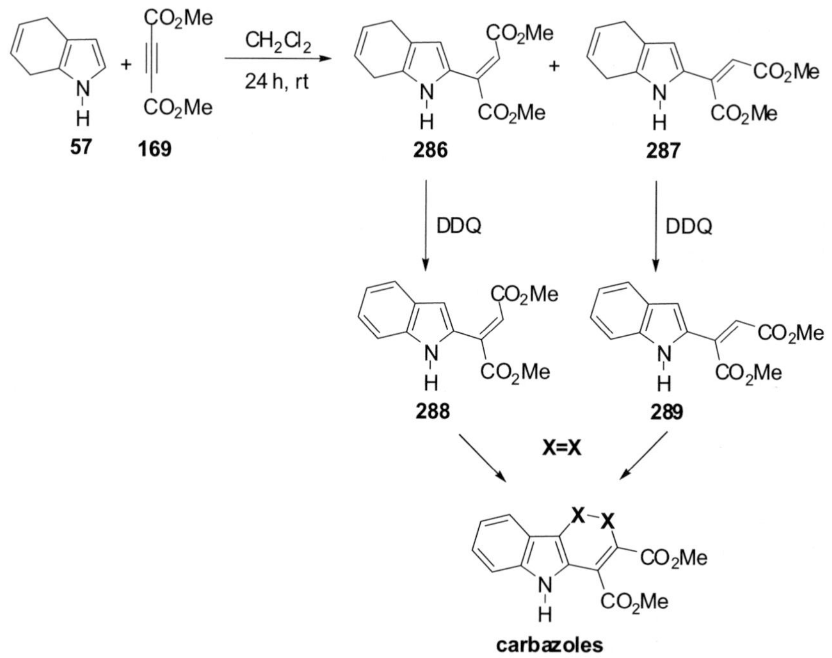

Scheme 63

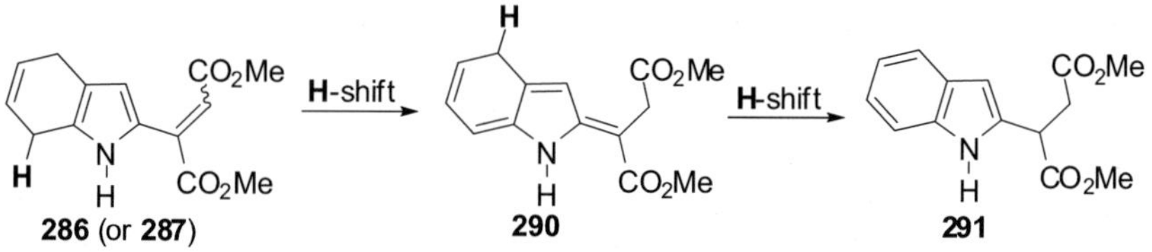

Scheme 64

range to the indole derivative **291** by two sequential hydrogen shifts, as shown in Scheme 64.

3
Michael Addition Applications for Pyrroles

3.1
Reactions Using Catalysts

Generally, indirect methods are used to prepare 2- and 3-alkyl pyrroles, which tend to polymerize under most reaction conditions. Recently, simple and direct methods for the synthesis of 2-alkyl pyrroles were reported via Michael additions catalyzed by various Lewis acids. InCl$_3$ is the first Lewis acid used to give 2-alkylated pyrroles via the conjugate addition with electron-deficient olefins of pyrrole (**1**) without polymerization [218]. The reactions proceeded smoothly at ambient temperature in excellent yields with high selectivity. The electron-deficient olefins, such as alkyl-, aryl- and benzyl vinyl ketone, 2-benzylidenemalononitrile, bis(benzylidene) ketone and β-nitrostyrenes afforded the corresponding 2-alkylated pyrroles (**292–299**) in 65–80% yields (Scheme 65). By increasing the reaction time and changing the molar ratio of reactants, the reaction of unsaturated ketones with pyrroles afforded 2,5-dialkylated products (**300–305**) in high yields.

Also, aluminum dodecyl sulfate trihydrate [Al(DS)$_3$]·3H$_2$O is a mild catalyst for Michael addition reactions of the pyrrole, which is an acid-sensitive compound, producing high yields without any polymerization reaction [70]. A rapid entry to C-alkyl pyrroles was supplied through microwave-assisted addition of pyrroles to electron-deficient olefins using silica gel supported reagent. This method is fast, efficient, environmentally friendly and solvent-free. However, the use of a catalytic-amount of BiCl$_3$ under the same microwave irradiation facilitates both the production of 2-alkylpyrroles from the more hindered electron-deficient olefins and the reaction of 2-alkylpyrroles to give 2,5-dialkylpyrroles such as **306** and **307** (Scheme 66). In fact, 2-alkylpyrroles show much more reactivity than pyrrole in the presence of 5 mole% BiCl$_3$ under MW [219]. Furthermore, BiCl$_3$-catalyzed Michael addition of pyrroles with unsaturated ketones, 2-benzylidenemalononitrile, diethyl 2-benzylidenemalonate and nitroolefins generated the corresponding adducts in high yields [220].

Molecular iodine-promoted Michael addition is a simple and efficient method for generating 2-pyrrolyl-2-phenyl-1-nitroalkanes in good yields (Scheme 67) [86]. Cr^{+3}-Catsan (Cr^{+3} exchanged commercially available montmorillonite clay) and ZnCl$_2$, which were first used as Lewis acids for Michael reactions of pyrrole, showed different selectivity under the same conditions [221]. In general, while the reactions catalyzed by Cr^{+3}-Catsan

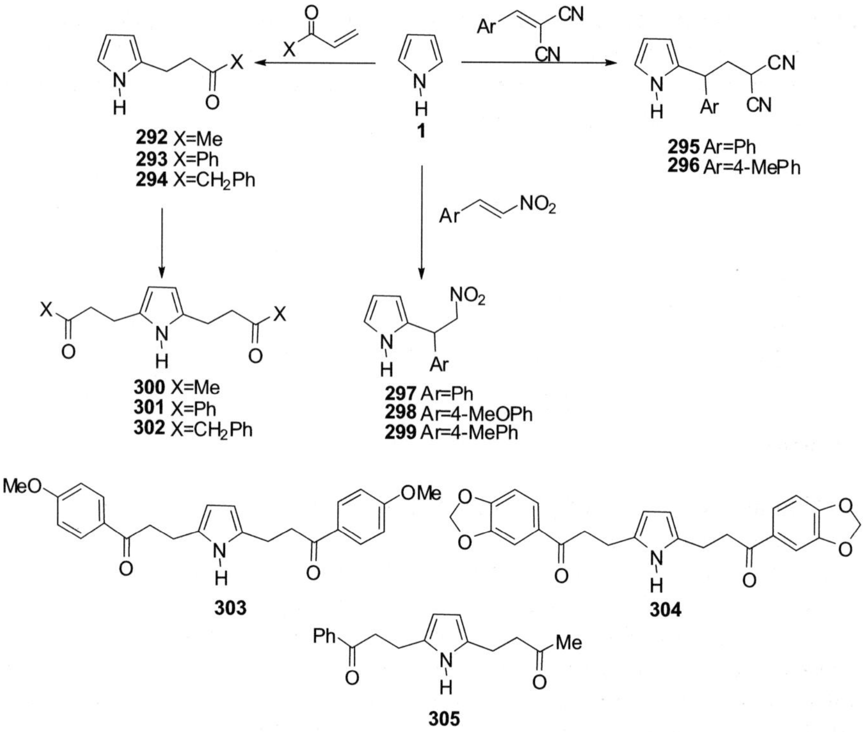

O
X
Ar
CN
CN
N
H
1
X
O
292 X=Me
293 X=Ph
294 X=CH2Ph
CN
Ar
CN
295 Ar=Ph
296 Ar=4-MePh
Ar
NO2
X
X
O
O
300 X=Me
301 X=Ph
302 X=CH2Ph
NO2
Ar
297 Ar=Ph
298 Ar=4-MeOPh
299 Ar=4-MePh
MeO
OMe
303
304
Ph
Me
305

Scheme 65

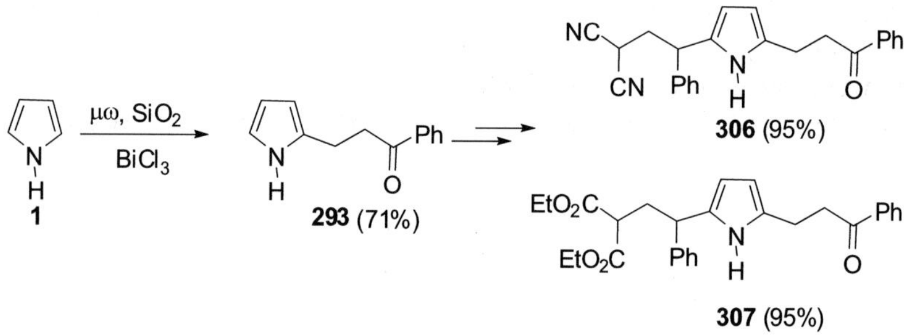

μω, SiO2
BiCl3
1
Ph
293 (71%)
NC
CN
Ph
306 (95%)
EtO2C
EtO2C
307 (95%)

Scheme 66

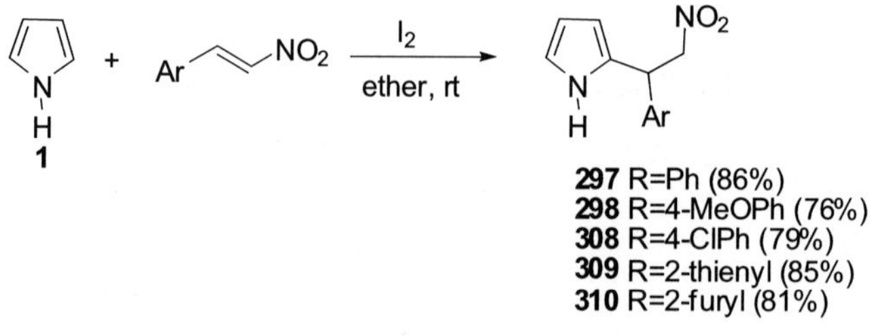

1
+
Ar
NO2
I2
ether, rt
NO2
Ar
297 R=Ph (86%)
298 R=4-MeOPh (76%)
308 R=4-ClPh (79%)
309 R=2-thienyl (85%)
310 R=2-furyl (81%)

Scheme 67

afforded predominantly the 2-alkylpyrrole derivative, $ZnCl_2$ gave the 2,5-dialkylpyrrole derivatives as the main product. $ZrCl_4$ has emerged as a new, safe, economical, air- and moisture-tolerant catalyst for Michael additions (311–314) of 1-(p-anisyl)-2,5-dimethyl-1H-pyrrole (Scheme 68) [79].

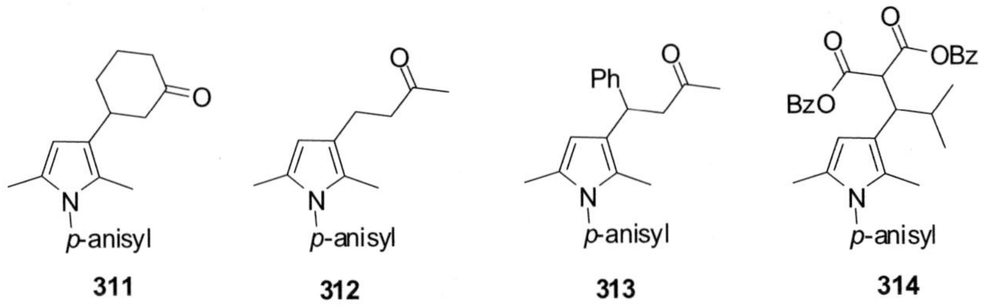

311   **312**   **313**   **314**

Scheme 68

Although N-substituted pyrroles are usually obtained from pyrrolyl anion, this anion exhibits ambident behavior as a nucleophile (Scheme 69) [222–224]. When the pyrrole anion is alkylated, 2- and 3-alkylated pyrroles may form, as well as N-alkylpyrroles. Pyrrole N-alkylation procedures show limited or no interference from C-alkylation. These limited methods do, however, require harsh reaction conditions, long reaction time, use of toxic solvents or catalyst and give low yield. Thus, the development of a mild, efficient and regioselective pyrrole (1) N-alkylation process is still much sought. Recently, the reaction of pyrrole (1) with some olefins such as methyl acrylate, acrylonitrile and methyl vinyl ketone in the ionic liquid, 1-butyl-3-methylimidazolium hexafluorophosphate ([Bmim][PF$_6$]), in the presence of KOH were performed and efficiently produced N-Michael adducts 316–318 of pyrrole as the single isomer. KF/Al$_2$O$_3$ is also a versatile and green cata-

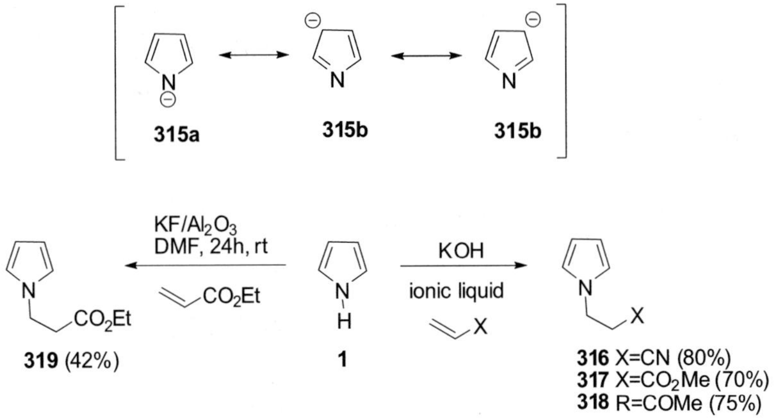

315a   **315b**   **315b**

319 (42%)   **1**   **316** X=CN (80%)
317 X=CO$_2$Me (70%)
318 R=COMe (75%)

Scheme 69

lyst system, which catalyzes the reaction between pyrrole and ethyl acrylate to form aza-Michael adducts **319** (Scheme 69) [225, 226].

3.2
Enantioselective Syntheses

The first example of an enantioselective catalytic conjugate addition for pyrrole was the reaction of *N*-methyl pyrrole with (*E*)-cinnamaldehyde in the presence of a series of benzyl imidazolidinone · HX salts **320** (Scheme 70) [227]. This reaction provided the desired Michael adduct **323** with excellent enantioselectivity (80–93% ee). The highest yield and enantiocontrol is achieved with catalyst **320d** at − 30 °C in 93% ee and 87% yield. The scope of the organocatalyst has been expanded by modifying both pyrrole and phenyl substituent on the olefin. A further illustration of the utility of this organocatalytic conjugate alkylation is the reaction of *N*-methyl pyrrole with excess crotonaladehyde, which controls the alkylation at both C-2 and C-5 of pyrrole. Thus, the synthesis of the 2,5-disubstituted pyrrole **324** is accomplished in 98% ee and 90 : 10 selectivity for the C_2-isomer. The non-symmetrical disubstituted pyrrole **325** could be provided with two variations of unsaturated aldehydes in 99% ee and 90 : 10 *anti*-selectivity.

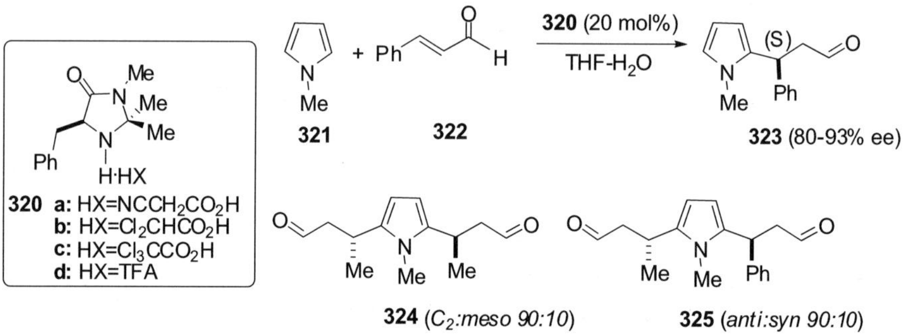

Scheme 70

The reaction of both pyrrole and *N*-methylpyrrole (**321**) with dimethyl *p*-nitrobenzylidene malonate (**326**) in the presence of the catalyst (*S*)-**93** gave the Michael adducts **327–328** in excellent yields (99%) [98], but the enantioselectivity of the products was quite low (28–36% ee), respectively (Scheme 71). Regarding catalyst **93**, the Michael adduct **329** was obtained from *N*-methylpyrrole (**321**) and the alkylidene malonate in moderate yield (62%) and low enantioselectivity 18% ee (Scheme 71) [100]. But, the bis(oxazoline) **93**-catalyzed reaction of both pyrrole (**1**) and *N*-methylpyrrole (**321**) with various α′-hydroxy enones **95** as Michael acceptor worked perfectly (Scheme 71) [99]. The elaboration of these adducts through sequen-

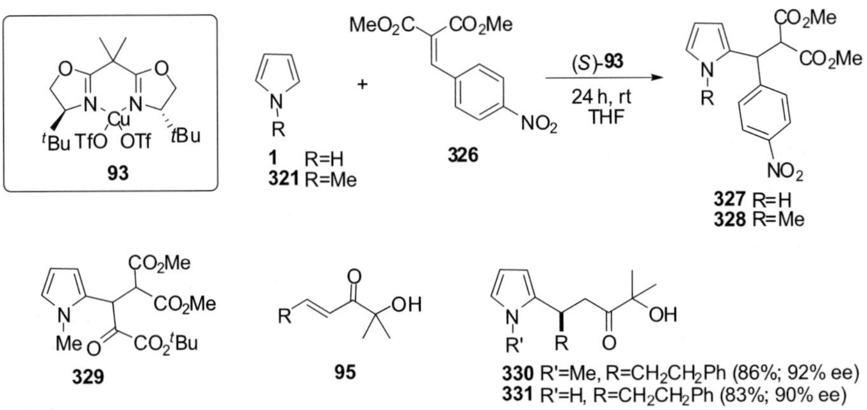

Scheme 71

tial reduction and oxidative diol cleavage afforded the aldehydes **333–334** (Scheme 72) [100]. Enantioselective Michael additions of pyrrole and indoles organocatalyzed by chiral imidazolidinones **118** and **320** were explored with B3LYP/6-31G(d) density functional theory and the enantioselectivities observed with these two catalysts have been explained [228].

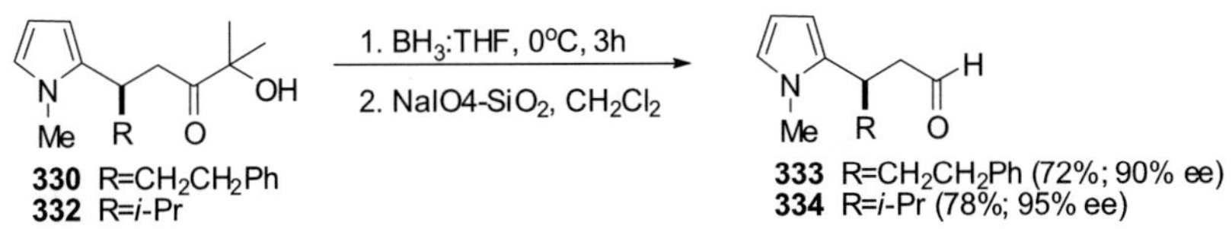

Scheme 72

Michael addition reactions of racemic 2-(arylsulfinyl)-1,4-benzoquinones **335a–b** and enantiomerically pure **335c** with N-(*tert*-butoxycarbonyl)-2-(*tert*-butyldimethylsiloxy)pyrrole (**336**) have been studied under thermal conditions (in CH$_2$Cl$_2$ at – 90 °C) in the presence of ZnBr$_2$, BF$_3$ – OEt$_2$, Eu(fod)$_3$ and SnCl$_4$ as the catalysts [229]. Under ZnBr$_2$ and BF$_3$ – OEt$_2$, the reactions completely afforded the Michael adducts **337** and **338**, whereas in the presence of Eu(fod)$_3$ and SnCl$_4$ the pyrrole[3,2-*b*]benzofuranes **339–340** were obtained from a tandem process involving the intramolecular Michael reaction by a cyclization of intermediate **337** and **338** in quantitative yields. The best results were obtained for the stereoselective formation of **337c** (100% de; 67% yield) by using BF$_3$ – OEt$_2$ as the catalyst and the acceptor **335c**. Although the reactions of **335b** and **335c** with Eu(fod)$_3$ directly yielded the pyrrole[3,2-*b*]benzofuranes **339b** (86% de; 70%) and **339c** (80% de; 40%) in a highly stereoselective manner, the mixture of the diastereoisomers **339** and **340** were obtained under SnCl$_4$ (Scheme 73).

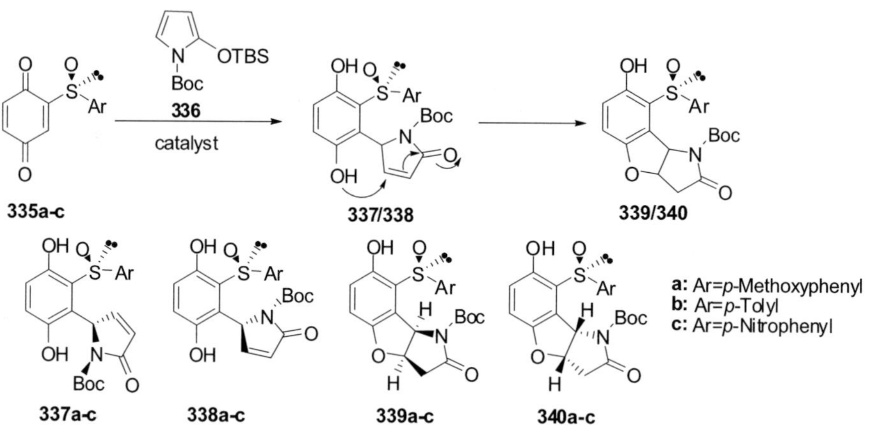

Scheme 73

3.3
Synthesis of Natural Products

Peramine (352) is a pyrrole alkaloid identified as a major insect feeding deterrent isolated from perennial ryegrass infected with the entophyte *Acremonium lolii*. Due to the interesting heterocyclic ring system and biological activity, the synthesis of peramine is attractive [230].

For the synthesis of peramine (352), the key step is the Michael addition of the potassium salt of the nitroolefin 341 and methyl pyrrole-2-carboxylate (342) [231]. After the nitro-group of Michael adduct 343 was reduced using NaBH$_4$ – CoCl$_2$, the amine 344 cyclized to the lactam 345 under reflux in toluene in 88% yield. The elimination of the ethanol with excess of potassium hydride in THF at room temperature gave the unsaturated lactam 346. After N-methylation, the amine compound 351 was synthesized using a four-step reaction. The amine 351 was converted to peramine (352) with excess of S-methylthiouronium hydrogen sulfate in NaOH (2 M) at room temperature for 48 h (Scheme 74).

Many bromopyrrole alkaloids displaying interesting biological properties have been isolated from various marine organisms [232]. Longamide (353) was obtained from the Caribbean sponge *Agelas longissima* and a *Homoxinella* species [233]. Longamide B methyl ester (356) exhibits weak cytotoxic activity in vitro against some leukemia cell lines [233]. While longamide B (357) shows modest antibiotic activity against several strains of gram-positive bacteria, hanishin (358) is cytotoxic towards human non-small-cell lung carcinoma (NSCLC) [234]. Starting from pyrrole, (±)-longamide (353) was synthesized in equilibrium with aldehyde 354 [235]. The reaction of longamide (353) with the sodium salt of methyl diethylphosphonoacetate afforded longamide B methyl ester (356) upon a Wadsworth–Horner–Emmons olefination and a facile in-

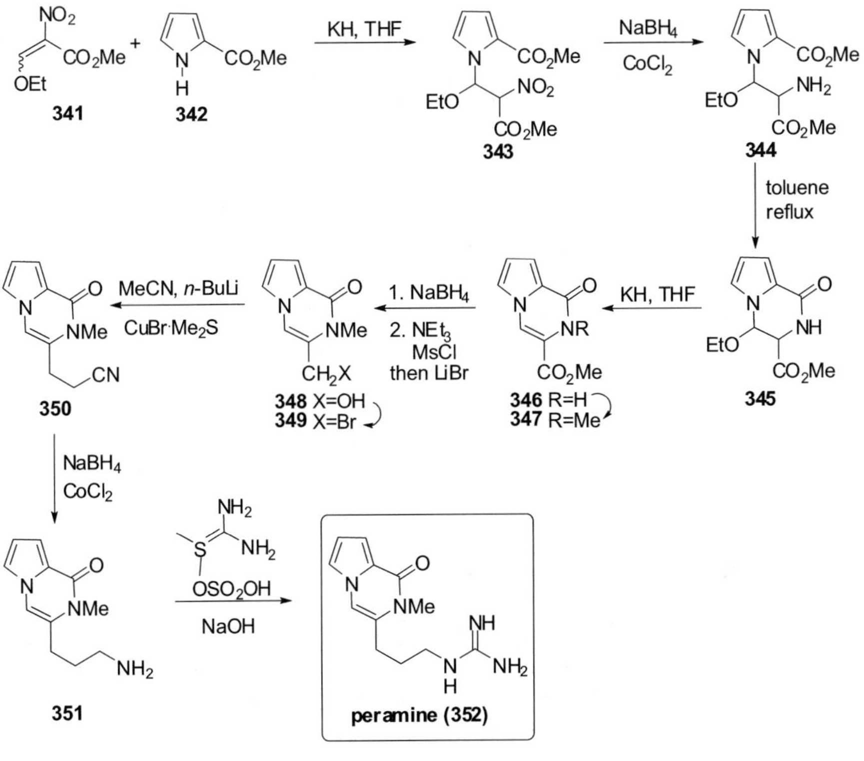

Scheme 74

tramolecular Michael addition, respectively. While the saponification of ester **356** provided longamide B (**357**), the racemic form of hanishin (±)-**358** is readily prepared by reacting acid **357** with acidic ethanol (Scheme 75).

The *Aspidosperma* family of indole alkaloids has inspired many synthetic strategies for the construction of their pentacyclic framework of the parent compound aspidospermidine (**366**), since the initial clinical success of two derivatives, vinblastine (**10**) and vincristine, as anticancer agents. The alkaloids such as (–)-rhazinal (**369**) and (–)-rhazinilam (**6**) have been identified as novel leads for the development of new generation anticancer agents [10, 11]. Bis-lactams (–)-leucunolam (**370**) and (+)-*epi*-leucunolam (**371**) have biogenetic and structural relationships with these compounds [236]. Recently, enantioselective or racemic total syntheses of some of the these natural product were achieved. One successful synthesis was the preparation of the tricyclic ketone **365**, an advanced intermediate in the synthesis of aspidospermidine (**366**), from pyrrole (**1**) (Scheme 76) [14]. The key step is the construction of the indolizidine **360**, which represents the first example of the equivalent intramolecular Michael addition process [14, 237, 238]. The DIBAL-H mediated reduction product was subject to mesylation under the Crossland–

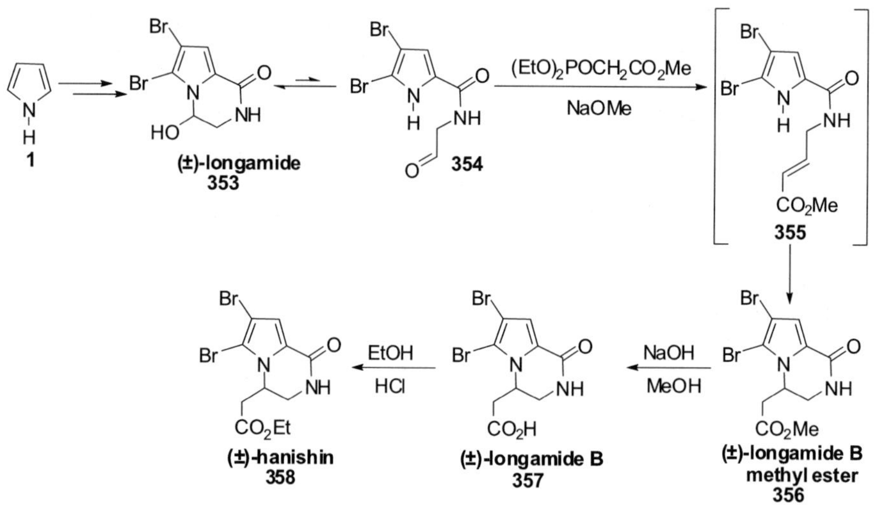

Scheme 75

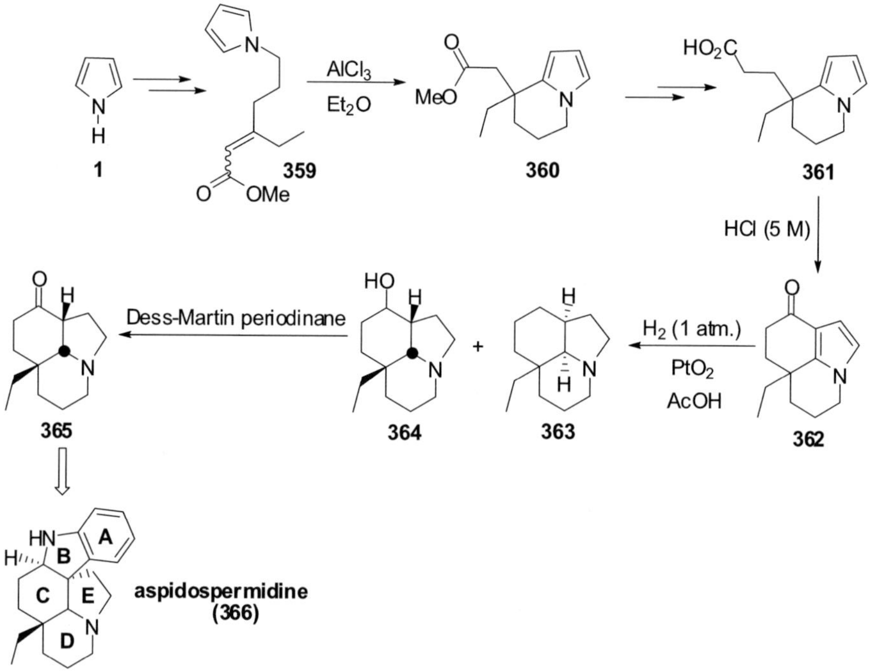

Scheme 76

Servis conditions. The resulting mesylate reacted with excess sodium cyanide in N,N'-dimethylproplenurea (DPMU) to give the nitrile. The intramolecular Friedel–Crafts type cyclization of the hydrolysis product acid **361** provided the ketone **362**, which incorporates the CDE-ring substructure associated with aspidospermidine (**366**). The hydrogenation of the ketone **362** in the presence of PtO_2 at 18 °C in acetic acid gave the mixture (1 : 1 ratio) of the full-reduction product **363** and the alcohol **364**. The oxidation of this mixture with Dess–Martin periodinane gave the known ketone **365** (28% from **362**), which could be separated from the saturated amine **363** by flash chromatography. The conversion of the ketone **365** to (±)-aspidospermidine (**366**) in one-step was a known process beforehand.

The synthesis of the enantiomerically enriched (74% ee) tetrahydroindolizine **368** is the most crucial step for synthesis of (−)-rhazinal (**369**), (−)-rhazinilam (**6**), (−)-leucunolam (**370**) and (+)-*epi*-leucunolam (**371**) alkaloids [239]. The selective intramolecular conjugate additions of pyrrole to N-tethered Michael acceptors were achieved by using chiral organocatalyst **320c** (Scheme 77).

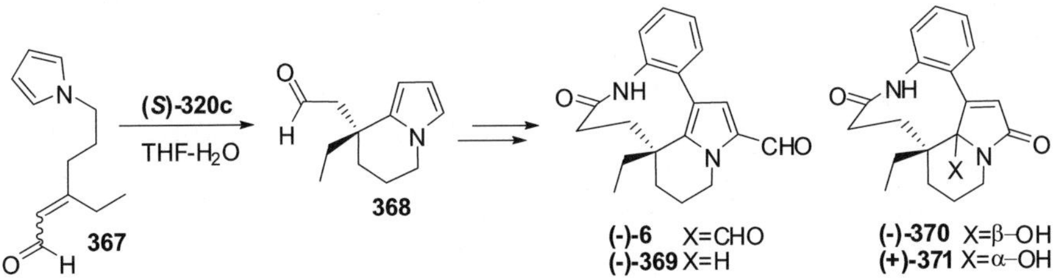

Scheme 77

Enantioselective Michael additions of pyrroles to α,β-unsaturated 2-acyl imidazoles were accomplished by Sc(III)triflate-bis(oxazolinyl)pyridine com-

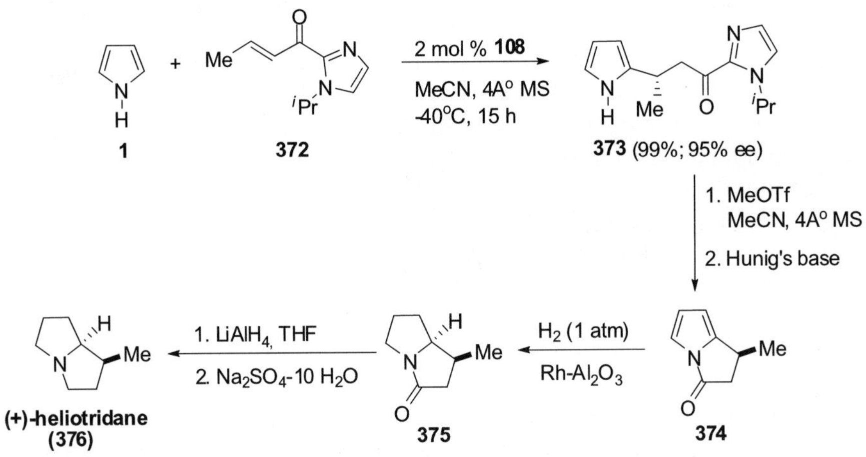

Scheme 78

plex **108** catalysis in optimal yield (98%) and excellent enantioselectivity (86–95% ee) [105, 106]. The 2-acyl imidazole moiety could be transformed into a range of carbonyl derivatives. This advantageous has been applied to a short and enantioselective synthesis of the alkaloid (+)-heliotridane (**376**). One-pot methylation and cyclization of Michael addition product **373** to 2,3-dihydro-1*H*-pyrrolizine **374** was performed by using a slight excess (1.1 equiv.) of methyl triflate and DMAP or *N,N*-diisopropylethylamine (Hunig's base), respectively (Scheme 78). The synthesis of (+)-heliotridane (**376**) was completed with the hydrogenation of **374** with Rh-Al$_2$O$_3$ to afford hexhydro-pyrrolizin-3-one **374** in a quantitative yield and subsequent LiAlH$_4$ reduction [106].

3.4
Miscellaneous Michael Additions for Pyrroles

The synthesis of β-substituted pyrroles from a pyrrolic precursor is quite difficult, but, there are some primary methods to synthesize β-substituted pyrroles [240]. For example, a removable electron-withdrawing group at the α-position or placement of a bulky substituent on the nitrogen orient electrophilic addition to the β-position. Another one is isomerization of an α-substituted pyrrole to the corresponding β-substituted pyrrole. In an alternative route, pentaammineosmium η^2-pyrrole [Os(NH$_3$)$_5$(4,5-η^2-pyrrole)] complexes undergo alkylation with Michael acceptors predominantly at the β-position [241, 242]. Depending on pyrrole, electrophile, solvent, temperature, the presence of Lewis acids and concentration, the resulting products are either Michael addition or 1,3-dipolar cycloaddition. The pyrrole complexes ([Os(NH$_3$)$_5$(4,5-η^2-pyrrole)](OTf)$_2$) are prepared from various pyrrole and [Os(NH$_3$)$_5$(OTf)](OTf)$_2$ starting materials in DMAc or in a cosolvent mixture of DMAc (*N,N*-dimethyl acetamide) and DME (1,2-dimethoxyethane). Reaction of the 1-methylpyrrole complex **377** with 1 equiv. of methyl vinyl ketone (MVK) (or acrolein) in methanol gave β-alkylation products **378** (or **379**) in high yields. Treatment of pyrrole complexes **380** and **381** with 2 equiv. of MVK afforded the corresponding 1,3-dialkylated 1*H*-pyrrole complexes **382**–**383**. The conjugate addition of 2,5-dimethylpyrrole complex **384** with MVK in acetonitrile gave a 1 : 1 ratio of β-alkylated 3H-tautomer **385** along with a second alkylation product. While the reactions of pyrrole **380** and 1-methyl pyrrole **377** complexes with 3-butyn-2-one in methanol underwent conjugate addition to form products **386** and **387**, the reaction of the alkyne with **377** in DMSO gave a new product **388**, which follows a retro-Mannich reaction to generate **386** quantitatively when treated with a protic solvent or undried reagent grade acetone. When the 2,5-dimethylpyrrole complex **384** reacted with MVK in acetonitrile, two addition products **390** and **385** were isolated as a 1 : 1 mixture. When the reaction is monitored in CD$_3$CN, the compound **390** has been observed as arising from a retro-Mannich reaction

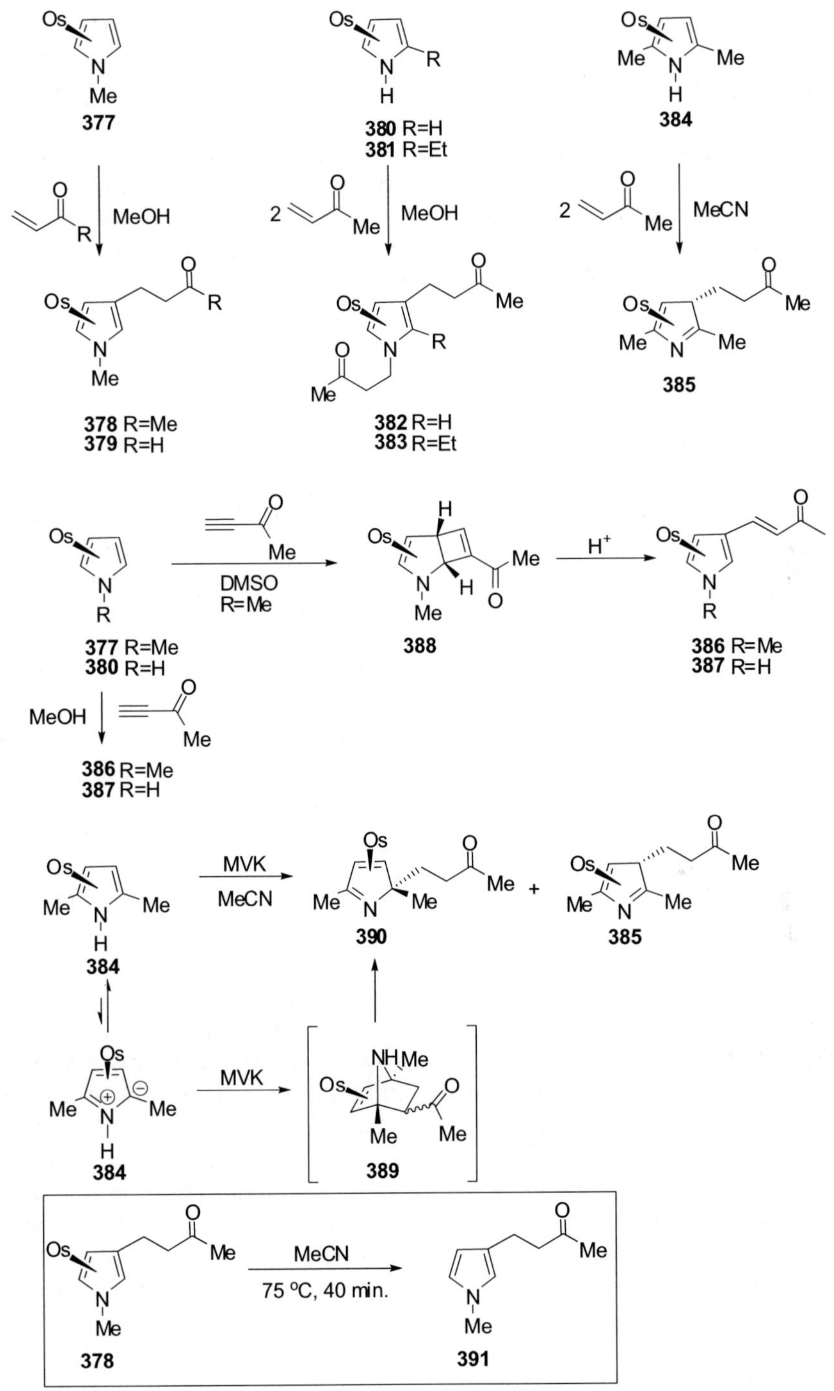

Os
N
Me
377
Os
R
N
H
380 R=H
381 R=Et
Os
Me
N
Me
H
384
O
R
MeOH
2
O
Me
MeOH
2
O
Me
MeCN
Os
O
R
N
Me
378 R=Me
379 R=H
Os
O
Me
O
N
R
Me
382 R=H
383 R=Et
Os
O
Me
Me
N
Me
385
Os
N
R
377 R=Me
380 R=H
O
Me
DMSO
R=Me
H
Os
Me
N
H
O
Me
388
H+
Os
O
N
R
386 R=Me
387 R=H
MeOH
O
Me
386 R=Me
387 R=H
Os
Me
N
Me
H
384
MVK
MeCN
Os
O
Me
N
Me
Me
390
+
Os
O
Me
Me
N
Me
385
Os
Me
N
Me
H
384
MVK
NH Me
Os
O
Me
Me
389
Os
O
Me
N
Me
378
MeCN
75 °C, 40 min.
O
Me
N
Me
391

Scheme 79

of 1,3-dipolar cycloaddition product **389**, which has been characterized as a 7-azabicyclo[2.2.1]heptane complex, followed by proton transfer. Decomplexation of Michael adduct **378** to give **391** could be achieved in 77% yield by heating (Scheme 79).

N-Bonded pyrrolyl complex **392** is prepared from the reaction of $(PMe_2Ph)_3ReCl_3$ with excess pyrrolyllithium in ether at room temperature [243]. The reaction of **392** with DMAD (**169**) is the first Michael type reaction observed for the N-bonded pyrrole complex. The reaction proceeds at room temperature in the presence of a catalytic amount of acetic acid in toluene for 3 days (Scheme 80).

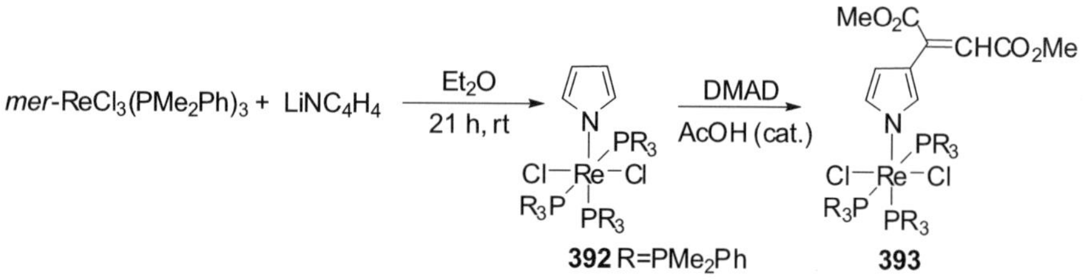

$$mer\text{-}ReCl_3(PMe_2Ph)_3 + LiNC_4H_4 \xrightarrow[21\,h,\,rt]{Et_2O} \quad \xrightarrow[AcOH\,(cat.)]{DMAD}$$

392 R=PMe$_2$Ph   **393**

Scheme 80

Pyrrole-substituted β-amino ester **394** was synthesized via aza-Michael conjugate addition (Scheme 81) [149].

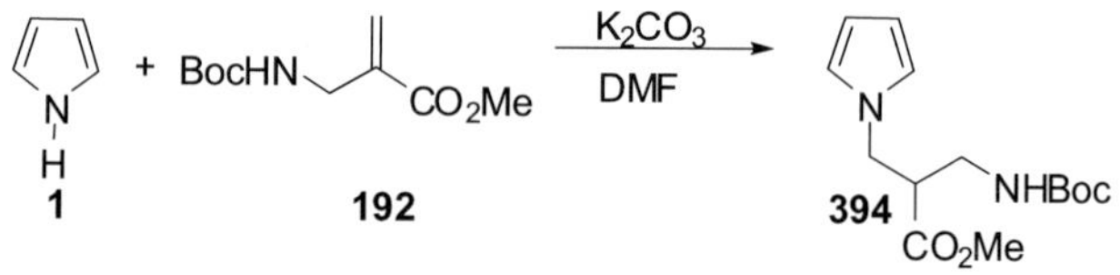

$$\xrightarrow[DMF]{K_2CO_3}$$

1   **192**   **394**

Scheme 81

The Michael addition of methyl α-acetamidoacrylate (**196**) with pyrrole (**1**) under silica-supported Lewis acid (Si(M) : Si(Zn), Si(Al) and Si(Ti)) assisted by microwave irradiation (MW) afforded the alanine derivatives **395** and **396** dependent on the reaction conditions (Scheme 81) [153]. Both MW and thermal activation for pyrrole gave only Michael product **396**, whereas alanine derivatives **395**, which are the α-Michael addition product, and **396** were observed with Al and Ti-catalyst. This behavior shows that aluminium and titanium Lewis acids can form a new acceptor in an irreversible way. The Si(M) or p-TsOH catalyzed reactions of N-benzylpyrrole **397** with the acrylate **196** under MW gave the product **398** as sole product. The reaction yield has been increased by using a catalytic amount of p-TsOH (Scheme 82).

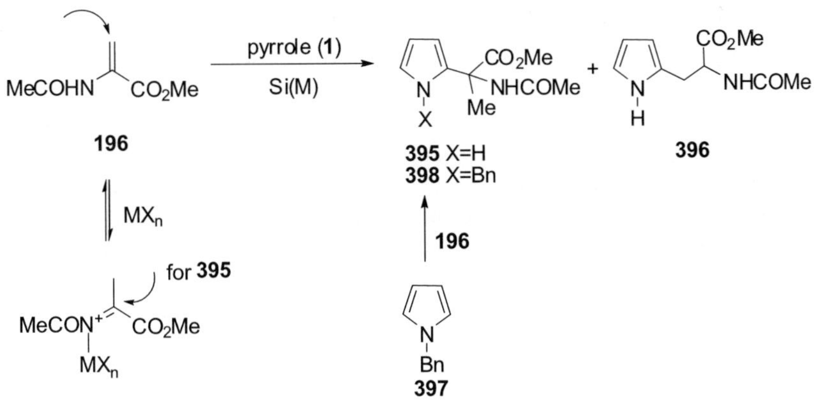

Scheme 82

4
Conclusion

The indole and pyrrole rings are incorporated into many biologically active molecules. Therefore, the functionalization of indole and pyrrole cores via Michael-type additions has been discussed. This chapter especially focuses on studies of the last 10 years on catalyst systems, enantioselective synthesis and the design of natural products or biological active molecules as related to Michael additions of indole and pyrrole.

References

1. O'Hagan D (2000) Nat Prod Rep 17:435
2. Black D StC (1996) In: Bird CW (ed) Comprehensive heterocyclic chemistry II, vol 2. Elsevier, Oxford UK, p 39
3. Banwell M, Goodwin TE, Ng S, Smith JA, Wong DJ (2006) Eur J Org Chem 3043
4. Hesse M (1974) Indolalkaloide. Verlag Chemie, Weinheim
5. Sundberg RJ (1970) In: Blomquist AT (ed) The chemistry of indoles. Academic, New York
6. Remers WA (1973) In: Houlihan WJ (ed) The chemistry of heterocyclic compounds, vol 25, chap 1. Wiley, New York
7. Bosch J, Luisa Bennasar M-L, Amat M (1996) Pure Appl Chem 68:557
8. Hibino S, Choshi T (2001) Nat Prod Rep 18:66
9. Knölker H-J, Reddy KR (2002) Chem Rev 102:4303
10. Banwell M, Edwards A, Smith J, Hamel E, Verdier-Pinard P (2000) J Chem Soc Perkin Trans I p 1497
11. Baudoin O, Cesario M, Guenard D, Gueritte F (2002) J Org Chem 67:1199
12. Abraham DJ, Rosenstein RD (1972) Tetrahedron Lett 13:909
13. De Silva KT, Ratcliffe AH, Smith GF, Smith GN (1972) Tetrahedron Lett 13:913
14. Banwell M, Smith J (2002) J Chem Soc Perkin Trans I, p 2613

15. Joule JA, Mills K, Smith GF (eds) (1995) Heterocyclic chemistry, 3rd edn. Chapman and Hall, London
16. Speeter ME, Anthony WC (1954) J Am Chem Soc 76:6208
17. Noble RL, Beer CT, Cutts JH (1958) Ann NY Acad Sci 76:882
18. Svoboda GH, Neuss N, Gorman MJ (1959) Am Pharm Assoc Sci Ed 48:659
19. Yokoshima S, Ueda T, Kobayashi S, Sato A, Kuboyama T, Tokuyama H, Fukuyama T (2003) Pure Appl Chem 75:29
20. Chen ZH, Muller P, Swager TM (2006) Org Lett 8:273
21. Sha C-K (1996) Adv Nitrogen Heterocycl 2:147
22. LeHoullier CS, Gribble GW (1983) J Org Chem 48:2364
23. Bonnett R, North SA (1981) Adv Heterocycl Chem 29:341
24. Chen Y-L, Lee M-H, Wong W-Y, Lee AWM (2006) Synlett, p 2510
25. Fleming I (1976) In: Frontier orbitals and organic chemical reactions. Wiley, New York, p 58
26. Fukui K, Yonezawa T, Nagata C, Shingu H (1954) J Chem Phys 22:1433
27. Bandini M, Melloni A, Umani-Ronchi A (2004) Angew Chem Int Ed 43:550
28. Bandini M, Melloni A, Tommasi S, Umani-Ronchi A (2005) Synlett, p 1199
29. Trost BM, Fleming I (eds) (1991) Comprehensive organic synthesis, vol 4. Pergamon, Oxford
30. Perlmutter P (1992) Conjugate addition reactions in organic synthesis. Pergamon, Oxford
31. Mukaiyama T, Kobayashi S (1994) Org React 46:1
32. Little RD, Masjedizadeh MR, Wallquist O, McLoughlin JI (1995) Org React 47:315
33. Johnson JS, Evan DA (2000) Acc Chem Res 33:325
34. Christoffers J (2001) Synlett, p 723
35. Leonard J (1994) Contemp Org Synth 1:387
36. Rossiter BE, Swingle NM (1992) Chem Rev 92:771
37. Lewandowska E (2006) Tetrahedron 63:2107
38. Wabnitz TC, Yu Q-J, Spencer JB (2004) Chem Eur J 10:484
39. Harrington PE, Kerr MA (1998) Can J Chem 76:1256
40. Harrington PE, Kerr MA (1996) Synlett, p 1047
41. Loh TP, Wei LL (1998) Synlett, p 975
42. Loh TP, Pei J, Lin M (1996) Chem Commun, p 2315
43. Babu G, Perumal PT (2000) Aldrichimica Acta 33:16
44. Yadav JS, Abraham S, Reddy BVS, Sabitha G (2001) Synthesis, p 2165
45. Bandini M, Melchiorre P, Melloni A, Umani-Ronchi A (2002) Synthesis, p 1110
46. Cox ED, Cook JM (1995) Chem Rev 95:1797
47. Cox ED, Diaz-Harauzo H, Huang Q, Reddy MS, Ma C, Harris B, McKernan R, Skolnick P, Cook JM (1998) J Med Chem 41:2537
48. Mobilio D, Humber LG, Katz AH, Demerson CA, Hughes P, Brigance R, Conway K, Shah U, Williams G, Labbadia F, De Lange B, Asselin A, Schmid J, Newburger J, Jensen NP, Weichman BM, Chau T, Neuman G, Wood DD, Van Engen D, Taylor N (1988) J Med Chem 31:2211
49. Humber LG (1987) Med Res Rev 7:1
50. Agnusdei M, Bandini M, Melloni A, Umani-Ronchi A (2003) J Org Chem 68:7126
51. Zhao S, Liao X, Cook JM (2002) Org Lett 4:687
52. Faulkner DJ (2001) J Nat Prod Rep 18:1
53. Miyake FY, Yakushijin K, Horne DA (2002) Org Lett 4:941
54. Jiang B, Yang C-G, Wang J (2002) J Org Chem 67:1396
55. Bandini M, Fagioli M, Melloni A, Umani-Ronchi A (2003) Synthesis, p 397

56. Bandini M, Cozzi PG, Giacomini M, Melchiorre P, Selva S, Umani-Ronchi A (2002) J Org Chem 67:3700
57. Arai K, Yamamoto Y (1990) Chem Pharm Bull 38:2929
58. Kaji A, Saito R, Nomura M, Miyamoto K, Kiryama N (1998) Biol Pharm Bull 21:945
59. Kaji A, Saito R, Hata Y, Kiriyama N (1999) Chem Pharm Bull 47:77
60. Kaji A, Iwata T, Kiriyama N, Wakusawa S, Miyamoto K (1994) Chem Pharm Bull 42:1682
61. Alvi KA, Pu H, Luche M, Rice A, App H, McMahon G, Dare H, Margolis B (1999) J Antibiot 52:215
62. Kaji A, Saito R, Nomura M, Miyamoto K, Kiriyama N (1997) Anticancer Res 17:3675
63. Zhang B, Salituro G, Szalkowski D, Li Z, Zhang Y, Royo I, Vilella D, Diez MT, Pelaez F, Ruby C, Kendall RL, Mao X, Griffin P, Calaycay J, Zierath JR, Heck JV, Smith RG, Moller DE (1999) Science 284:794
64. Yadav JS, Reddy BVS, Swamy T (2004) Synthesis, p 106
65. Lopez-Alvarado P, Alonso MA, Carmen A, Menendez JC (2001) Tetrahedron Lett 42:7971
66. Bartoli G, Bosco M, Giuli S, Giuliani A, Lucarelli L, Marcantoni E, Sambri L, Torregiani E (2005) J Org Chem 70:1941
67. Bartoli G, Bartolacci M, Bosco M, Foglia G, Giuliani A, Marcantoni E, Sambri L, Torregiani E (2003) J Org Chem 68:4594
68. Bandini M, Fagioli M, Umani-Ronchi A (2004) Adv Synth Catal 346:545
69. Alam MM, Varala R, Adapa SR (2003) Tetrahedron Lett 44:5115
70. Fiouzabadi H, Iranpoor N, Nowrouzi F (2005) Chem Commun, p 789
71. Li W-J, Lin X-F, Wang J, Li G-L, Wang Y-G (2005) Synlett, p 2003
72. Zhan Z-P, Yang R-F, Lang K (2005) Tetrahedron Lett 46:3859
73. Murugan R, Karthikeyan M, Perumal PT, Reddy BSR (2005) Tetrahedron 61:12275
74. Rajabi F, Saidi R (2005) J Sulf Chem 26:251
75. Fiouzabadi H, Iranpoor N, Jafari AA (2006) J Mol Catal A Chem, p 168
76. Tahir R, Banert K, Solhy A, Sebti S (2006) J Mol Catal A Chem, p 39
77. Singh DU, Singh PR, Samant SD (2006) Synth Commun 36:1265
78. Nayak SK (2006) Synth Commun 36:1307
79. Firouzabadi H, Iranpoor N, Jafarpour M, Ghaderi A (2006) J Mol Catal A Chem, p 150
80. Kumar V, Kaur S, Kumar S (2006) Tetrahedron Lett 47:7001
81. Arumugam P, Perumal P (2006) Chem Lett 35:632
82. Ye M-C, Yang Y-Y, Tang Y, Sun X-L, Ma Z, Qin W-M (2006) Synlett, p 1240
83. Srivastava N, Banik BK (2003) J Org Chem 68:2109
84. Cavdar H, Saracoglu N (2005) Tetrahedron 61:2401
85. Wang S-Y, Ji S-J, Loh T-P (2003) Synlett, p 2377
86. Banik BM, Fernandez M, Alvarez C (2005) Tetrahedron Lett 46:2479
87. Lin C, Hsu J, Sastry MNV, Fang H, Tu Z, Liu J-T, Ching-Fa Y (2005) Tetrahedron 61:11751
88. Ji S-J, Wang S-Y (2003) Synlett, p 2074
89. Ji S-J, Wang S-Y (2005) Ultrason Sonochem 12:339
90. Li J-T, Dai H-G, Xu W-Z, Li T-S (2006) J Chem Res, p 41
91. Quin LD (2000) A guide to organophosphorus chemistry. Wiley, New York
92. Corbridge DEC (1990) Phosphorus, an outline of its chemistry, biochemistry, and technology, 4th edn. Elsevier, Amsterdam
93. Yavari I, Norouzi-Arasi H (2002) Phosphor Sulfur Silicon Relat Elem 177:87
94. Krawczyk H, Sliwinski M (2002) Synthesis, p 1351

95. Couthon-Gourves H, Simon G, Haelters J-P, Corbel B (2006) Synthesis, p 81
96. Moran J, Suen T, Beauchemin AM (2006) J Org Chem 71:676
97. Kusurkar RS, Alkobati NAH, Gokule AS, Chaudhari PM, Waghchaure PB (2006) Synth Commun, p 1075
98. Jensen KB, Thorhauge J, Hazell RG, Jorgensen KA (2001) Angew Chem Int Ed 41:160
99. Zhuang W, Hansen T, Jorgensen KA (2001) Chem Commun, p 347
100. Palomo C, Oiarbide M, Kardak BG, Garcia JM, Linden A (2005) J Am Chem Soc 127:4154
101. Yamazaki S, Iwata Y (2006) J Org Chem 71:739
102. Zhou J, Tang Y (2004) Chem Commun, p 432
103. Jia Y-X, Zhu S-F, Yang Y, Zhou Q-L (2006) J Org Chem 71:75
104. Lu S-F, Du D-M, Xu J (2006) Org Lett 8:2115
105. Evans DA, Frandrick KR, Song H-J (2005) J Am Chem Soc 127:8942
106. Evans DA, Frandrick KR (2006) Org Lett 8:2249
107. Evans DA, Scheidt KA, Frandrick KR, Lam HW, Wu J (2003) J Am Chem Soc 125:10780
108. Zhou J, Tang Y (2002) J Am Chem Soc 124:9030
109. Zhou J, Ye M-C, Huang Z-Z, Tang Y (2004) J Org Chem 69:1309
110. Ye M-C, Li B, Zhou J, Sun X-L, Tang Y (2005) J Org Chem 70:6108
111. Austin JF, MacMillan DWC (2002) J Am Chem Soc 124:1172
112. Paras NA, MacMillan DWC (2002) J Am Chem Soc 124:7894
113. Taylor MS, Jacobsen EN (2003) J Am Chem Soc 125:11204
114. Bandini M, Fagioli M, Melchiorre P, Melloni A, Umani-Ronchi A (2003) Tetrahedron Lett 44:5843
115. Bandini M, Fagioli M, Garavelli M, Melloni A, Trigari V, Umani-Ronchi A (2004) J Org Chem 69:7511
116. Bandini M, Garelli A, Rovinetti M, Tommasi S, Umani-Ronchi A (2005) Chirality 17:522
117. Angeli M, Bandin M, Garell A, Piccinnell F, Tommas S, Umani-Ronch A (2006) Org Biomol Chem, p 3291
118. Bandini M, Melloni A, Tommasi S, Umani-Ronchi A (2003) Helv Chim Acta 86:3753
119. Herrera RP, Sgarzani V, Bernardi L, Ricci A (2005) Angew Chem Int Ed 44:6576
120. Fleming EM, McCabe T, Connon SJ (2006) Tetrahedron Lett 47:7037
121. Zhuang W, Hazell RG, Jorgensen KA (2005) Org Biomol Chem, p 2566
122. Sarstedt B, Winkerfeldt E (1983) Heterocycles 20:469
123. Hughes I, Raphael RA (1983) Tetrahedron Lett 24:1441
124. Magnus PD, Sear NL (1984) Tetrahedron 40:2795
125. Gribble GW, Berthel SJ (1992) Tetrahedron 48:8869
126. Bergman J, Pelcman B (2004) J Org Chem 69:7511
127. Kaneko T, Wong H, Okamoto KT, Clardy J (1985) Tetrahedron Lett 26:4015
128. Rahman A (ed) (1988) Studies in natural products chemistry. vol 1, Part A. Elsevier, Amsterdam
129. Kase H, Iwahashi K, Matsuda Y (1986) J Antibiot 39:1059
130. Sezaki M, Sasak T, Nakazaw T, Taked U, Iwat M, Watanab T, Koyam M, Ka F, Shomur T, Kojim M (1985) J Antibiot 38:1437
131. Hirayama N, Lida T, Shirahata K (1986) Acta Crystallogr 42:1402
132. Funato N, Takayanagi H, Konda Y, Toda Y, Harigaya Y, Iwai Y, Omura S (1994) Tetrahedron Lett 35:1251
133. Barry JF, Wallace TW, Walshe NDA (1993) Tetrahedron Lett 34:5329
134. Barry JF, Wallace TW, Walshe NDA (1995) Tetrahedron 51:12797

135. Link JT, Raghavan S, Danishefsky SJ (1995) J Am Chem Soc 117:552
136. Omura S, Iwai Y, Hirano A, Nakagawa A, Awaya J, Tsuchya H, Takahashi Y, Masuma R (1977) J Antibiot 30:275
137. Pindur U, Kim Y-S (1996) J Heterocycl Chem 33:623
138. Ikegami F, Murakoshi I (1994) Phytochemistry 5:1089
139. Dunnill M, Fowd L (1965) Phytochemistry 4:935
140. Shinozaki H, Konishi S (1974) Neuropharmacology 13:665
141. Evans RH, Francis AA, Hunt K, Martin MR, Watkins JC (1978) J Pharm Pharmac 30:364
142. Ikegami F, Komada Y, Kobori M, Hawkins DR, Murakoshi I (1990) Phytochemistry 29:2507
143. Juaristi E (ed) (1997) Enantioselective synthesis of β-amino acids. Wiley, New York
144. Cardillo G, Tomasini C (1996) Chem Soc Rev 25:117
145. Shinagawa S, Kanamaru T, Harada S, Asai M, Okazaki M (1987) J Med Chem 30:1458
146. Kabawata N, Inamoto T, Hashimoto S (1992) J Antibiot 45:513
147. Matsuura F, Hamada Y, Shiouri T (1994) Tetrahedron 50:11303
148. Ferreira PMT, Maia HLS, Monteiro LS (1999) Tetrahedron Lett 40:4099
149. Huck J, Duru C, Roumestant ML, Martinez J (2003) Synthesis, p 2165
150. Rhodes CN, Brown DR (1993) J Chem Soc Faraday Trans 89:1387
151. Cativiela C, Figureras F, Garcia JI, Mayoral JA, Pires E, Royo AJ (1993) Tetrahedron: Asymmetry 4:621
152. Cativiela C, Fraile JM, Garcia JI, Mayoral JA, Pires E, Royo AJ, Figureras F (1993) Tetrahedron 43:4073
153. de la Hoz A, Diaz-Ortiz A, Gomez MV, Mayoral JA, Moreno A, Sanchez-Migallon AM, Vazquez E (2001) Tetrahedron 57:5428
154. Asselin AA, Humber LG, Voith K, Metcalf G (1986) J Med Chem 29:648
155. Nichols DE, Cassady JM, Persons PE, Yeung MC, Clemens JA, Smalstig EB (1989) J Med Chem 32:2128
156. Mewshaw RE, Marquis KL, Shi X, McGaughey G, Stack G, Webb MB, Abou-Gharbia M, Wasik T, Scerni R, Spangler T, Brennan JA, Mazandarani H, Coupet J, Andree TH (1998) Tetrahedron 54:7081
157. Daukshas VK, Martinkus RS, Gineitite VL, Urbonene SL (1982) Chem Heterocycl Compd (Engl Transl) 18:932
158. Ennis MD, Baze ME, Smith MW, Lawson CF, McCall RB, Lahti RA, Piercey MF (1992) J Med Chem 35:3058
159. Andrew M, Birch AM, Bradley PA (1999) Synthesis, p 1181
160. Partsvaniya DA, Akhvlediani RN, Zhigachev VE, Gordeev EN, Kuleshova LN, Suvorov NN, Vigdorchik MM, Mashkovskii MD (1986) Chem Heterocycl Compd (Engl Transl) 22:1311
161. Mayer S, Merour J-Y, Joseph B, Guillaumet G (2002) Eur J Org Chem, p 1646
162. Kebabian JW, Calne DB (1979) Nature (London) 277:93
163. Barnett A (1986) Drugs Fut 11:49
164. Setler PE, Sarau HM, Zirkle CL, Saunders HL (1978) Eur Pharmacol Sci 50:419
165. Kraxner J, Hübner H, Gmenier P (2000) Arch Pharm Pharm Med Chem 333:287
166. Yokoyama Y, Matsumoto T, Murakami Y (1995) J Org Chem 60:1486
167. Efange SMN, Mash DC, Khare AB, Quyang Q (1998) J Med Chem 41:4486
168. Gmeiner P, Sommer J, Höfner G (1995) Arc Pharm 328:329
169. Osawa T, Namiki M (1983) Tetrahedron Lett 24:4719
170. Fahy E, Potts BCM, Faulkner DJ, Smith K (1991) J Nat Prod 54:564
171. Bifulco G, Bruno I, Riccio R, Lavayre J, Bourdy G (1995) J Nat Prod 58:1254

172. Bell R, Carmeli S, Sar N (1994) J Nat Prod 57:1587
173. Garbe TR, Kobayashi M, Shimizu N, Takesue N, Ozawa M, Yukawa H (2000) J Nat Prod 63:596
174. Chakrabarty M, Basak R, Ghosh N (2001) Tetrahedron Lett 42:3913
175. Denis J-N, Mauger H, Valle'e Y (1997) Tetrahedron Lett 38:8515
176. Chalaye-Mauger H, Denis J-N, Averbuch-Pouchot M-T, Vallee Y (2000) Tetrahedron 56:791
177. Chakrabarty M, Basak R, Ghosh N, Harigaya Y (2004) Tetrahedron 60:1941
178. Spadoni G, Balsamini C, Bedini A, Duranti E, Tontini A (1992) J Heterocycl Chem 29:305
179. Bonjoch J, Sole D (2000) Chem Rev 100:3455
180. Ohshima T, Xu Y, Takita R, Shimizu S, Zhong D, Shibasaki M (2002) J Am Chem Soc 124:14546
181. Mori M, Nakanishi M, Kajishima DA, Sato Y (2003) J Am Chem Soc 125:9801
182. Gorman M, Neuss N, Biemann K (1962) J Am Chem Soc 84:1058
183. Moza BK, Trojanek J (1963) Collect Czech Chem Commun 28:1427
184. Buchi G, Matsumoto K, Nishimura H (1971) J Am Chem Soc 93:3299
185. Ando M, Buchi G, Ohnuma T (1975) J Am Chem Soc 97:6880
186. Heureux N, Wouters J, Marko IE (2005) Org Lett 7:5245
187. Hsu D-S, Rao PD, Liao C-C (1998) Chem Commun, p 1795
188. Rao PD, Chen C-H (1998) Chem Commun, p 155
189. Hsu P-Y, Lee Y-C, Liao C-C (1998) Tetrahedron Lett 39:659
190. Liu W-C, Liao C-C (1998) Synlett, p 912
191. Carlini R, Higgs K, Older C, Randhawa S, Rodrigo R (1997) J Org Chem 62:2330
192. Coleman RS, Grant EB (1995) J Am Chem Soc 117:10889
193. Hsieh M-F, Rao PD, Liao C-C (1999) Chem Commun, p 1441
194. Singh NB, Singh NP (1994) Tetrahedron 50:6441
195. Desiraju GR (ed) (1987) Organic solid state chemistry. Elsevier, Amsterdam
196. Toda F, Akai H (1990) J Org Chem 55:3446
197. Li X-L, Wang Y-M, Matsuura T, Meng J-B (1999) J Heterocycl Chem 36:697
198. Bodwell GJ, Li J, Miller OD (1999) Tetrahedron 55:12956
199. Wong DT, Bymaster FP, Engelman EA (1995) Life Sci 57:411
200. Brodfuehrer PR, Chen B, Sattleberg TR, Smith PR, Reddy JP, Stark DR, Quinlan SL, Reid JG, Thottathil JK, Wang S (1997) J Org Chem 62:9192
201. Denhart DJ, Mattson RJ, Ditta JL, Macor JE (2004) Tetrahedron Lett 45:3803
202. Belokon YN, Harutyunyan S, Vorontsov EV, Peregudov AS, Chrustalev VN, Kochetkov KA, Pripadchev D, Saygan AS, Beck AK, Seebach D (2004) ARKIVOC iii:132
203. Sabitha G, Kumar MR, Reddy MSK, Yadav JS, Krishna KVSR, Kunwar AC (2005) Tetrahedron Lett 46:1659
204. Pindur U (1988) Heterocycles 27:1253
205. Pindur U (1995) In: Moody CJ (ed) Advances in nitrogen heterocycles, vol 1. JAI, Greenwich, p 121
206. Sundberg R (1996) In: Meth-Cohn O (ed) Best synthetic methods, sub-series key systems and functional groups indoles. Academic, London
207. Saxton JE (ed) (1994) The chemistry of heterocyclic compounds, vol 25, Part IV. Wiley, Chichester
208. Macor JE (1990) Heterocycles 31:993
209. Madalengoitia JS, Macdonald TL (1993) Tetrahedron Lett 34:6237
210. Blechert S, Wirth T (1992) Tetrahedron Lett 33:6621
211. Wiest O, Steckhan E (1993) Angew Chem Int Ed Engl 32:901

212. Elango S, Srinivasan PC (1993) Tetrahedron Lett 34:1347
213. Balasubramanian T, Balasubramanian KK (1994) Chem Commun, p 1237
214. Dufour B, Motorina I, Fowler FW, Grierson DS (1994) Heterocycles 37:1455
215. Rodriguez-Salvador L, Zaballos-Garcia E, Gonzalez-Rosende E, Sidi MD, Sepulveda-Arques J, Jones RA (1997) Synth Commun 27:1439
216. Joseph B, Da Costa H, Merour J-Y, Leonce S (2000) Tetrahedron 56:3189
217. Cavdar H, Saracoglu N (2006) J Org Chem 71:7793
218. Yadav JS, Abraham S, Reddy BVS, Sabitha G (2001) Tetrahedron Lett 42:8063
219. Zhan Z-P, Yang W-Z, Yang R-F (2005) Synlett, p 2425
220. Zhan Z-P, Yu J-L, Yang W-Z (2006) Synth Commun 36:1373
221. Zhang C-X, Wang Y-Q, Duan Y-S, Ge Z-M, Cheng T-M, Li R-T (2006) Catal Commun 7:534
222. Guida WC, Mathre DJJ (1980) Org Chem 45:3172
223. Hamaide T (1990) Synth Commun 20:2193
224. Wang N, Teo K, Anderson HJ (1977) Can J Chem 55:4112
225. Le Z-G, Chen Z-C, Hu Y, Zheng Q-G (2004) Synthesis, p 1951
226. Yang L, Xu L-W, Xia C-G (2005) Tetrahedron Lett 46:3279
227. Paras NA, MacMillan WC (2001) J Am Chem Soc 123:4370
228. Gordillo R, Carter J, Houk KN (2004) Adv Synth Catal 346:1175
229. Arroyo Y, de Paz M, Rodriguez JF, Sanz-Tejedor MA, Ruano JLG (2002) J Org Chem 67:5638
230. Rowan DD, Hunt MB, Gaynor DL (1986) J Chem Soc Chem Commun, p 935
231. Brimble MA (1990) J Chem Soc Perkin Trans I, p 311
232. Christophersen C (1985) In: Brossi A (ed) The alkaloids, vol 24. Academic, Orlando, chap 2
233. Umeyama A, Ito S, Yuasa E, Arihara S, Yamad T (1998) J Nat Prod 61:1433
234. Mancini I, Guella G, Amade P, Roussakis C, Pietra F (1997) Tetrahedron Lett 38:6271
235. Banwell MG, Bray AM, Wills AC, Wong D (1999) J New J Chem 23:687
236. Goh SH, Ali RMA, Wong WH (1989) Tetrahedron 45:7899
237. Banwell MG, Edwards AJ, Jolliffe KA, Smith JA, Hamel E (2003) Org Biomol Chem 296
238. Banwell MG, Beck DAS, Smith JA (2004) Org Biomol Chem 2:157
239. Banwell MG, Beck DAS, Willis AC (2006) ARKIVOC iii:163
240. Anderson HJ, Loader CE (1985) Synthesis, p 353
241. Hodges LM, Gonzalez J, Koontz JI, Myers WH, Harman WD (1993) J Org Chem 58:4788
242. Hodges LM, Gonzalez J, Koontz JI, Myers WH, Harman WD (1995) J Org Chem 60:2125
243. DuBois MR, Vasquez LD, Peslherbe L, Noll BC (1999) Organometallics 18:2230

Top Heterocycl Chem (2007) 11: 63–101
DOI 10.1007/7081_2007_063
© Springer-Verlag Berlin Heidelberg
Published online: 4 July 2007

Chemistry of the Welwitindolinones

J. Carlos Menéndez

Universidad Complutense, Departamento de Química Orgánica y Farmacéutica,
Facultad de Farmacia, Plaza de Ramón y Cajal, s.n., 28040 Madrid, Spain
josecm@farm.ucm.es

Abstract Marine cyanobacteria are a prolific source of bioactive natural products. The welwitindolinones are a group of structurally unique metabolites isolated from cyanobacteria, containing 3,4-bridged or spiro cyclobutaneoxindole moieties. Due to their challenging structures and their broad spectrum of biological activities, these compounds have attracted the attention of synthetic chemists for some time, although so far their complexity has prevented the completion of total syntheses for most of these natural products. This chapter is devoted to a discussion of the current knowledge about the occurrence, biological properties, biosynthesis and synthetic approaches to the welwitindolinones.

Keywords Bicyclo[4.3.1]decanes · Biosynthesis · Indole alkaloids ·
Marine natural products · Spiro-fused cyclobutanes

Abbreviations

ABC	ATP-Binding Cassette
ATP	Adenosine triphosphate
BOC	*tert*-Butyloxycarbonyl
Brsm	Based on recovered starting material
CDMT	2-Chloro-4,6-dimethoxy[1,3,5]triazine
DDQ	Dichlorodicyanoquinone
DPPA	Diphenylphosphoryl azide
GTP	Guanosine triphosphate
HMPA	Hexamethylphosphoramide
LDA	Lithium diisopropylamide
LHMDS	Lithium hexamethyldisilazide
MDR	Multidrug resistance
NMM	N-Methylmorphioline
NBS	*N*-Bromosuccinimide
PCC	Pyridinium chlorochromate
P-gp 170	Glycoprotein P-170
T2IMDA	Type 2 intramolecular Diels–Alder
TBS	*tert*-Butyldimethylsilyl
TEMPO	2,2,6,6-Tetramethylpiperidine-*N*-oxyl
Tf	Triflate (trifluoromethylsulfonate)
TFA	Trifluoroacetate
TMS	Trimethylsilyl
TPAP	Tetrapropylammonium perruthenate
p-Ts	*p*-Tolylsulfonyl

1
Occurrence of the Welwitindolinones

Welwitindolinones are a family of structurally unusual oxindole alkaloids obtained from marine cyanobacteria (Fig. 1). The first members of this group were isolated in 1994 by Moore and coworkers, who were studying the lipophilic extracts of the blue green algae *Hapalosiphon welwitschii* and *Westiella intricata* [1]. These extracts exhibited reversing activity towards multidrug resistance (MDR), insecticidal activity against blowfish larvae, and antifungal activity, the first two properties being largely associated with *N*-methylwelwitindolinone C isothiocyanate, also known as welwistatin (1). Some other related 3,4-bridged oxindoles, such as welwitindolinone C isothiocyanate 2, *N*-methylwelwitindolinone B isothiocyanate (3) and its norderivative (4), and *N*-methylwelwitindolinone C isonitrile (5) were also isolated. Besides these compounds, a structurally novel spirocyclobutane oxindole derivative, welwitindolinone A isonitrile (6) was also identified. Subsequent work led to the isolation of some other oxidized welwitindolinones from several *Fischerella* species. These compounds contain a 3-hydroxy group (compounds 7 and 8) or an ether link between C-3 and C-14, accompanied by oxidation of C-13 (compound 9, *N*-methylwelwitindolinone D isonitrile) [2].

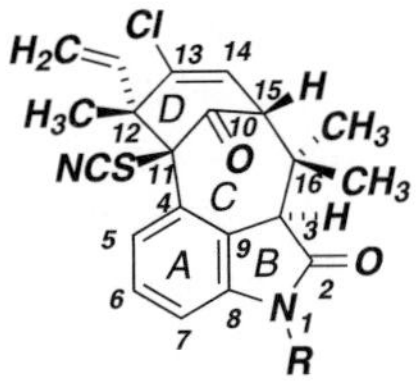

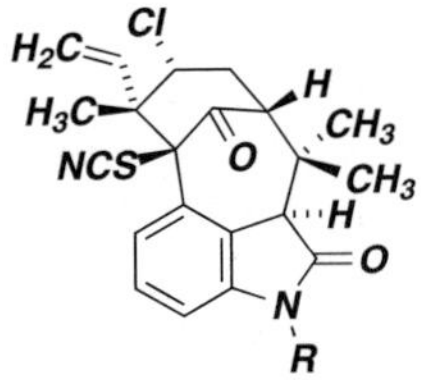

R	Cmpd.	Name
CH₃	1	N-methylwelwitindolinone C isothiocyanate (Welwistatin)
H	2	Welwitindolinone C isothiocyanate

R	Cmpd.	Name
CH₃	3	N-methylwelwitindolinone B isothiocyanate
H	4	Welwitindolinone B isothiocyanate

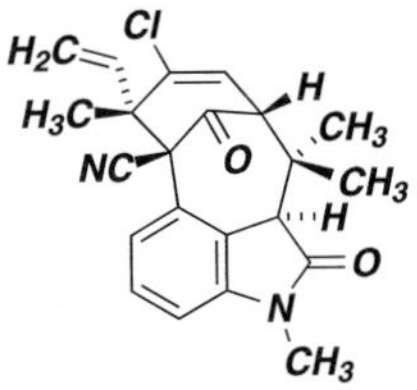

N-methylwelwitindolinone C isonitrile (5)

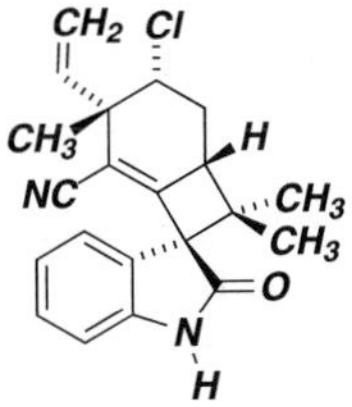

Welwitindolinone A isonitrile (6)

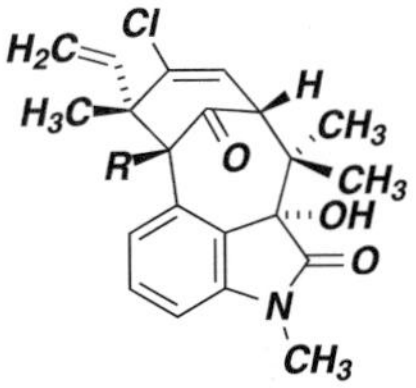

7 R = NC
8 R = NCS

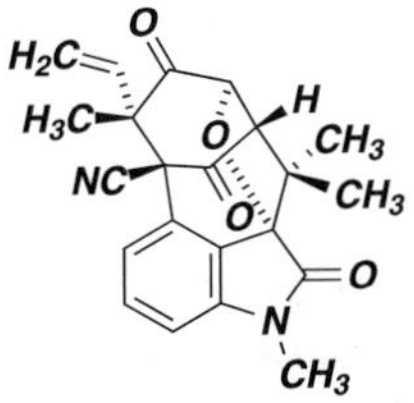

N-methylwelwitindolinone D isonitrile (9)

Fig. 1 Structures of the welwitindolinones

2
Biological Properties of the Welwitindolinones

The lipophilic extracts from *Hapalosiphon welwitschii* and *Westiella intricata* exhibited MDR reversing activity, insecticidal activity against blowfish larvae, and antifungal activity. The first two properties were largely associated with welwistatin, while welwitindolinone A isonitrile (**6**) accounted for most of the antifungal activity found in the extracts. When all these findings are taken into account, welwistatin (**1**) emerges as the biologically more relevant of the welwitindolinones.

Inhibition of multidrug resistance (MDR) to antitumor agents is one of the most interesting biological properties of welwistatin. In fact, marine cyanobacteria are very interesting in the search for new MDR modulators, since it has been estimated that about 1% of their extracts have this activity. In vitro models with the MDR phenotype have been mainly related to overexpression of a 170 kD membrane glycoprotein known as P-gp 170, which is a member of a superfamily of ATP-dependent transport proteins known as ABC (ATP-Binding Cassette) transporters. Welwistatin inhibits the P-gp 170-mediated resistance of MCF-7/ADR cells to lipophilic anticancer drugs such as vinblastine, taxol, actinomycin D, daunomycin, and colchicine (but not to the less lipophilic cisplatin) at 10^{-7} M doses [3], which represents a potency 20 to 100-fold higher than that of verapamil, the reference MDR reversal compound [4, 5]. Although structure-activity relationships have not been established for P-gp inhibition by welwistatin, the isothiocyanate group seems to play a very important role, because natural welwistatin analogues in which an isonitrile group replaced this moiety were inactive; indeed, some simple isothiocyanates have subsequently been shown to behave as MDR reversors [6, 7]. The interaction of welwistatin with P-gp 170 was corroborated by photoaffinity labeling of this transport protein with [^{3}H]azidopine in membranes from SK-VLB-1 cells [3].

Welwistatin also inhibits cell proliferation with reversible depletion of cellular microtubules in ovarian carcinoma cells and A-10 vascular smooth muscle cells by inhibiting the polymerization of tubulin, but it does not alter the ability of tubulin to bind [^{3}H]colchicine or to hydrolyze GTP [8]. Due to the cytotoxicity associated with the inhibition of tubulin polymerization, which is the main mechanism of action of antitumor drugs such as vincristine and vinblastine, and because P-gp-overexpressing cells show virtually no resistance to welwistatin due to its MDR reversal properties, this natural product could be a good candidate in the chemotherapy of drug-resistant tumors.

3
Biosynthesis of the Welwitindolinones

The welwitindolinones are biogenetically related to other families of natural products known as fischerindoles and hapalindoles, as shown in the biosynthetic proposal made by Moore [1] that is summarized in Schemes 1 and 2. A chloronium ion-induced condensation between isonitrile 10, derived from L-tryptophan, and polyene 11, derived from geranyl pyrophosphate, would give the hypothetical intermediate 12, known as 12-epihapalindole E isonitrile, which has been proposed as the common precursor of all these chlorine-containing alkaloids [1, 9]. An enzyme-controlled, acid-catalyzed condensation of the isopropenyl group onto the indole C(2) or C(4) positions would give the fischerindole 13 or the hapalindole 14, respectively (Scheme 1).

On the other hand, oxidation of the indole system in 12 would give 15, a direct precursor of the spiro compound 6 (welwitindolinone A isonitrile).

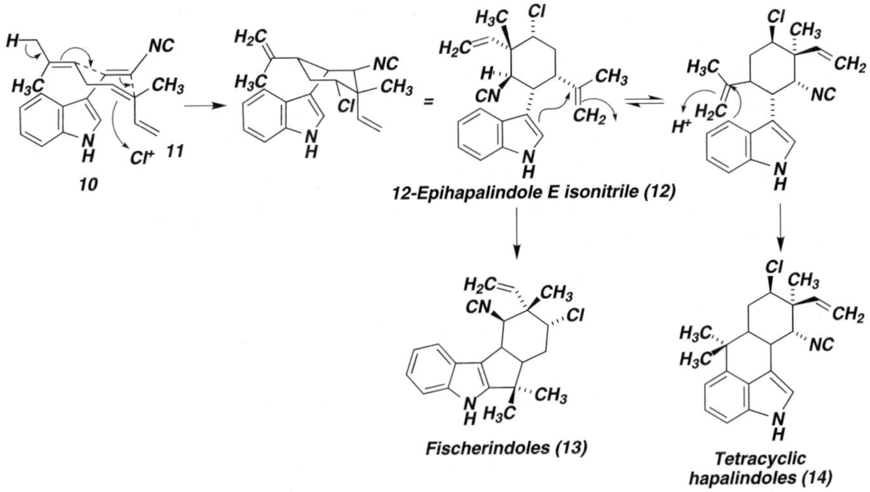

Scheme 1 Biosynthesis of the fischerindoles and hapalindoles according to Moore

Epoxidation of the cyclohexene bond of the latter to give **16**, followed by intramolecular Friedel-Crafts-type chemistry would finally afford compound **4**, containing the welwitindolinone skeleton [1]. The origin of the isothiocyanate unit in compounds such as **1** is less clear and this group has been proposed

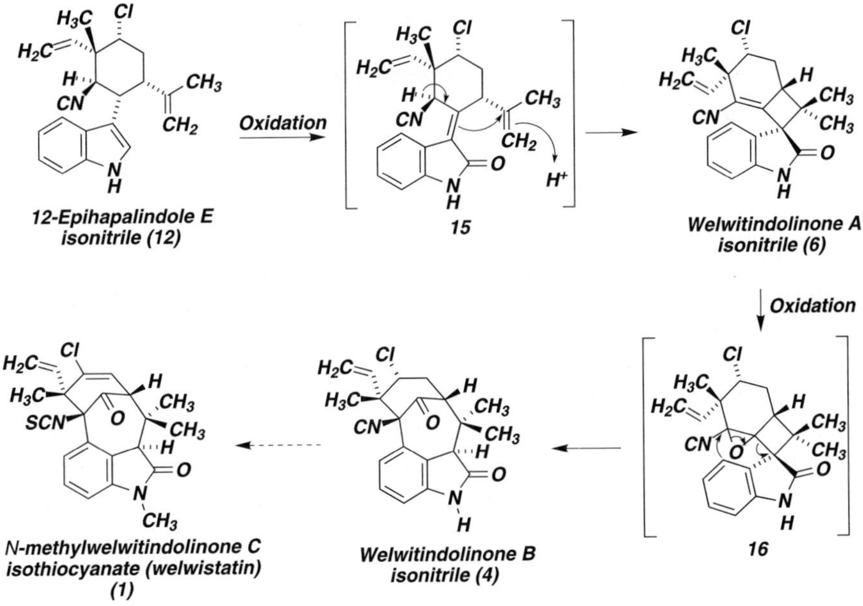

Scheme 2 Biosynthesis of the welwitindolinones according to Moore

to arise directly from inorganic thiocyanate or by introduction of sulfur into isonitrile intermediates [10] (Scheme 2).

Some crucial details of this biosynthetic proposal have recently been challenged. In particular, it has been proposed that the role of 12 as an intermediate is unlikely because of the absence of architectures similar to that of 12 among the known natural members of the hapalindole-fischerindole-welwitindolinone family. Instead, fischerindole I has been proposed as the actual precursor to welwitindolinone A 6, a hypothesis that is supported by synthetic studies (Sect. 5.2) [11].

4
Synthetic Efforts Towards Welwitindolinone C Isothiocyanate (Welwistatin)

4.1
Introduction

In addition to its biological interest, the structure of welwistatin poses a significant synthetic challenge because of its four-ring compact structure in which an oxindole system and a cyclohexanone are linked through a seven-membered ring. It also contains four stereogenic centers, including two adjacent quaternary centers, a third quaternary carbon containing the gem-dimethyl substituent and two unusual and reactive moieties, such as the isothiocyanate and vinyl chloride functions (Fig. 2).

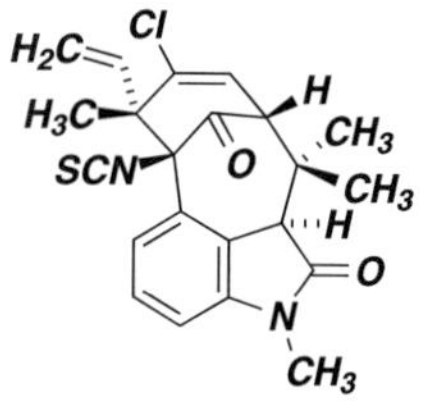

- Compact, unprecedented ring system
- Bicyclo[4.3.1]decane system about an oxindole
- Four stereocenters, two of them quaternary
- Three quaternary carbons, two of them adjacent
- Gem-dimetyl substituent
- Bridgehead nitrogen function
- Unusual, sensitive functions: isothiocyanate, vinyl chloride

Welwistatin (1)

Fig. 2 Main challenges in the construction of welwistatin

A good deal of effort has been devoted to the preparation of derivatives of the cyclohepta[cd]indole and bicyclo[4.3.1]decane ring systems, that can be considered related to the ABC and CD ring systems of welwistatin, respectively, but much of this work cannot be translated easily to methods relevant to welwistatin synthesis. For instance, many successful approaches to the cyclohepta[cd]indole system are based on intramolecular radical cyclizations [12] or Heck reactions [13]. In many published examples, a substituent at the indole C-2 avoids the competing reaction at this position, but this device renders the reaction products inadequate to achieve the oxindole structure of welwistatin. Because this work on potential welwistatin fragments has been recently reviewed [14], it will not be discussed here.

No total synthesis of welwistatin has been completed so far, although a number of groups have completed syntheses of the complete tetracyclic framework.

4.2
Failed Approaches to the Welwistatin Skeleton

Several groups have reported the preparation of advanced synthetic intermediates that they were then unable to cyclize to compounds containing the complete welwistatin framework. In the first published approach to a welwistatin fragment, Konopelski planned the formation of the welwistatin C4–C11 bond through a coupling reaction with a lead (IV) 4-indolyltriacetate derivative [15, 16] and a subsequent aldol-type condensation to form the C15–C16 bond of the natural compounds. The model cyclohexanone derivative 18 was arylated very efficiently with the indolyllead reagent 17 to give the 4-substituted indole derivative 19. It is interesting to note that the excellent diastereoselectivity observed in the formation of 19 was attributed to an attractive interaction between the carbanionic center derived from the intermediate β-ketoester and the distal trialkylsilyloxy group. Compound 19 was then

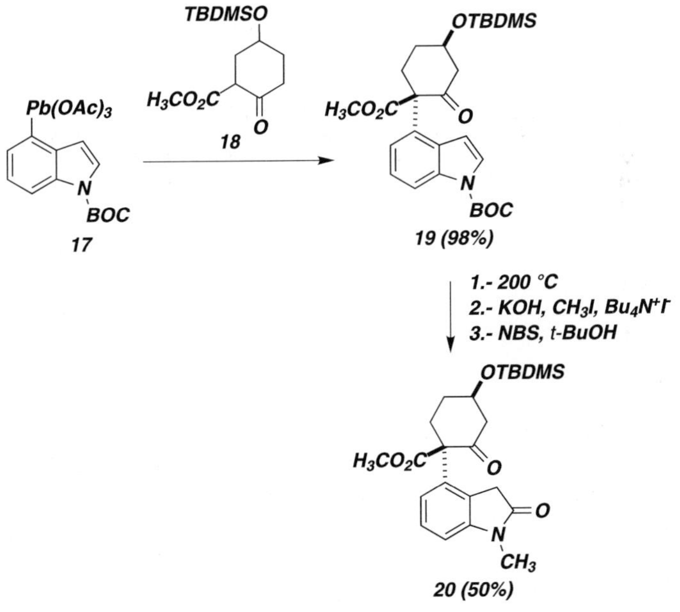

Scheme 3 Creation of the C15–C16 bond of a welwistatin model compound by C-arylation of a cyclohexanone derivative

transformed into oxindole **20** by thermal elimination of the BOC protection, followed by *N*-methylation and transformation of the indole ring into oxindole in the presence of NBS and *tert*-butyl alcohol (Scheme 3), but this compound could not be used as a precursor of the welwistatin skeleton because of the failure of all attempts to create the *C* ring by aldol chemistry [17, 18].

This strategy also required the preparation of a suitable precursor of the welwistatin *D* ring, as shown in Scheme 4. The starting material for this endeavor was 4-hydroxycyclohexanone **21**, which was transformed into the corresponding acetal by reaction with the enantiomerically pure hydrobenzoin **22** as a chiral auxiliary to generate the required stereochemistry at the welwistatin C(12)-position, leading to compound **25** after subsequent TEMPO oxidation of the hydroxyl group. Methoxycarbonylation of the **23** anion by its exposure to methyl carbonate led to the β-dicarbonyl derivative **24**, which was submitted to a standard selenation-oxidation-elimination protocol to create a double bond conjugated with both carbonyls (compound **25**). Lewis acid-catalyzed conjugate addition of methylmagnesium bromide to this conjugated

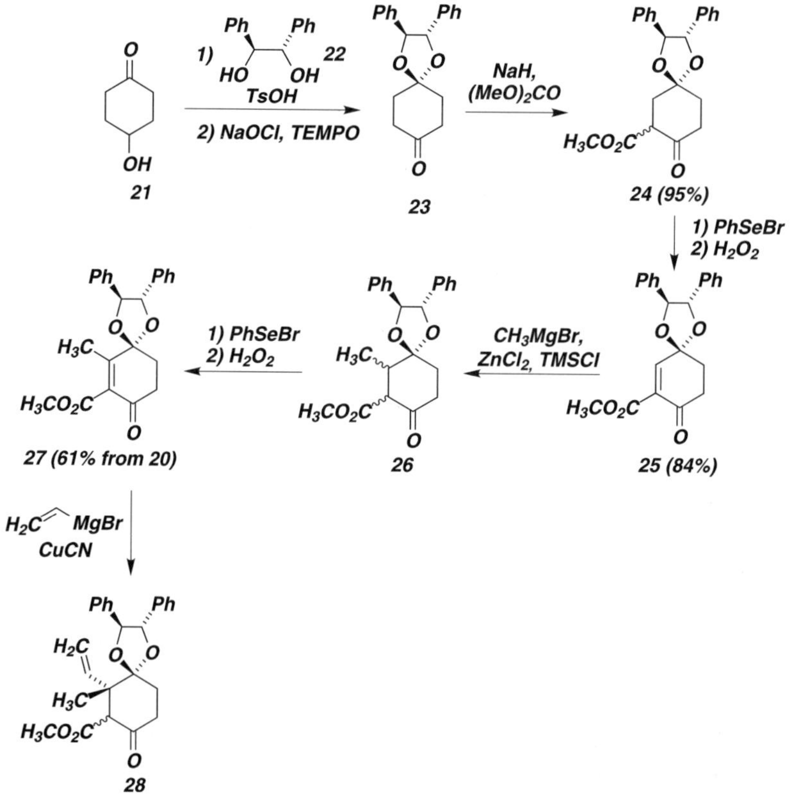

Scheme 4 Synthesis of the welwistatin *D* ring by Konopelski

system afforded **26**, which contains the welwistatin C-12 methyl substituent. A selenium-assisted dehydrogenation similar to the one previously applied to **24** gave compound **27**, which was then vinylated by diastereoselective conjugate addition of vinylmagnesium bromide to give compound **28** [19].

With welwistatin precursor **28** in hand, the synthetic plan involved its arylation with indolyllead triacetate **17** in conditions similar to those established for the model system **18**. However, this reaction failed to give the expected product **29** (Scheme 5), probably because of the steric hindrance imposed by the presence of the methyl and vinyl substituents, as shown by the fact that cyclohexanone derivative **24** gave the reaction in quantitative yield, affording compound **30** [20].

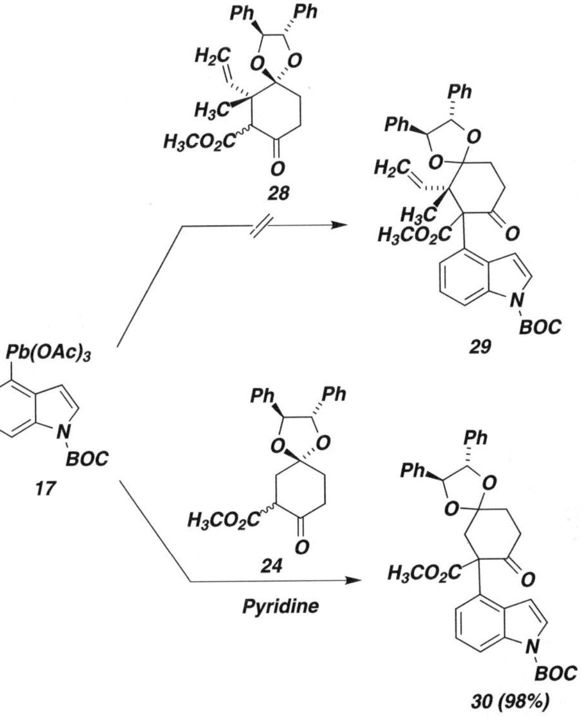

Scheme 5 Failed approach to welwistatin based on a *C*-arylation reaction

In another failed approach, addition of 1-pyrrolidinocyclohexene **32** to (*E*)-(1-methyl-2-oxoindolin-3-ylidene)acetophenone **31**, followed by acid hydrolysis, could be controlled simply by changes in the reaction temperature, and was found to give diastereoselectively compounds **33** or **34**. The latter compound was shown by NMR and X-ray diffraction data to have the suitable geometry to allow the creation of the crucial bond between the position adjacent to the cyclohexanone carbonyl and the oxindole C-4 position.

However, all attempts to activate the α-position of the cyclohexanone ring in order to facilitate a subsequent diazotization, which was to be followed by rhodium carbenoid-mediated aryl C–H insertion onto C-4 [21], were unsatisfactory [22]. One of the approaches was based on the generation of the silyl enol ether 35, but attempts to achieve its α-acylation led only to the formation of Paal–Knorr-type cyclization products 36. Chemoselective formylation of 34 to 37 was possible by reaction with ethyl formate in the presence of a large excess of sodium ethoxide, but in situ oxidation of the desired compound 37 to 38, which was the major isolated product, made the reaction unpractical (Scheme 6).

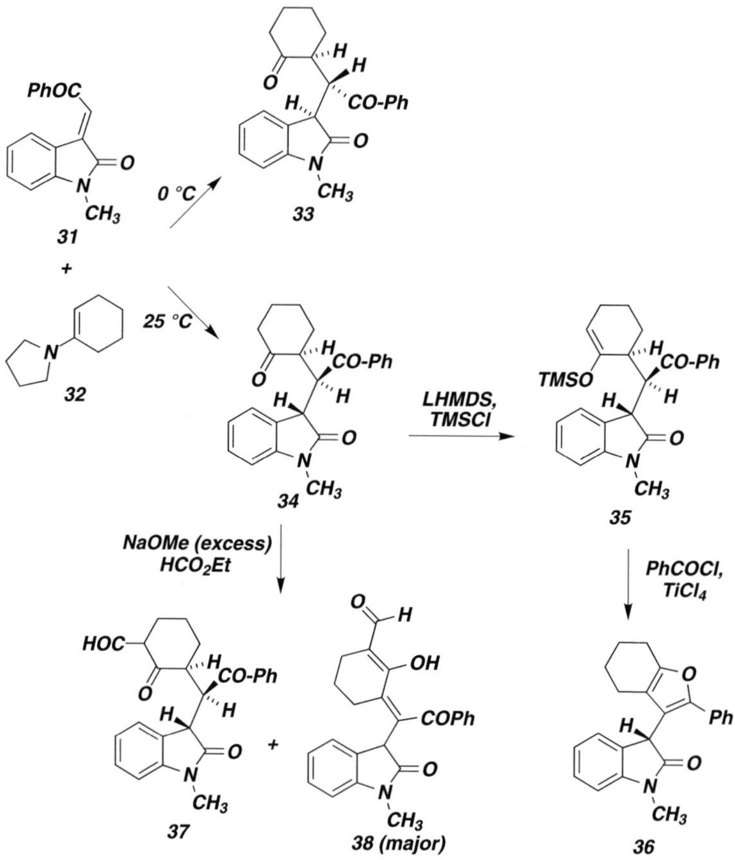

Scheme 6 Studies on an approach to welwistatin based on an enamine addition onto a (2-oxoindolin-3-ylidene)acetophenone derivative

In a related example, the desired rhodium carbenoid could be generated from 3-substituted indole or oxindole derivatives, which are readily accessible by ytterbium triflate-catalyzed Michael reactions, but several problems were found that prevented the formation of the cycloheptane ring via an aryl CH insertion at C-4. For instance, oxindole substrate **39** gave readily the rhodium carbenoid **41** in two steps, but this intermediate failed to give the desired CH insertion to **42**. Instead, it was attacked by the oxygen of the oxindole to give the zwitterion **43**, which evolved either to the indolooxepine **44** (*a*, Scheme 7), or by intramolecular cyclization (*b*) to give epoxide **45** that then rearranged to **46**.

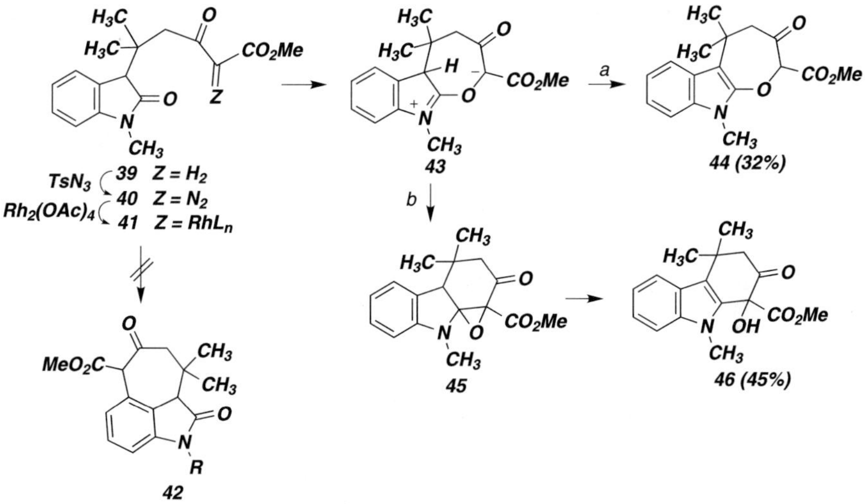

Scheme 7 Failed attempts at the construction of the cyclohepta[*cd*]indole system based on a rhodium carbenoid insertion

The attempted cyclization of rhodium carbenoid **47** onto the indole C-4 position was also unsuccessful, and the only observed products were compounds **48** and **49**, arising from insertion onto the indole C-2 position (Scheme 8). An attempt to force the desired mode of cyclization by suppression of the indole C2-C3 double bond also failed, and thus 2,3-dihydro carbenoid **50** gave only the intramolecular dismutation derivative **51** [23].

4.3
Syntheses of the Complete Welwistatin Core

4.3.1
Synthesis of the Welwistatin Core by Wood

The preparation of compound **52** represented the first synthesis of an intermediate containing the complete welwistatin carbon framework [24]. The synthetic planning is shown in Scheme 9, and is based on the construction of

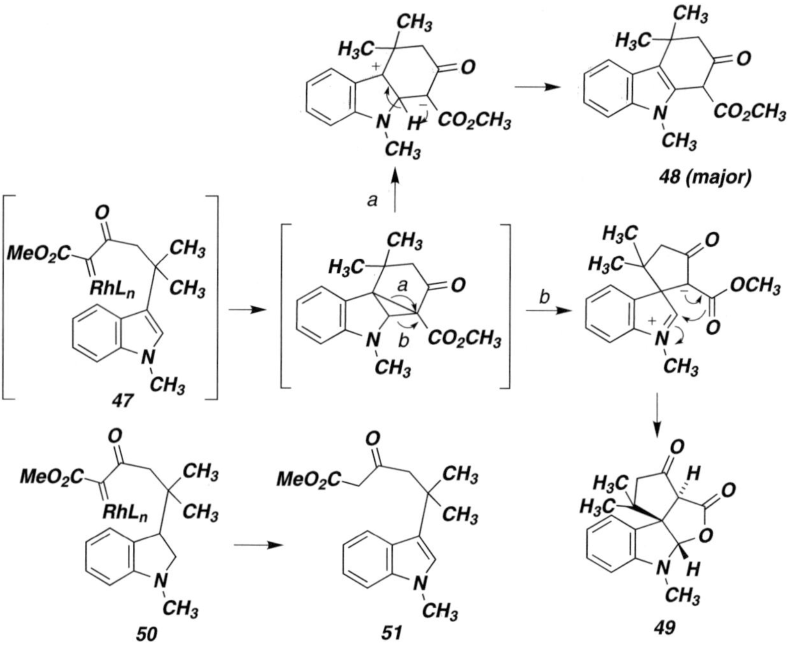

Scheme 8 Further attempts at the construction of the cyclohepta[*cd*]indole system using rhodium carbenoid chemistry

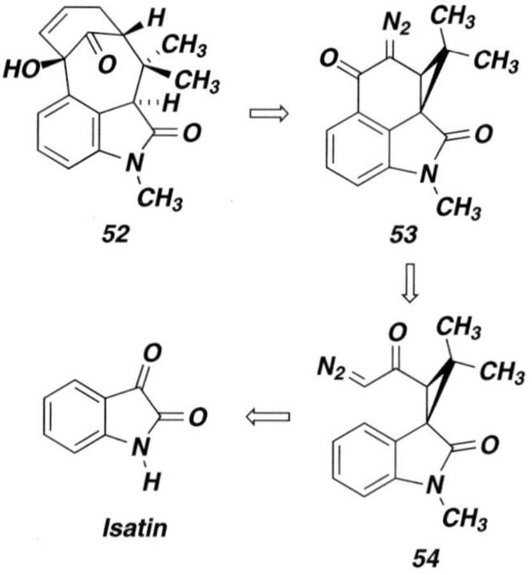

Scheme 9 Retrosynthetic analysis of the welwistatin core by Wood

the seven-membered ring of the 3,4-bridged oxindole core through opening of the cyclopropane ring of diazo ketone **53**, which was prepared by rhodium carbenoid-mediated aryl C-H insertion in **54**.

Wood's synthesis started with a Wittig reaction between isatin and ethyl triphenylphosphoranylidene acetate to give compound **55**, which was submitted to a cyclopropanation reaction through sequential treatment with isopropyl triphenylphosphorane and methyl iodide [25], giving **56**. This compound was transformed into the α-diazoketone **54** through a four-step sequence, comprising *N*-methylation and saponification, followed by conversion of the acid thus obtained into to the corresponding acid chloride and final reaction with diazomethane. The initial studies on the planned intramolecular aryl C–H insertion in compound **54**, using Rh₂(TFA)₄ as a catalyst, gave disappointing results, since the reaction gave low yields of a mixture of compounds **58** and **59**, both presumably arising from the common intermediate **57**. After extensive optimization work, it was found that addition to the reaction mixture of the mildly Lewis-acidic clay montmorillonite K allowed the preparation of the desired **59** in a reproducible 57% yield, which allowed to proceed with the synthetic plan. Compound **59** was oxidized with PCC to give the corresponding α-diketone, which, upon treatment with tosyl azide and base, underwent regioselective diazotization to

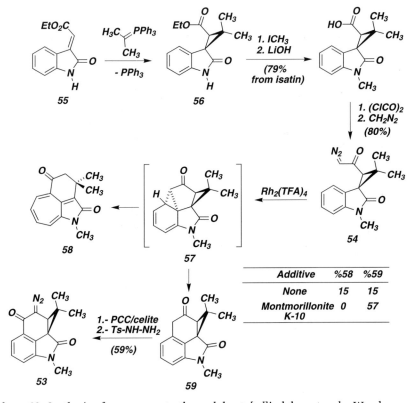

Additive	%58	%59
None	15	15
Montmorillonite K-10	0	57

Scheme 10 Synthesis of a precursor to the cyclohepta[*cd*]indole system by Wood

give **53** (Scheme 10). It is relevant to mention at this point that a previous attempt at the preparation of a cyclohepta[*cd*]indole derivative from a C-3 substituted oxindole failed at the aryl C–H insertion stage, as previously mentioned [23].

Compound **53**, via generation of its rhodium carbenoid, was coupled with allyl alcohol to give the non-isolated enol intermediate **60** which underwent a cyclopropane ring-opening to give **61**, containing the welwistatin seven-membered ring. Ethynylation of **61** by addition of an ethynyl Grignard reagent to its keto group afforded compound **62**, where the stage was set for a Claisen rearrangement intended to induce migration of the allyl group to the position where it would allow creating ring *D* through a metathesis reaction. Thus, thermal treatment of **62** followed by separation of the major (98 : 2) diastereoisomer, bearing a α-allyl substituent, gave compound **63**. Partial reduction of the carbon–carbon triple bond with hydrogen in the presence of Lindlar's catalyst gave a suitable substrate for a ring-closing metathesis that was finally transformed into **52** by treatment with the first-generation Grubbs catalyst (Scheme 11). The tandem OH-insertion-Claisen rearrangement used in this sequence proved to be of general value [26].

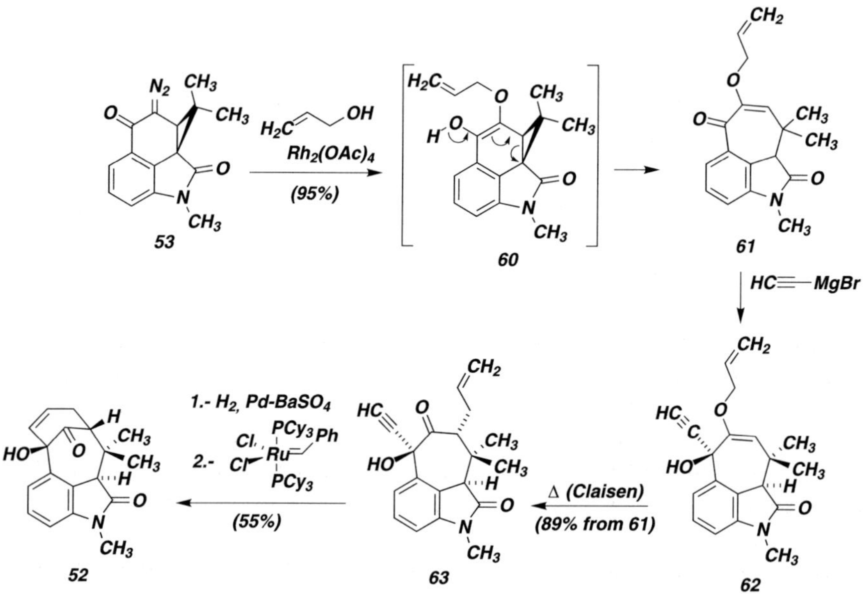

Scheme 11 Final stages of the preparation of the welwistatin core by Wood

4.3.2
Synthesis of the Welwistatin Core by Rawal

The preparation of compound **64** and the subsequent installation of a bridge-head nitrogen function by the Rawal group [27] represented the second successful synthesis of the complete welwistatin carbon framework. Their strategy (Scheme 12) was based on the creation of the welwistatin C4–C11 bond through a palladium-catalyzed intramolecular coupling in compound **65**, which would be available using a Lewis acid-catalyzed displacement of the tertiary hydroxyl in compound **66** by a cyclohexenyl silyl enol ether. The starting material for this route was 4-bromoindole.

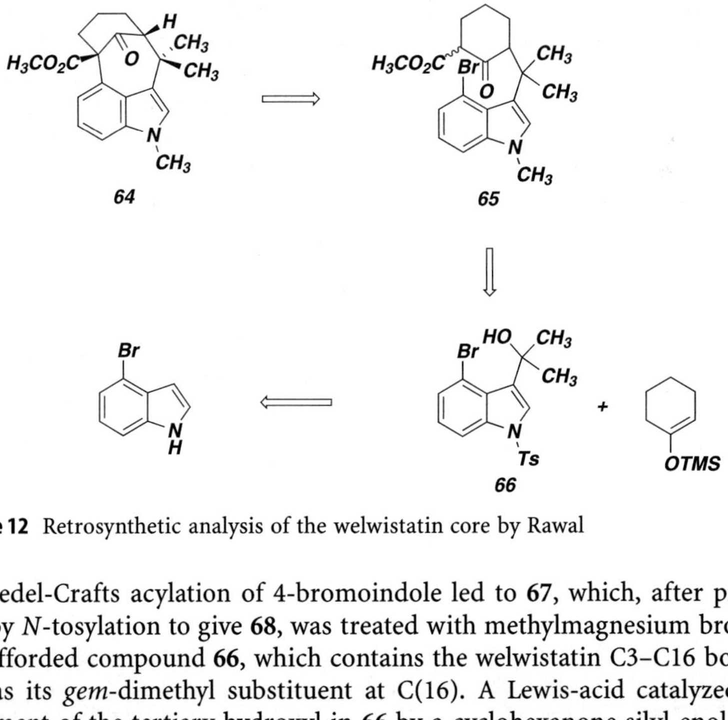

Scheme 12 Retrosynthetic analysis of the welwistatin core by Rawal

Friedel-Crafts acylation of 4-bromoindole led to **67**, which, after protection by N-tosylation to give **68**, was treated with methylmagnesium bromide and afforded compound **66**, which contains the welwistatin C3–C16 bond as well as its *gem*-dimethyl substituent at C(16). A Lewis-acid catalyzed displacement of the tertiary hydroxyl in **66** by a cyclohexanone silyl enol ether afforded intermediate **69**, which was then deprotected and N-methylated to **70** (Scheme 13).

In order to facilitate the desired palladium-catalyzed creation of the welwistatin C4–C11 bond, compound **70** was regioselectively transformed into β-ketoester **65** by treatment with LDA in the presence of HMPA, followed by addition of Mander's reagent. The crucial palladium-catalyzed cross-coupling reaction took place in the presence of palladium acetate, tri-*tert*-butylphosphine and potassium *tert*-butoxide, and afforded compound **64**, which contains the complete welwistatin skeleton. The installation of a nitro-

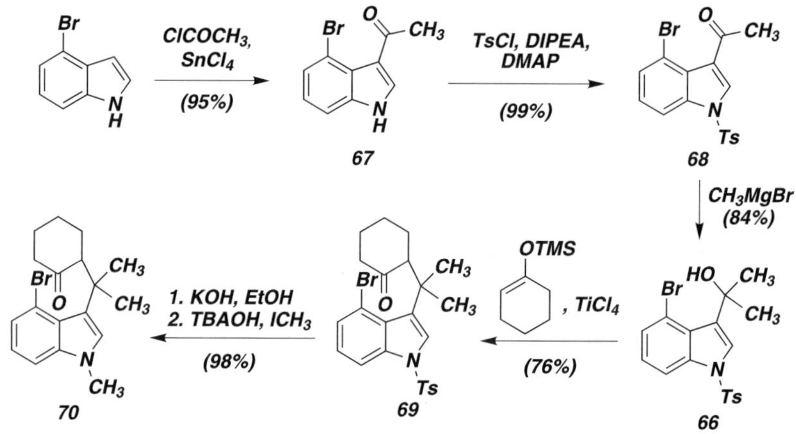

Scheme 13 Initial stages of the Rawal route to the welwistatin core

gen function at the bridgehead position of **64**, which seems to be essential for biological activity, was also studied. To this end, the ester group had to be hydrolyzed, in order to use the carboxyl function as a starting point for a Curtius rearrangement. This task proved more difficult than expected, since **64** was quite resistant to hydrolysis, even under forcing conditions. This low reactivity was explained considering that the conformation of the bicyclic system, as studied by X-ray diffraction, imposes severe steric constraints that force the ester group into a position where both faces of the carbonyl are inaccessible to

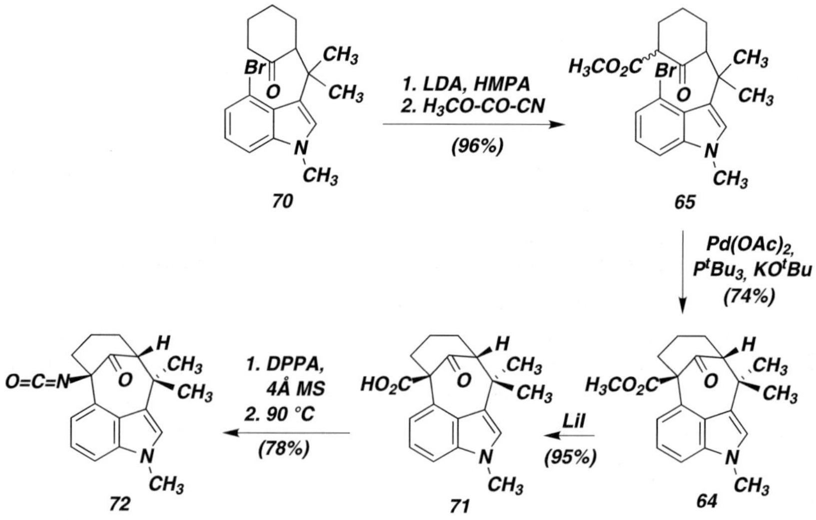

Scheme 14 Synthesis of the welwistatin core by Rawal

hydrolytic attack. This problem was solved by the use of nucleophilic dealkylation conditions, and thus refluxing **64** in pyridine with an excess of lithium iodide afforded acid **71** in 95% yield. Treatment of **71** with diphenylphosphoryl azide (DPPA) gave isocyanate **72**, which was found to be remarkably resistant towards nucleophiles such as benzyl alcohol (Scheme 14).

4.3.3
Synthesis of the Welwistatin Core by Simpkins

As part of a project studying bridgehead metalation chemistry, the Simpkins group became interested in the disconnections for the welwistatin skeleton shown in Scheme 15. After the failure of all efforts to translate disconnection (a) into practice, they focused their work on route (b), in spite of its similarity to the route unsuccessfully attempted by Konopelski [28] (Sect. 4.2).

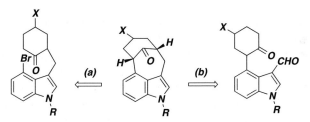

Scheme 15 Retrosynthetic analysis of the welwistatin core by Simpkins

Palladium-catalyzed coupling of cyclohexanone with 4-bromo-1-methylindole, under Buchwald conditions, allowed accessing the starting material **73**. Compound **73** was then formylated to **74** using the Vilsmeier–Haak reaction. All attempts to induce an intramolecular base-catalyzed aldol cyclization of compound **74** were unsuccessful, but acidic conditions led to an equimolecular mixture of two bridged structures, namely compounds **75** and **76**, which can formally be considered as dismutation products of the expected aldol product. Their formation can be explained by assuming that the initial aldol reaction leading to **77** is followed by hydroxide elimination to generate the iminium species **78**. Hydride transfer from **77** to **78**, as shown in Scheme 16, would easily explain the formation of the observed products.

In subsequent work, it was found that the reaction conditions could be modified to give a single product. Thus, carrying out the acid-catalyzed cyclization in the presence of an external source of hydride, such as triethylsilane, led to compound **75** in 92% yield, while the presence of an hydride acceptor such as DDQ resulted in the formation of **76** as the sole product, albeit in 45% yield. Compound **77** was finally transformed into the oxindole derivative **79** by oxidation with *N*-bromosuccinimide in the presence of *tert*-butyl alcohol (Scheme 17).

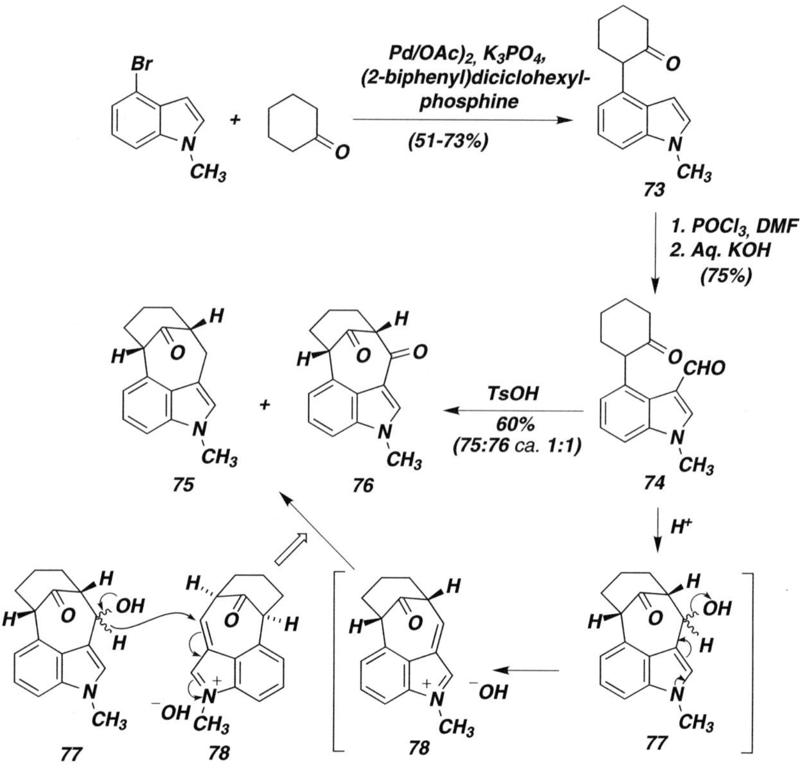

Scheme 16 Synthesis of the welwistatin core by Simpkins

4.3.4
Synthesis of the Welwistatin Core by Funk

Greshock and Funk have recently disclosed the synthesis of compound **80**, the most advanced welwistatin intermediate prepared to date [29]. Their approach differs from all others in that it does not start from an indole or oxindole derivative, but rather is deigned to build the indole ring system at a late stage, using an in-house developed [30] annelation sequence from intermediate **81**, which would be available from **82** via an electrocyclic ring closure reaction that can also be viewed as an intramolecular vinylogous Michael addition. The bicyclic structure **83** was to be built at the early stages of the synthesis, also using synthetic methodology developed by the Funk group [31] (Scheme 18).

Cyclohexanone derivative **85** was chosen as the starting material because of its ready availability from 3-methylanisole, both in racemic and enantiomerically pure forms [32]. Because of the previously established synthetic equivalence between 6-bromomethyl-4H-1,3-dioxine and bromomethyl vinyl

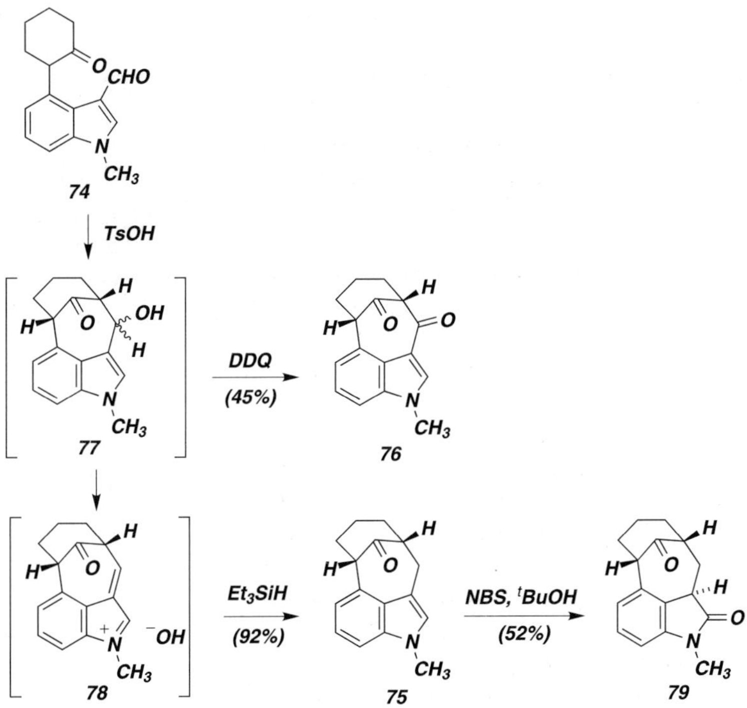

Scheme 17 Intramolecular aldol route to the welwistatin core

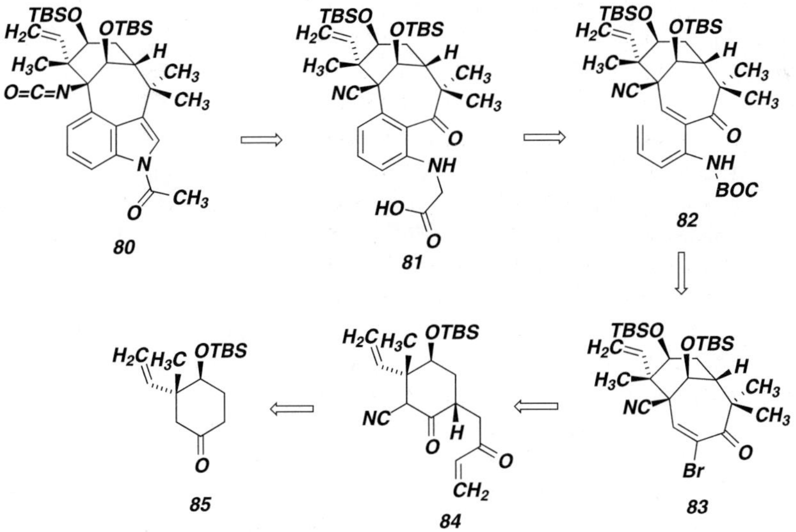

Scheme 18 Retrosynthetic analysis of the welwistatin core by Funk

ketone [31], its reaction with the kinetic enolate derived from **85** was studied. It was hoped that this reaction would proceed by axial alkylation from the α face through a chair conformer possessing equatorial vinyl and trialkylsilyloxy substituents, since alkylation from the β face would be hampered by a 1,3-diaxial interaction with the silyloxy substituent. This expectation was fulfilled in practice, since the reaction of **85** with the bromomethyl dioxine derivative in the presence of lithium hexamethyl disilazide afforded compound **86** with higher than 10 : 1 diastereoselectivity. Regioselective generation of the less hindered enolate derived from **86** followed by its cyanation by treatment with tosyl cyanide afforded compound **87**. Thermal degradation of the 1,3-

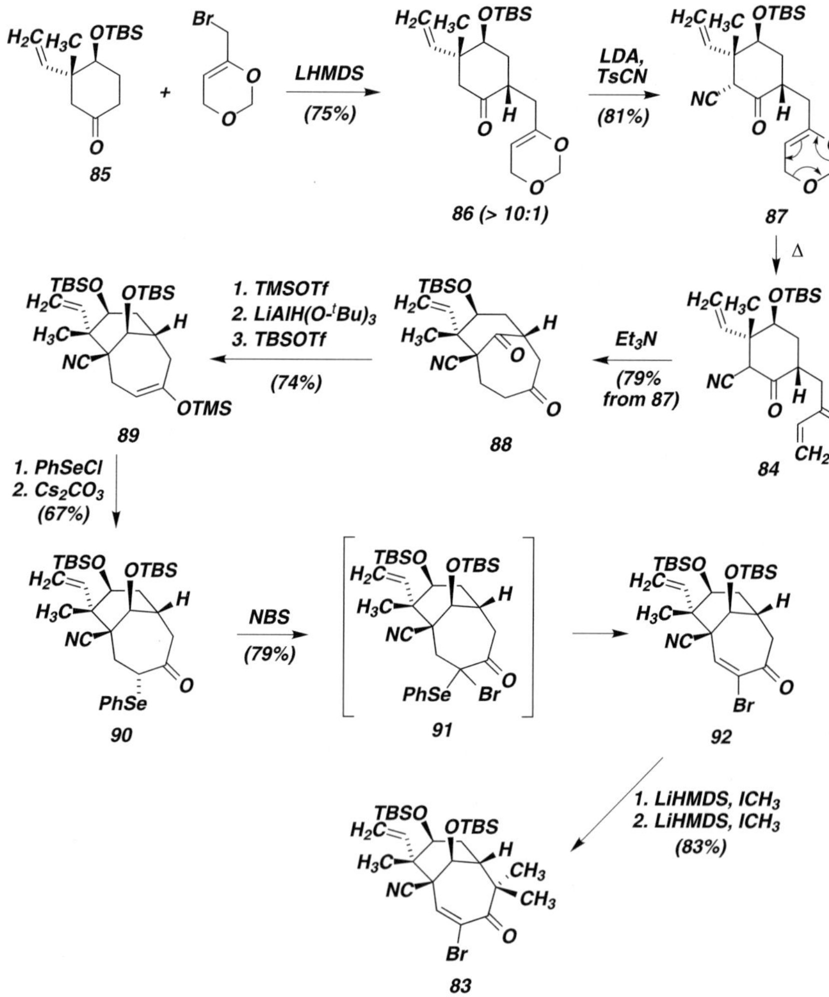

Scheme 19 Synthesis of a welwistatin CD fragment by Funk

dioxine ring in **87** through a retro-hetero Diels–Alder pathway afforded **84**, having an unmasked α, β-unsaturated ketone moiety. This set the stage for the generation of the bicyclo[4.3.1]decane system through an intramolecular Michael reaction, which was triggered by simple addition of triethylamine to **84** to give compound **88**, with no epimerization at the bridgehead position. Parallel work had shown that the presence of the carbonyl function at the one-carbon bridge would be troublesome during the electrocyclic ring closure leading to the indole ring (see Scheme 20). For this reason, chemoselective activation of the more reactive carbonyl group of **88** by its transformation into the corresponding silyl enol ether was followed by stereoselective reduction of the troublesome ketone carbonyl with $LiAlH(O - tBu)_3$, and the resulting alcohol was protected as a TBS ether to give **89**. This compound was transformed into **90** after its reaction with phenylselenyl chloride followed by equilibration of the epimeric mixture of selenides in the presence of cesium carbonate. The α-bromoenone functionality required for the subsequent cross-coupling step was secured by treatment of **90** with N-bromosuccinimide, which afforded compound **92** in a single step, presumably through the intermediacy of the undetected α-bromo-α-phenylselenylketone **91**. Interestingly, the selenide epimeric to **90** afforded only desbromo-**92** under the same conditions. The *gem*-dimethyl substituent was introduced at this stage by two sequential enolate methylations, leading to compound **83** (Scheme 19).

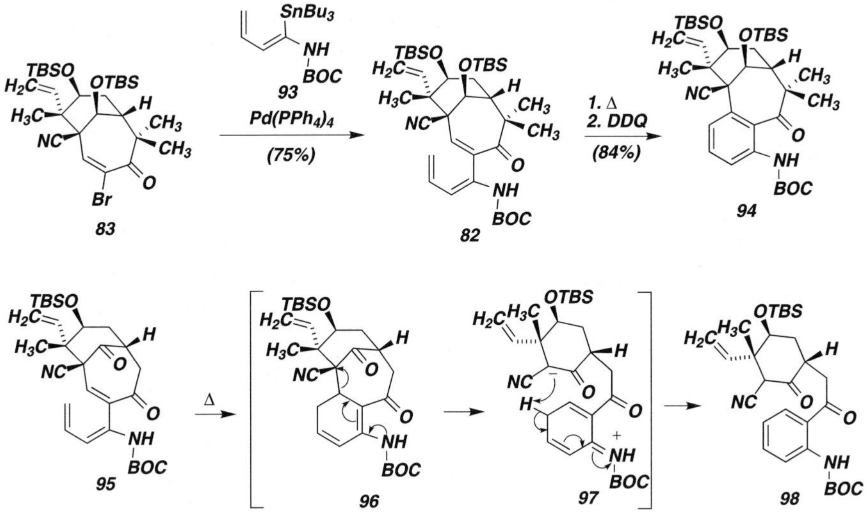

Scheme 20 Synthesis of a welwistatin ACD fragment by Funk

The Stille coupling of α-bromoenone **83** with the previously known [30] stannane **93** proceeded uneventfully and gave amidotriene **82**, the key precursor for the planned electrocyclization reaction. Heating **82** in refluxing toluene afforded the expected cyclohexadiene derivative, which was dehydrogenated in situ with DDQ to give compound **94**. As previously mentioned, the product of the electrocyclic ring closure of compound **95** containing a second carbonyl group at the one-carbon bridge proved to be unstable under thermal conditions and evolved through a retro-Michael reaction to give **98** (Scheme 20).

Removal of the *N*-protecting group in **94** with trifluoroacetic acid followed by reductive amination of the resultant aniline with glyoxylic acid afforded compound **81**, which was transformed into indole derivative **99** using the conditions previously established in methodological studies by the same group [30]. Compound **99** was employed to verify the feasibility of introducing the bridgehead nitrogen function through a Hofmann rearrangement. To this end, the nitrile function was hydrated in the presence of the Parkins catalyst **100** [33] to give amide **101**, which underwent a Hofmann-type rearrangement in the presence of lead tetraacetate [34] that afforded isocyanate **80** (Scheme 21).

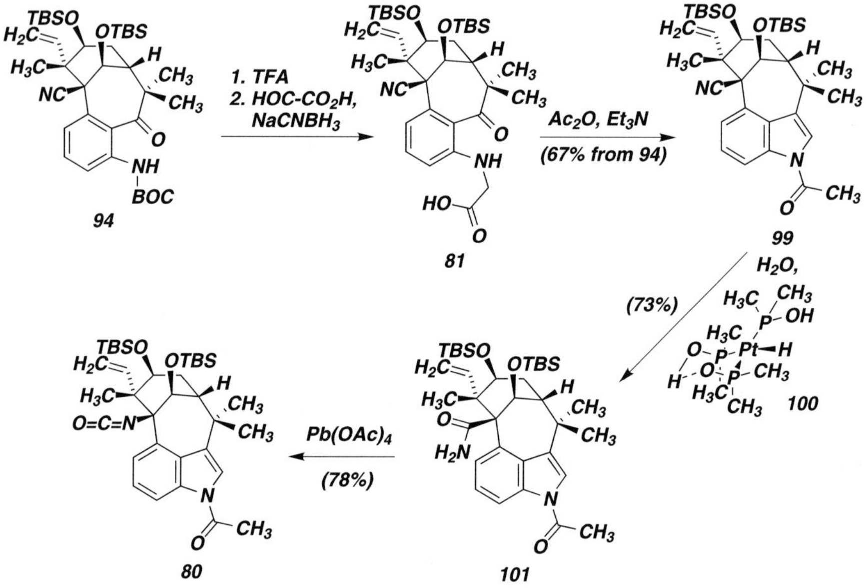

Scheme 21 Final stages of the synthesis of the welwistatin core by Funk

4.3.5
Synthesis of the Welwistatin Core by Shea

Lauchli and Shea have recently published a synthesis of the welwistatin core based on the use of a type 2 intramolecular Diels–Alder (T2IMDA) reaction to generate the fused bicyclo[4.3.1]system [35]. Thus, the preparation of compound **102** was planned from precursor **103**, where the indole framework is employed to tether the diene and dienophile components. The synthesis of **103** was planned from alkyne **104**, which should be available from 4-bromoindole via Sonogashira chemistry (Scheme 22).

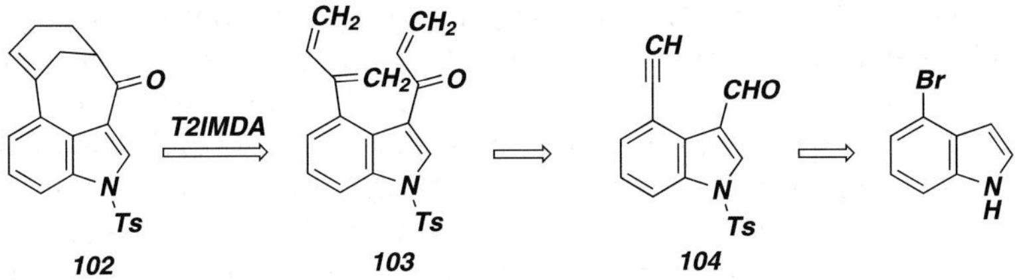

Scheme 22 Retrosynthetic analysis of the welwistatin core by Shea

Once again, a 4-bromoindole derivative was used as the starting material. Compound **105**, prepared by Vilsmeier–Haak formylation of 4-bromoindole,

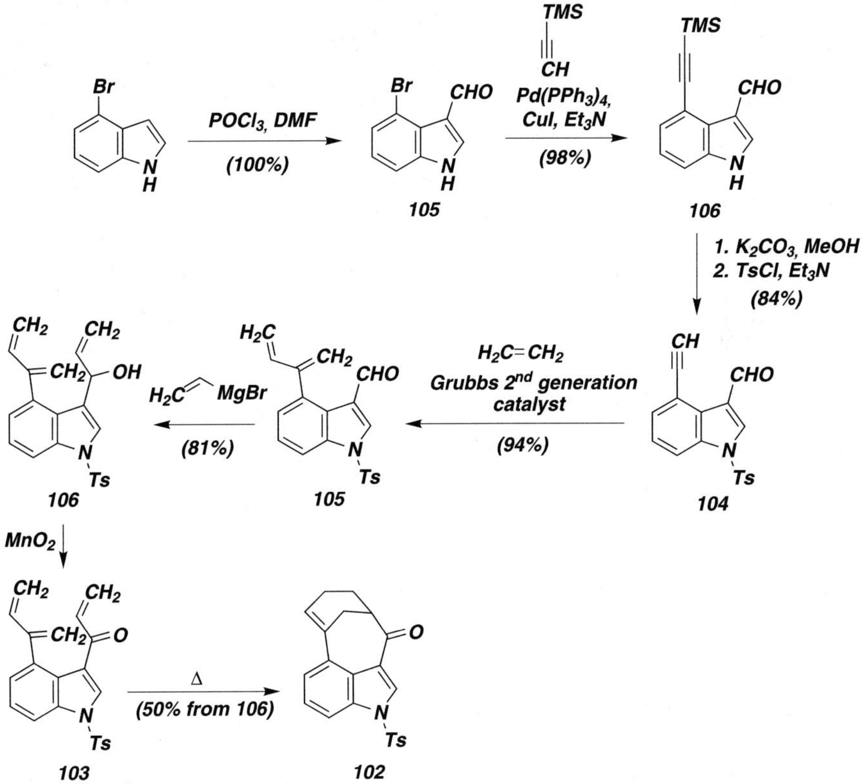

Scheme 23 The Shea intramolecular Diels–Alder route to the welwistatin core

was used as the substrate for a Sonogashira coupling with trimethylsily-lacetylene, which afforded **106**. Deprotection of the acetylene under basic conditions proceeded with no noticeable problem, and was followed by *N*-tosylation to **104**. This *N*-protection step was necessary for improving the solubility of the material in toluene for the next reaction, which consisted of an ene-yne metathesis reaction with ethylene in the presence of the second-generation Grubbs catalyst, leading to compound **105**. Addition of vinylmag-nesium bromide to **105** gave the allylic alcohol **106**, which was found to be more stable than its analogue bearing a methyl group at the indole nitro-gen, which decomposed rapidly, presumably via a carbocation whose forma-tion was hampered by the electron-withdrawing tosyl substituent. Manganese dioxide oxidation afforded vinyl ketone **103**, the substrate for the IMDA reac-tion, which proceeded under remarkably mild conditions and furnished the desired compound **102** (Scheme 23).

In order to introduce functionality at key positions, an intramolecular Diels–Alder reaction with furan as the diene component was planned. Suzuki coupling of the previously mentioned **105** with furan-3-boronic acid **106** gave compound **107**, which was *N*-tosylated to **108**. Construction of the dienophile portion was performed as in the previous case, by addition of vinylmagne-sium bromide to give **109**, followed by MnO$_2$ oxidation to give the cyclization precursor, which was immediately heated at 120 °C in toluene and afforded compound **110**, which incorporates an oxygen atom at both the ketone and vinyl chloride positions of welwistatin (Scheme 24).

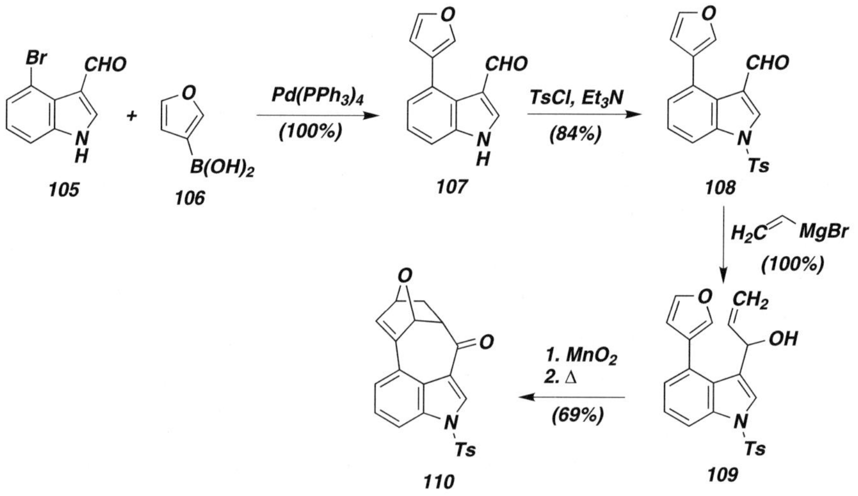

Scheme 24 Further studies on the IMDA route to the welwistatin core

4.3.6
Synthesis of the Welwistatin Core by Menéndez

Our group is working on a different approach to the welwistatin core that involves the initial preparation of a suitable tricyclic cyclohepta[cd]indole as a welwistatin ABC fragment, followed by construction of the D ring at a later stage (Scheme 25). Such an approach has the advantage of allowing the study of simplified welwistatin fragments in order to contribute to the establishment of structure-activity relationships. Thus, the synthesis of 111, a highly functionalized derivative of the welwistatin core, was planned involving as the key step a Michael-intramolecular aldol sequence from compound 112, which would be prepared by ring expansion from 113, the well-known Kornfeld's ketone. This starting material is readily available through an intramolecular Friedel-Crafts cyclization of 3-(1-pivaloyl-3-indolyl)propionic acid, where the pivaloyl group plays a crucial role in the regioselectivity of the reaction in favor of the less hindered C-4 position.

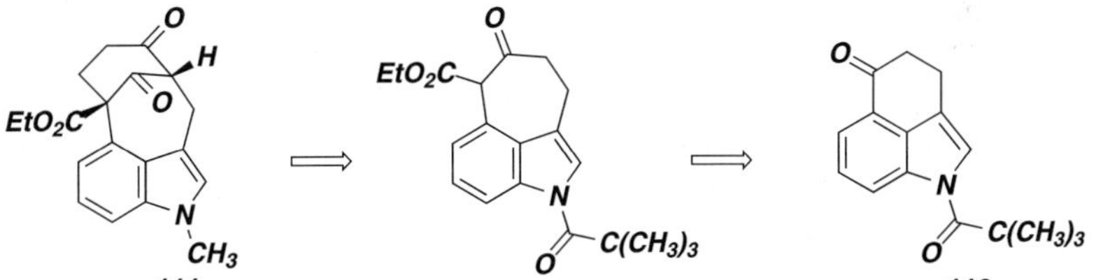

Scheme 25 Retrosynthetic analysis of the welwistatin core by Menéndez

As summarized in Scheme 26, transformation of compound 113 to the desired cyclohepta[cd]indole derivative 112 was achieved in a regioselective fashion by exposure to triethyloxonium tetrafluoroborate. Treatment of 112 with acrolein in the presence of potassium carbonate afforded the Michael adduct 114, which was cyclized to 115 in the presence of DBU. In the course of these studies, we found that it was possible to achieve hydrolysis of the pivaloyl protection in the same synthetic operation to give 116 by addition of water and an additional amount of DBU. The Michael addition-intramolecular aldol sequence could be performed in one operation in the presence of DBU, and this anionic domino process could also be combined with the hydrolysis of the pivaloyl group in the presence of water-DBU to give a 93% yield of 116 directly from 112. Although the reaction lacked diastereoselectivity, this is of no consequence since the stereocenter adjacent to the hydroxyl group is lost during subsequent stages of the sequence. Further elaboration of 116 to 111 involved chemoselective N-methylation under phase-transfer catalysis and oxidation of the hydroxyl group in the presence of TPAP [37].

EtO2C N2
BF4 Et3O
90%
CH2
K2CO3, THF
90%
1. Acrolein, DBU
2. H2O, DBU
93%
DBU, THF
1. ICH3, Bu4N+ HSO4-,
KOH, H2O-CH2Cl2
2. TPAP, NMO
65%
H2O,
DBU
90%
113
112
114
111
116
115

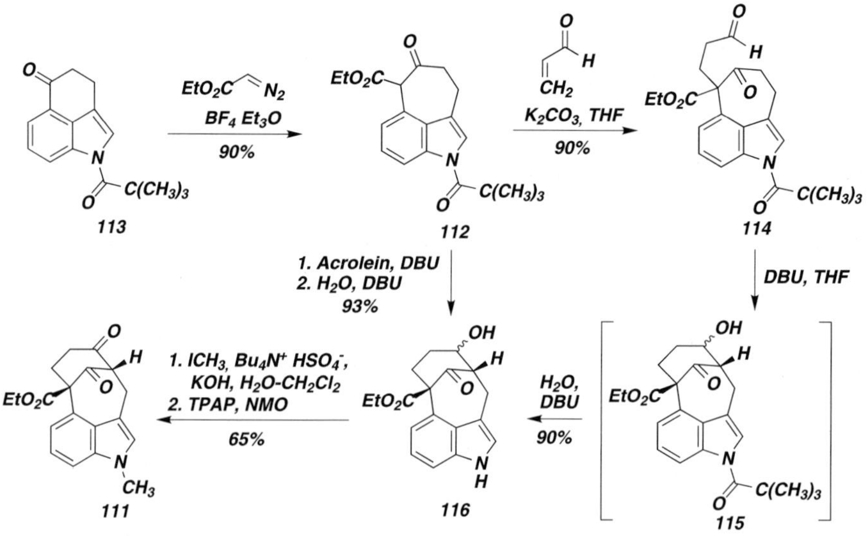

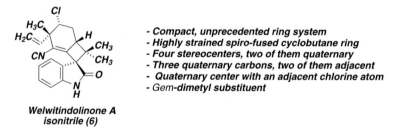

Scheme 26 The domino Michael-aldol route to the welwistatin core by Menéndez

5 Synthesis of Welwitindolinone A Isonitrile

5.1 Introduction

Welwitindolinone A isonitrile also has a number of structural features that make it a very challenging target, most notably the sterically crowded spiro-fused cyclobutane ring, which at first sight might seem thermodynamically unstable (Fig. 3).

Welwitindolinone A
isonitrile (6)
- Compact, unprecedented ring system
- Highly strained spiro-fused cyclobutane ring
- Four stereocenters, two of them quaternary
- Three quaternary carbons, two of them adjacent
- Quaternary center with an adjacent chlorine atom
- Gem-dimetyl substituent

Fig. 3 Main challenges in the construction of welwitindolinone A

Two total syntheses of welwitindolinone A isonitrile have been completed so far. The first of them, due to Baran, uses a ring-contraction strategy to generate the cyclobutane ring, and is remarkable for the absence of protecting groups. In the second (racemic) synthesis, the Wood group installed the cyclobutane ring early in the sequence and then created the spirooxindole system using a new samarium-mediated procedure.

5.2
Synthesis of Welwitindolinone A Isonitrile by Baran

Work on the total synthesis of welwitindolinone A isonitrile (6) by the Baran group [11] was inspired by the notion that its biosynthesis does not proceed from 12-epihapalindole E isonitrile 12, the hypothetical biosynthetic intermediate postulated by Moore (Scheme 2), but instead fischerindole I (117) is the actual precursor to 6 by an oxidative ring contraction. Compound 117 could arise by dehydrogenation of the cyclohexane ring in the enantiomer of fischerindole G (118). The preparation of the latter compound would involve a Friedel-Crafts cyclization of intermediate 119, which should be available from (R)-carvone oxide 120 using a direct indole coupling methodology developed by the same group [39, 40]. Thus, the successful implementation of the retrosynthetic plan outlined in Scheme 27 would provide access to three different natural products or their enantiomers, two of them belonging to the fischerindole class and the third one a welwitindolinone.

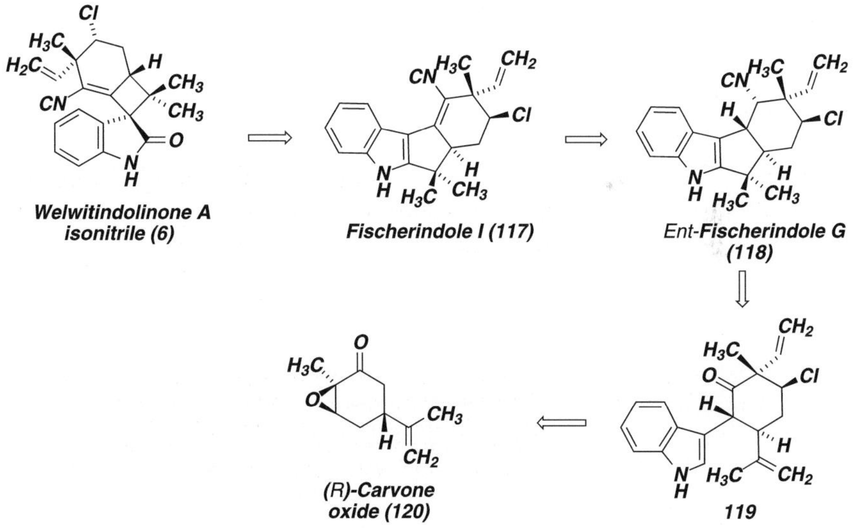

Scheme 27 Retrosynthetic analysis of welwitindolinone A isonitrile by Baran

The studies that eventually led to the synthesis of **6** started by installation of one of the all-carbon stereocenters by addition of vinylmagnesium bromide to the lithium enolate of (R)-carvone oxide **120**. This reaction afforded compound **121** as a single isomer, albeit in only 30% isolated yield because of a competing reaction derived from attack of the negatively charged carbon atom of the enolate via an S_N2' pathway. Chlorination of **121** with *N*-chlorosuccinimide and triphenylphosphine afforded **122**, which has three of the four stereocenters of the target molecule. Although the overall yield of this sequence is modest, it is still higher than that achieved through a ten-step route previously employed for the preparation of an epimer of **122** [38]; besides, the method is very fast and can be performed in a multigram scale. Coupling of **122** with indole using previously established methodology [39, 40] gave the key intermediate **119** in 55% yield as a single diastereoisomer. The Friedel-Crafts cyclization of **119** to **123** proved problematic, and a number of acidic catalysts failed to give a yield higher than 20% due to the formation of several side products. Eventually, a workable solution to this problem was found, consisting of the use of Montmorillonite K-10 under microwave irradiation. These conditions gave 40% isolated yield of **123** after one recycling of recovered starting material (57% brsm) (Scheme 28).

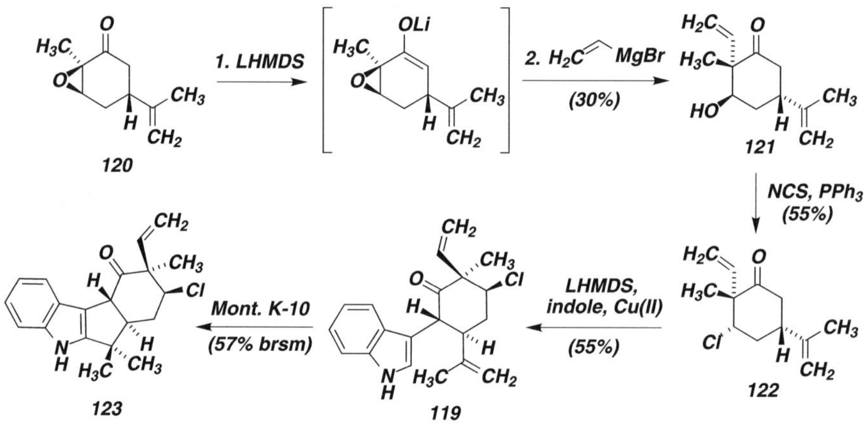

Scheme 28 Transformation of (*R*)-carvone oxide into a fischerindole precursor

Reductive amination of compound **123** proceeded with complete diastereoselectivity, which was opposed to the one previously found for the compound lacking the chloro substituent. Compound **124** thus obtained was initially considered unsuitable for later stages of the synthesis on stereochemical grounds, and for this reason access to its epimer **125** was sought using a four-step sequence [38], comprising carbonyl reduction, mesylation of the resulting alcohol, S_N2 displacement with lithium azide and final reduction. Compound **125** was successfully transformed into the corresponding isonitrile (**118**) by generation of the corresponding formamide derivative followed by dehydration with Burgess reagent. Compound **118** is the enantiomer of fischerindole G, and access to the naturally occurring isomer should be straightforward by this sequence using (*S*)-carvone oxide as the starting material. However, neither **118** nor its formamide precursor could be oxidized to fischerindole I **117** although a number of oxidants were tested. An attempt to use an indirect procedure also failed, since the reaction of **118** or its for-mamide precursor with *tert*-butyl hypochlorite or dimethyldioxirane gave the corresponding chloro- or hydroxy derivatives (e.g., **126**), but none of them could be transformed into **117**, presumably because the desired reaction in-volves a *syn* elimination (Scheme 29).

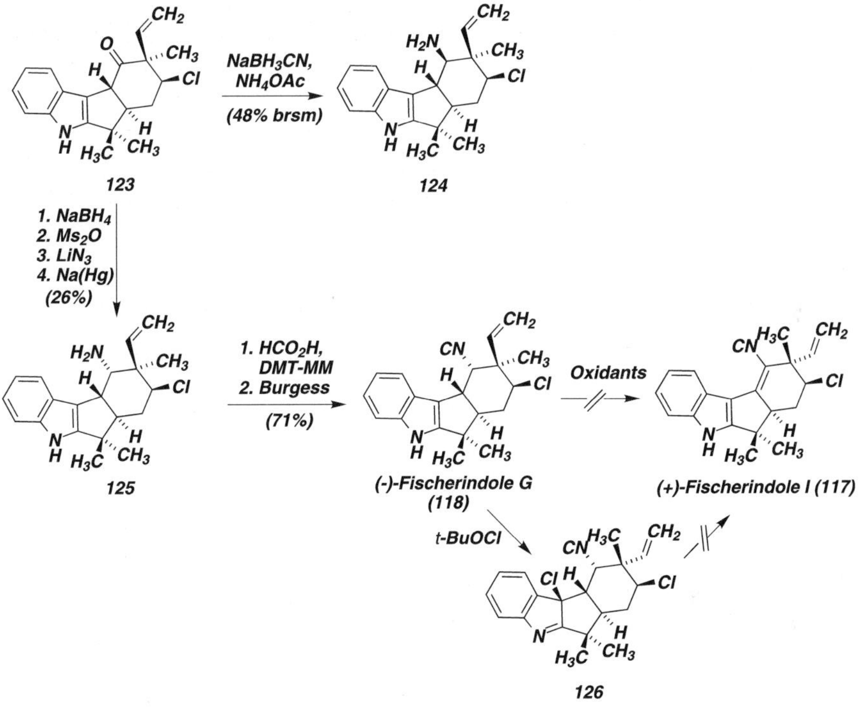

Scheme 29 Synthesis of (–)-fischerindole G by Baran

In view of these results, attention was turned to amine **124**, whose deriva-tives could be hoped to react with electrophiles from the opposite face. In-deed, formamide **127** gave (+)-fischerindole I (**117**) in 47% overall yield after treatment with *tert*-butyl hypochlorite and triethylamine followed by add-ition of silica gel deactivated with triethylamine and subsequent exposure to the Burgess reagent. This transformation presumably takes place by gener-ation of chloroindolenine **128**, *anti* elimination to give **129**, tautomerism to **130** and final dehydration (Scheme 30).

Finally, this chemistry was employed to access welwitindolinone A. This required the preparation of (–)-fischerindole I, which was achieved using the same route previously described starting from (*S*)-carvone. This starting

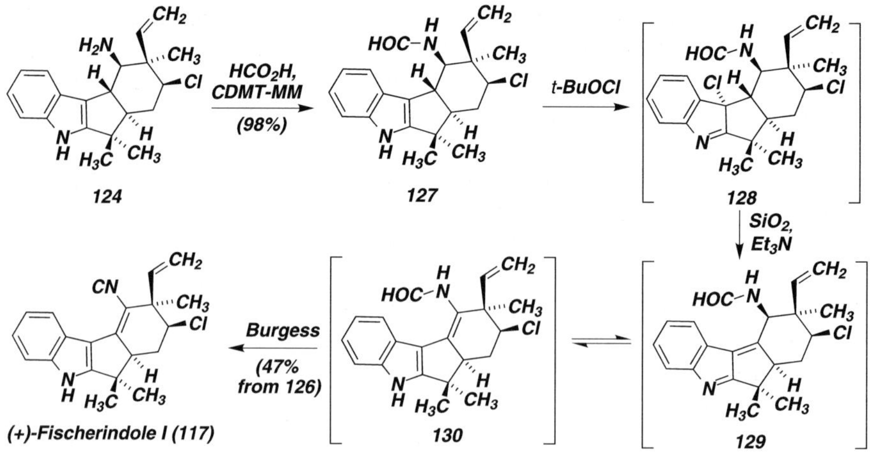

Scheme 30 Synthesis of (+)-fischerindole by Baran

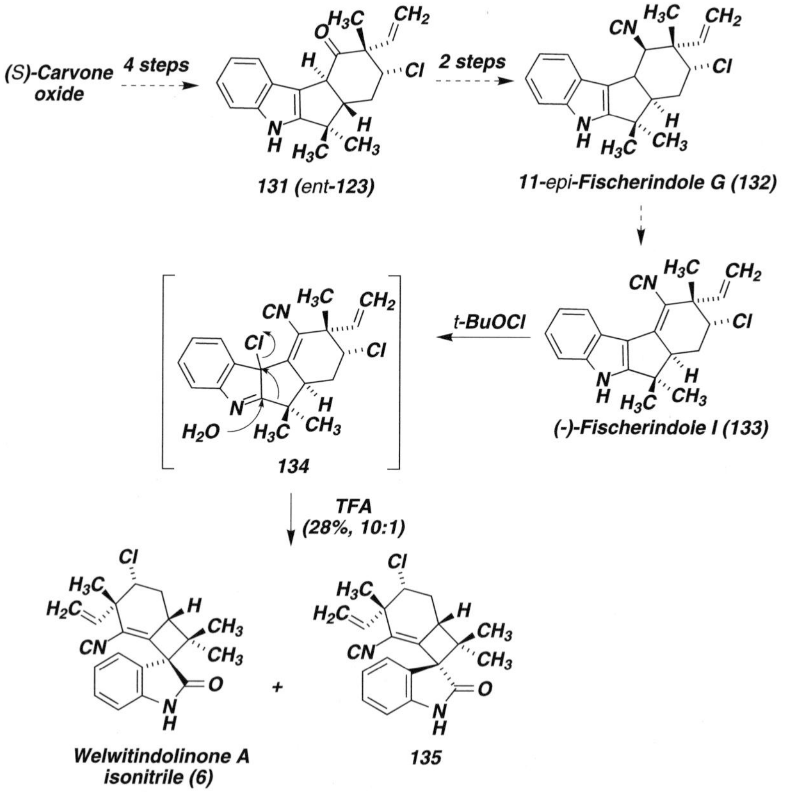

Scheme 31 Initial synthesis of welwitindolinone A isonitrile from (S)-carvone oxide by Baran

material was transformed into **131** and then to (+)-fischerindole I isonitrile **132**, which gave the target welwitindolinone A isonitrile by brief exposure to freshly prepared *tert*-butyl hypochlorite followed by acidic treatment through the intermediacy of **133** (Scheme 31). Besides its brevity and the use of some new synthetic methodology, this route is remarkable for the absence of any protecting groups. It allowed proposing a new biosynthetic pathway for the welwitindolinones, and it also served to confirm the absolute configuration of the natural product **6**, which had not been rigorously established.

A revised version of this synthesis has been subsequently developed that has a higher overall yield and is more easily scalable [42]. Reasoning that the yield and selectivity problems in the previous route were due to the use of *tert*-butyl hypochlorite, the transformation of 11-*epi*-fischerindole G **132** into (–)-fischerindole I **133** was carried out in excellent yield by exposure of **132** to DDQ in the presence of water, presumably through the intermediate unsaturated imine **136**. For the final ring contraction step, it was decided to replace the previously employed chlorohydroxylation by a hitherto unknown fluorohydroxylation, expecting that the increased hardness of fluorine over chlorine

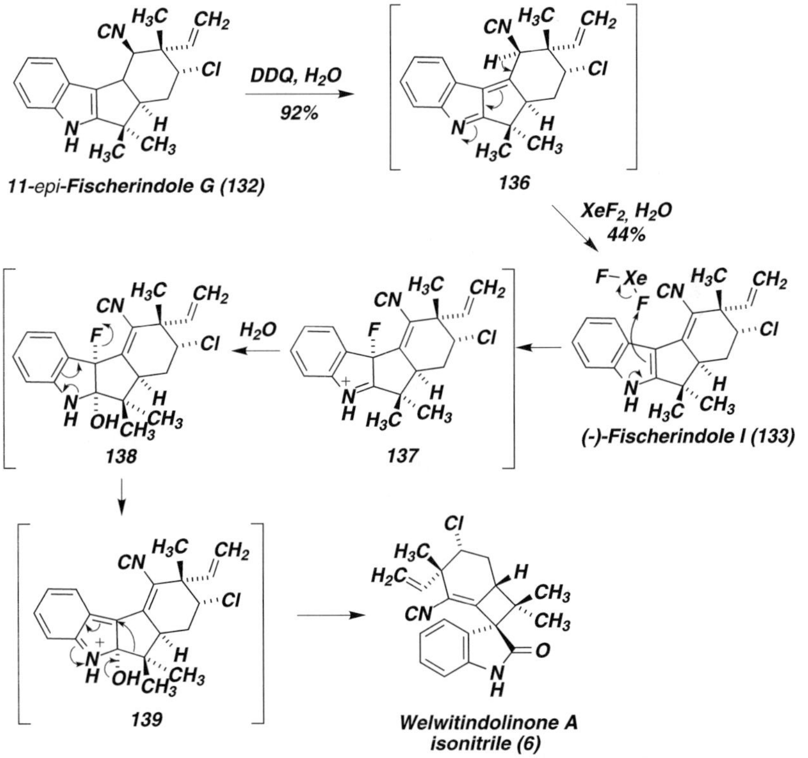

Scheme 32 Improved synthesis of welwitindolinone A isonitrile by Baran

would suppress side products derived from the isonitrile function. This was achieved by treatment with xenon difluoride and water in acetonitrile, presumably through a domino mechanism involving chemoselective fluorination of the indole ring to give **137**, which is trapped by water to give **138**. Elimination of fluoride would then give an azaorthoquinodimethane **139**, which would finally undergo a [1, 5] sigmatropic rearrangement leading to the diastereoselective generation of the spirocyclobutane unit of welwitindolinone A (Scheme 32).

5.3
Synthesis of Welwitindolinone A Isonitrile by Wood

The retrosynthetic approach to welwitindolinone A isonitrile (**6**) used by the Wood group is shown in Scheme 33. After recognition of the possibility of deriving the vinyl isonitrile fragment from a ketone, the disconnection of **6** to **140** was proposed. A literature report of a samarium (II) iodide-mediated reductive coupling of acrylates with isocyanates to give amides, which could be expected to lead to a new spirooxindole synthesis, prompted the disconnection of **140** to **141**. This compound was to be obtained from the readily available cyclohexadiene derivative **143**, by way of bicyclic ketone **142**.

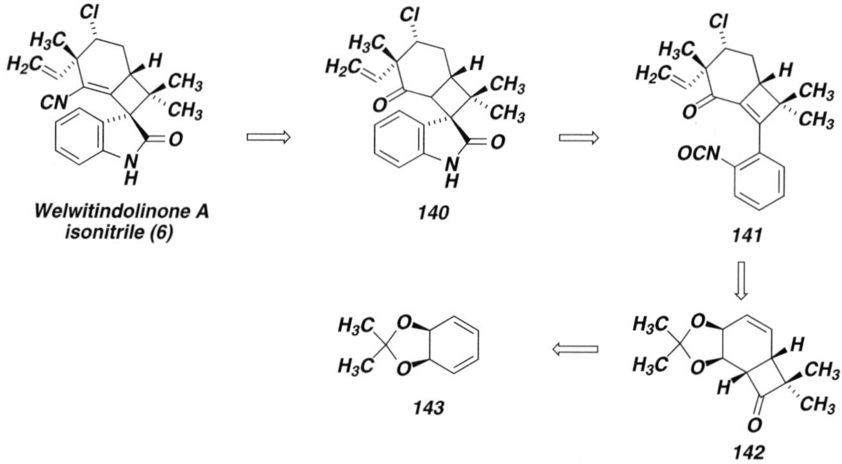

Scheme 33 Retrosynthetic analysis of welwitindolinone A isonitrile by Wood

A regio- and stereoselective [2+2] cycloaddition between cyclohexadiene acetonide **143** and ketene **144** afforded compound **142**. The required arylamine unit was introduced by means of the *ortho*-metallated aniline equivalent **145**, after the failure of many other aryl metal species to give the desired transformation. The triazene unit in compound **146** thus obtained was chemoselectively reduced to the corresponding amine by Ni Raney, and this was followed by simultaneous protection of the hydroxyl and amino groups by formation of the cyclic carbamate **147**. Acetonide removal followed by selective oxidation of the allylic alcohol furnished compound **148**. This compound was chosen as suitable substrate on which to prove the feasibility of generating the spirooxindole system by a new methodology developed for the purpose of this synthesis (see below). As the initial steps in this endeavor, **148** was first protected as a dithiolane and oxidized under Swern conditions to give compound **149** (Scheme 34).

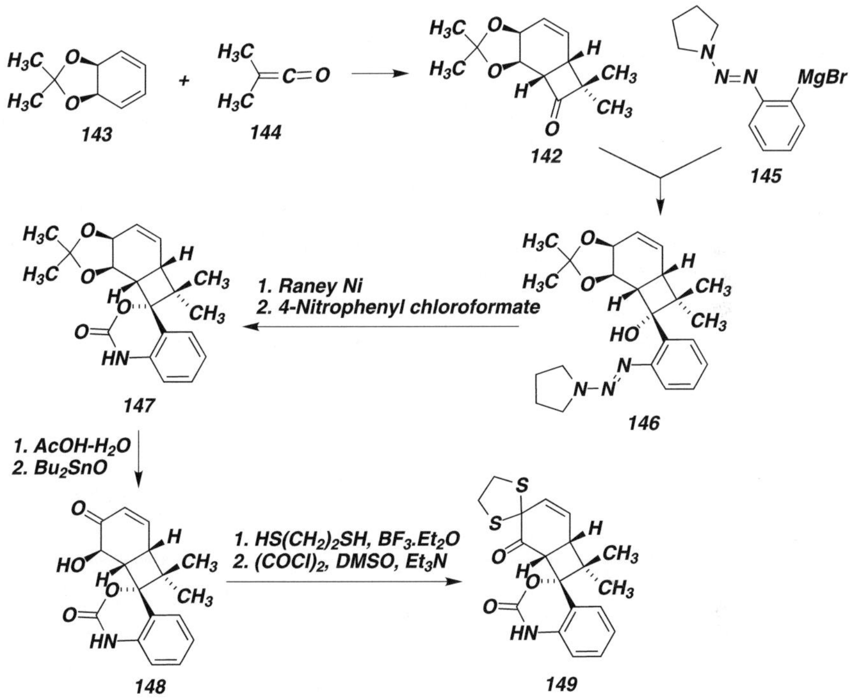

Scheme 34 Synthesis of spirooxindole precursors

Addition of DBU to a solution of **149** in THF induced an elimination reaction accompanied by loss of a molecule of CO_2 and provided the unstable amine **150**, which was converted in situ into isocyanate **151** by reaction with phosgene and triethylamine. After filtration to remove hydrochloride salts, the solution of **151** was treated with samarium (II) iodide in the presence of lithium chloride. These conditions, which had been previously determined to be optimal for spirooxindole generation on a model system, provided compound **152** as an inseparable 7 : 1 mixture of diastereoisomers [43]. The major component of this mixture was determined by NOE analysis to have the required configuration, which is consistent with bond formation from the less hindered, convex face of **151** (Scheme 35).

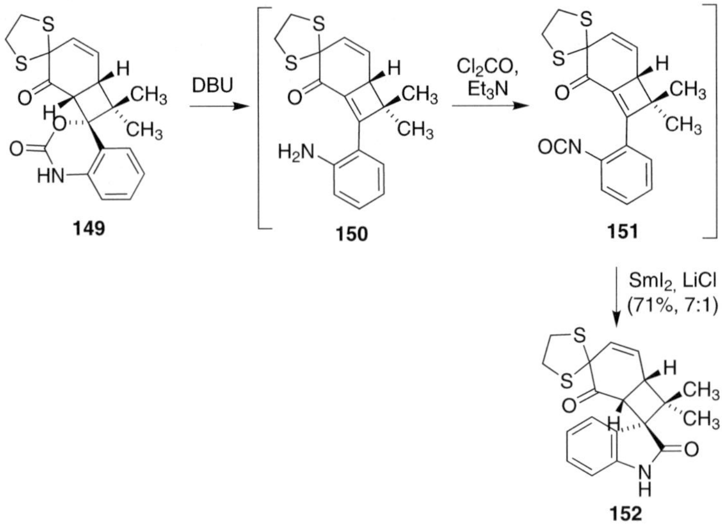

Scheme 35 Samarium iodide-promoted synthesis of spirooxindoles related to welwitindolinone A

With a suitable method for spirooxindole generation in hand, attention was turned to installing the quaternary center bearing the vinyl and methyl groups and also the adjacent chlorine-bearing stereocenter. An approach was devised that should address both issues simultaneously, based on the chloronium-induced semipinacol rearrangement from compound **153** shown in Scheme 36 [44]. It was hoped that the reaction mechanism would force methyl migration *anti* to the chloronium ion (intermediate **154**), leading to the correct relative stereochemical arrangement for the methyl and chlorine substituents in the reaction product **155**. In order to induce chlorination from the desired, more hindered concave face, the use of a bulky protection of the neighboring secondary hydroxyl (R group) was planned.

The tertiary alcohol **153** required as the starting material for this sequence was prepared from compound **148**, whose secondary hydroxyl was protected

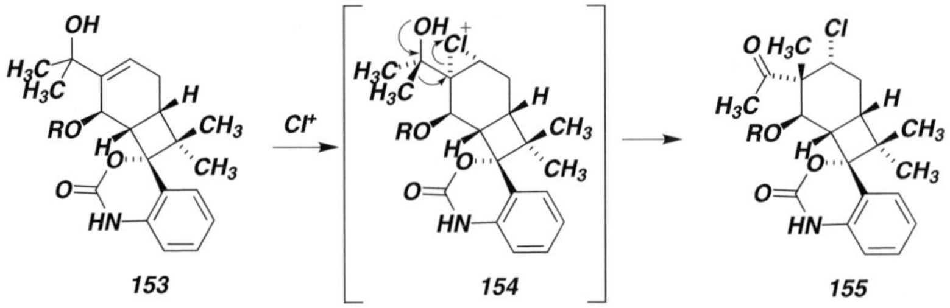

Scheme 36 Chloronium-induced semipinacol rearrangement for the installation of the quaternary stereocenter of welwitindolinone A. Model studies

as triisopropylsilyl ether. Enone **156** thus obtained was treated with lithium hexamethyldisilazide to protect the carbamate function as the corresponding lithium amide, and then with L-selectride and *N*-phenyltriflimide to give triflate **157**. This compound was submitted to a palladium-catalyzed insertion of carbon monoxide in the presence of methanol to give ester **158**, which afforded the desired tertiary alcohol **153** upon treatment with an excess of methylmagnesium bromide in the presence of cerium trichloride. The crucial reaction between **153** and a chloronium cation source could now be assayed. According to plan, treatment of **153** with sodium hypochlorite in the presence of cerium trichloride gave compound **155** in 78% yield and as a single diastereoisomer (Scheme 37).

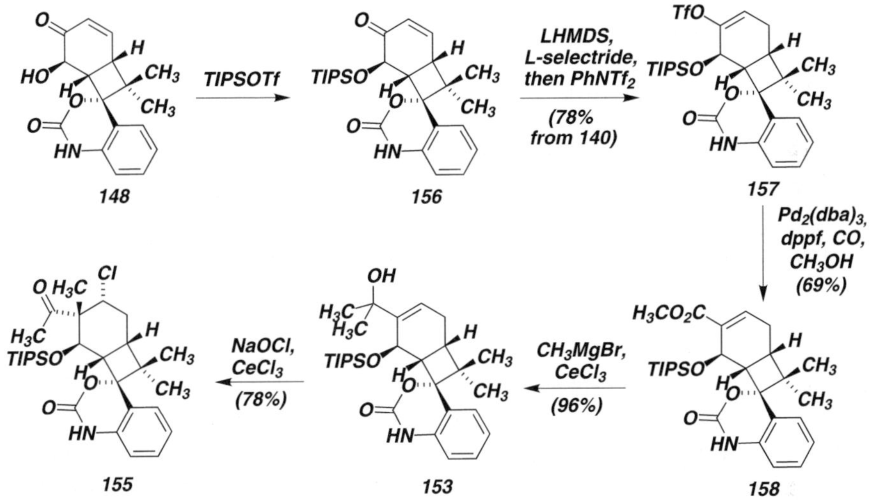

Scheme 37 Chloronium-induced semipinacol rearrangement for the installation of the quaternary stereocenter of welwitindolinone A

Deprotection of the secondary hydroxyl in compound **155** was carried out by exposure to fluorosilicic acid, and this step was followed by reduction of the resulting β-hydroxyketone moiety by tetramethylammonium triacetoxyborohydride to give diol **159** as a single diastereoisomer. Selective dehydration of the less hindered hydroxyl function by the Martin sulfurane reagent afforded the desired vinyl group, and the other hydroxyl was oxidized to ketone under Dess–Martin conditions to give compound **160**. Application of the method established in the preliminary studies afforded the desired spirooxindole derivative **140** as a single diastereoisomer. The only transformation that remained to complete the total synthesis was that of the ketone group of **140** into a vinyl isonitrile, but unfortunately all attempts to carry out this last step proved futile (Scheme 38).

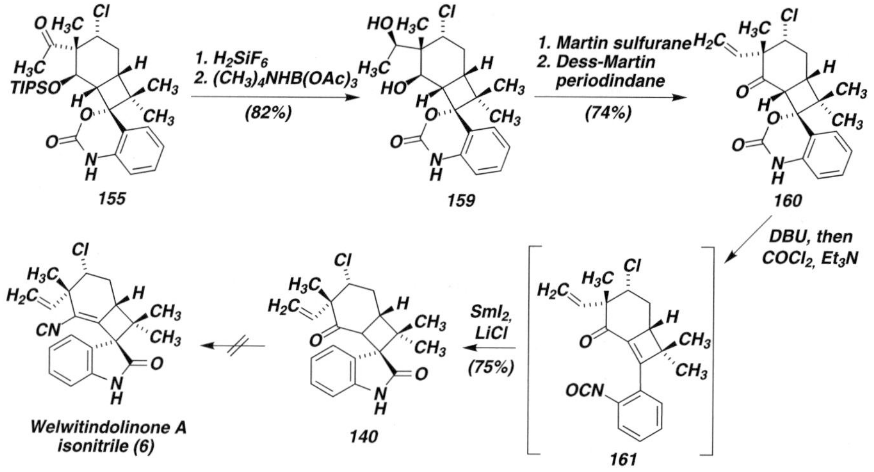

Scheme 38 Failed route to welwitindolinone A isonitrile by Wood

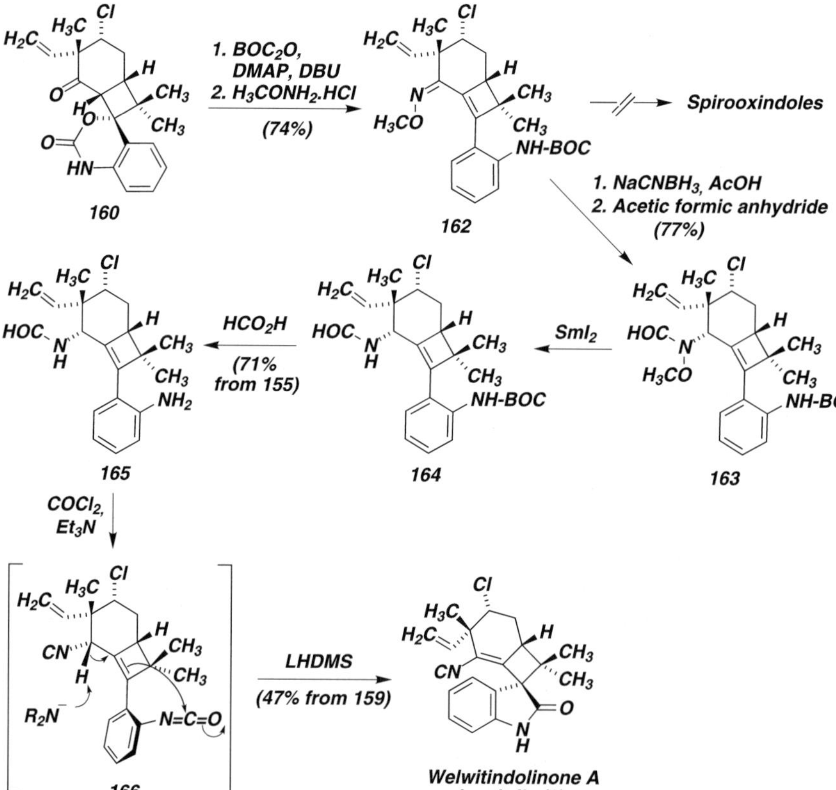

Scheme 39 Final steps of the succesful route to welwitindolinone A isonitrile by Wood

In order to overcome this last hurdle, the decision was taken to install, prior to oxindole formation, a nitrogen atom that could serve as a precursor to the isonitrile function. To this end, carbamate **160** was transformed into oxime **162** via a one-pot sequence involving decarboxylation and BOC protection that was followed by treatment of the enone thus obtained with *O*-methylhydroxylamine hydrochloride. However, compounds derived from **162** proved unreactive under the conditions of the previously developed SmI_2-mediated reductive cyclization, and hence an alternative method was necessary. Taking into account the possibility of deprotonating the position α to an isonitrile function using strong bases, the cyclization of isonitrile isocyanate **166** (Scheme 37) was planned. The preparation of this material started from **162**, which was reduced with sodium cyanoborohydride and then formylated to **163**. The reduction step occurred exclusively from the convex face, giving a compound with the correct stereochemistry for the final deprotonation step. Reductive cleavage of the N–O bond mediated by samarium (II) iodide furnished **164**, which gave **165** by acid-catalyzed BOC deprotection. Starting from **165**, the isonitrile and isocyanate functions could be obtained in a single synthetic operation by treatment with phosgene and triethylamine, leading to **166**. Exposure of the crude **166** to lithium hexamethyldisilazide gave the expected final cyclization and furnished the natural product in 47% yield and as a single diastereoisomer (Scheme 39).

6
Conclusion

The welwitindolinones have proved to be very challenging targets, with no total synthesis of welwitindolinone C (welwistatin) and only two of welwitindolinone A having being completed. Research into welwitindolinone synthesis has served as the stimulus to discover a number of new synthetic methodologies that will hopefully prove of value in other areas. It has also served to shed some light on the biosynthetic pathway of these fascinating and structurally complex molecules. Work in this field will certainly continue in the future, aimed at achieving the total synthesis of the elusive welwistatin, and also the preparation of its analogues for structure–activity relationship studies. It would also be of great interest to put to test the biosynthetic hypothesis according to which welwitindolinone A is the precursor to welwistatin.

Acknowledgements I would like to thank Professor Carmen Avendaño for her support and my coworkers Sonia Miranda, Juan Domingo Sánchez, Giorgio Giorgi, Miriam Ruiz, and Pilar López-Alvarado for their valuable contributions to the welwistatin project. Financial support of our research from MEC (grant CTQ2006-10930/BQU) and CAM-UCM (Grupos de Investigación, grant 920234) is gratefully acknowledged.

References

1. Stratmann K, Moore RE, Bonjouklian R, Deeter JB, Patterson GML, Shaffer S, Smith CD, Smitka TA (1994) J Am Chem Soc 116:9935
2. Jiménez JJ, Huber H, Moore RE, Patterson GML (1999) J Nat Prod 62:569
3. Smith CD, Zilfou JT, Stratmann K, Patterson GML, Moore RE (1995) Mol Pharmacol 47:241
4. Avendaño C, Menéndez JC (2002) Curr Med Chem 9:159
5. Avendaño C, Menéndez JC (2004) Med Chem Rev Online 1:249
6. Tseng E, Kamath A, Morris ME (2002) Pharm Res 19:1509
7. Kerr ID, Simpkins NS (2006) Lett Drug Des Discov 3:607
8. Zhang X, Smith CD (1996) Mol Pharmacol 49:288
9. Park A, Moore RE, Patterson GML (1992) Tetrahedron Lett 33:3257
10. Bornemann V, Patterson GML, Moore RE (1988) J Am Chem Soc 110:2339
11. Baran PS, Richter JM (2005) J Am Chem Soc 127:15394
12. Yokoyama Y, Matsushima H, Takashima M, Suzuki T, Murakami Y (1997) Heterocycles 46:133
13. Kaoudi T, Quiclet-Sire B, Seguin S, Zard SS (2000) Angew Chem Int Ed 39:731
14. Avendaño C, Menéndez JC (2004) Curr Org Synth 1:65
15. Elliott GI, Konopelski JP (2001) Tetrahedron 57:5683
16. Finet JP (1998) Ligand Coupling reactions with Heteroaromatic Compounds. Pergamon, Oxford (Tetrahedron Organic Chemistry Series, vol 18)
17. Konopelski JP, Lin J, Wenzel PJ, Deng H, Elliott GI, Gerstenberger BS (2002) Org Lett 4:4121
18. Elliott GI, Konopelski JP, Olmstead MM (1999) Org Lett 1:1867
19. Konopelski JP, Deng H, Schiemann K, Keane JM, Olmstead MM (1998) Synlett, p 1105
20. Deng H, Konopelski JP (2001) Org Lett 3:3001
21. Regitz M (1967) Angew Chem Int Ed 6:733
22. López-Alvarado P, García-Granda S, Álvarez-Rúa C, Avendaño C (2002) Eur J Org Chem, p 1702
23. Jung ME, Slowinski F (2001) Tetrahedron Lett 42:6835
24. Wood JL, Holubec AA, Stoltz BM, Weiss MM, Dixon JA, Doan BD, Shamji MF, Chen JM, Heffron TP (1999) J Am Chem Soc 121:6326
25. Krief A, Lecomte P (1993) Tetrahedron Lett 34:2695
26. Wood JL, Moniz GA, Pflum DA, Stoltz BM, Holubec AA, Dietrich HJ (1999) J Am Chem Soc 121:1748
27. Mackay A, Bishop RL, Rawal VH (2005) Org Lett 7:3421
28. Baudoux J, Blake AJ, Simpkins NS (2005) Org Lett 7:4087
29. Greshock TJ, Funk RL (2006) Org Lett 8:2643
30. Greshock TJ, Funk RL (2006) J Am Chem Soc 128:4946
31. Greshock TJ, Funk RL (2002) J Am Chem Soc 124:754
32. Frejd T, Polla M (1991) Tetrahedron 47:5883
33. Ghaffar T, Parkins AW (2000) J Mol Catal 160:249
34. Baumgarten HE, Smith HL, Staklis A (1975) J Org Chem 40:3554
35. Lauchli R, Shea KJ (2006) Org Lett 8:5287
36. Teranishi K, Hayashi S, Nakatsuka S, Goto T (1995) Synthesis p 506
37. Ruiz M (2007) DEA thesis. Universidad Complutense (Menéndez JC, López-Alvarado P supervisors)
38. Fukuyama T, Chen X (1994) J Am Chem Soc 116:3125
39. Baran PS, Richter JM (2004) J Am Chem Soc 126:7450

40. Baran PS, Richter JM, Lin DW (2005) Angew Chem Int Ed 44:609
41. Kim YH, Park HS, Kwon DW (1998) Synth Commun 28:4517
42. Baran PS, Maimone TJ, Richter JM (2007) Nature 446:404
43. Ready JM, Reisman SE, Hirata M, Weiss MM, Tamaki K, Ovaska TV, Wood JL (2004) Angew Chem Int Ed 43:1270
44. Reisman SE, Ready JM, Hasuoka A, Smith CJ, Wood JL (2006) J Am Chem Soc 128:1448

Top Heterocycl Chem (2007) 11: 103–144
DOI 10.1007/7081_2007_082
© Springer-Verlag Berlin Heidelberg
Published online: 19 September 2007

Lignans From *Taxus* Species

Gulacti Topcu[1] (✉) · Ozlem Demirkiran[2]

[1]Faculty of Science & Letters, Department of Chemistry, Istanbul Technical University,
Maslak, 34469 Istanbul, Turkey
topcugul@itu.edu.tr

[2]Faculty of Science & Letters, Department of Chemistry, Trakya University,
22030 Edirne, Turkey

Abstract Lignans are widely distributed in nature and exhibit various activities, including antitumor, antiviral, hepatoprotective, antioxidant, antiulcer, antiallergen, antiplatelet and antiosteoporotic activities. So far, *Taxus* species have received a great deal of interest in regard to their taxane diterpenes rather than their lignans. This chapter will review lignan biosynthesis and recent strategies in the synthesis of lignans and provide an overview of isolation and structural elucidation studies of all *Taxus* lignans, along with their biological activities. About 50 lignans, including neolignans and a few terpenolignans, isolated from eight *Taxus* species, are presented herein. Recent studies on the activities of lignans, particularly *Taxus* lignans, are outlined.

Keywords Biological Activity · Biosynthesis · Lignans · Synthesis · Taxaceae · *Taxus*

Abbreviations

AcOH	acetic acid
AIBN	2,2′-azabisisobutyronitrile
Bu	butyl
Bn	benzyl
DBU	1,8-diazabicyclo[2,2,2]octane
DIBAL-H	diisobutylaluminum hydride
DMAD	dimethyl acetylenedicarboxylate
DMSO	dimethylsulfoxide
LDA	lithium diisopropylamide
LHMDS	lithium hexamethyldisilazide
m-CPBA	*m*-chloroperoxybenzoic acid
Me	methyl
TBAF	tetrabutylammonium flouride
TBS	*t*-butyldimethylsilyl
Tf	trifluoromethanesulfonyl
THF	tetrahydrofuran
TFA	trifloroacetic acid
TFE	trifluoroethane
Ts	*p*-toluenesulfonyl
Ph	phenyl
PIDA	phenyliodinediacetate
PIFA	phenyliodinebistrifluoroacetic acid

1
Introduction

Lignans are a well-known class of widespread natural phenolic compounds that exhibit great structural and biological diversity and are commonly found in vascular plants from various families [1]. They are present at different levels of abundance in all plant parts, including roots, rhizomes, hardwood, bark, stems, leaves, flowers, fruits and seeds [2]. Lignans are of considerable pharmacological and clinical interest and are used in the treatment of cancer and other diseases [3]. The extensive pharmaceutical use of lignans is due to their antitumor, antiviral and hepatoprotective properties as well as many other beneficial activities.

Yew plants—Taxus species (Taxaceae)—contain lignans and are widely distributed across the Northern Hemisphere. The family Taxaceae comprises five genera consisting of *Amentotaxus* Pilger, Austro *Taxus* Compton, *Pseudotaxus* Cheng, *Taxus* Linn, and *Torreya* Arn [4]. It is common to indicate the region of occurrence of yew species when naming them, which results in names such as European, Canadian, American or Himalayan yew. However, there is some confusions over the true identities and nomenclature of the different species of genus *Taxus* [5]. Since all yews look very much alike, the presence of only one collective species, namely *T. baccata*, is often assumed. One approach used to classify the genus *Taxus* is based on macroscopic characters—their morphology (length and shape of needles, etc.)—while a second approach is based on microscopic characteristics of the plant. Interestingly, two botanical names are used for the Himalayan yew, *T. wallichiana* and *T. baccata*, and the Chinese yew is represented by five species, *T. chinensis* Rehd., *T. yunnanensis* Cheng et L.K. Fu, *T. mairei* Cheng et L.K. Fu, *T. cuspidata* Sieb et Zucc, and *T. wallichiana* Zucc [6]. The first three of these are considered to be synonymous. However, there is still uncertainty about which *Taxus* species are synonymous. Therefore, a better taxonomical approach may be to combine microscopic, macroscopic and phytochemical data.

Taxus species are well known due to the antitumor and anticancer activities of their taxane diterpenoids, which have attracted many researchers to them [5, 7, 8]. In *Taxus* species (Taxaceae), lignans have been shown to occur in different parts of the tree, extending from the needles to the roots [5, 8].

The chemical constituents of different *Taxus* species have been studied for over a hundred years. This genus has attracted a great deal of interest because they contain diterpenoids with taxane skeletons and lignans with various skeletons, as well as biflavonoids, steroids and sugar derivatives with remarkable biological and pharmacological properties. There are several reviews on *Taxus* species, but most cover only their taxane diterpenoids [9], although two recent reviews cover all types of compounds isolated from *Taxus* species [8, 10]. The present review may be the first to focus entirely on lignan constituents of *Taxus* species.

2
Chemistry and Classification of Lignans

2.1
Chemistry and Nomenclature

The lignans comprise a class of natural products, derived from cinnamic acid derivatives, which are related biochemically to phenylalanine metabolism.

In 1936, Haworth proposed [11] that the class of compounds derived from two $\beta\beta'$-linked C_6C_3 units, where the phenylpropane units are linked by the central carbon (C8) of their propyl side chains, should be called "lignans." In the 1970s, McCredie et al. [12] tried to extend the term "lignan" to include the compounds arising from the oxidative coupling of *p*-hydroxyphenylpropane units, and Gottlieb coined the term "neolignan" [13, 14] to refer to two phenylpropane units linked in a different way to C8–C8'. If the carbon skeleton of a lignan has an additional carboxylic ring which is formed by direct bonding between two atoms of the lignan skeleton, it is called a "cyclolignan." There are some definitions regarding lignans and neolignans [15].

However, IUPAC recently [16] recommended that the term "lignan" should be defined based on Haworth's 1936 definition and Gottlieb's 1972 and 1978 extensions, with some adaptations. The nomenclature of the diverse range of structures classified as lignans depends largely on trivial names and if necessary the appropriate numbering derived from the systematic IUPAC name. The term "neolignan" covers [14] lignans and related compounds where the two C_6C_3 units are linked by a different bond to the C-8 to C-8' bond. The neolignan linkage may include C-8 or C-8' so long as there are no direct carbon–carbon bonds between the C_6C_3 units, and if they are linked by an ether oxygen atom the compound is termed an "oxyneolignan." If there are multiple oxylinkages then bis-, tris-, etc., is used. Higher analogs of the lignans and neolignans are composed of three or more C_6C_3 units. The number of C_6C_3 units is indicated by the appropriate prefix placed before the term "neolignan;" thus there are three C_6C_3 units in sesquineolignan, four C_6C_3 units in dineolignan and five C_6C_3 units in sesterneolignan [16], and so on.

2.2
Classification of Lignans

Several classification schemes have been reported in the literature for lignans. Most of them divide lignans into eight or nine subgroups, including or excluding neolignans.

A general lignan classification scheme is illustrated in Fig. 1, which is very similar to that designed by Umezawa recently. He [1] classified the lignans into eight groups and twelve subgroups based on the way in which oxygen is incorporated into the skeleton and the cyclization pattern. Neolignans are not covered by his scheme. There are two additional lignan groups to those devised by Umezawa in Fig. 1: cyclobutane and benzofuran. In Fig. 1, furan lignans are depicted with and without 9(9')-oxygen, as shown by Umezawa. Likewise, dibenzylbutyrolactols are shown separately from dibenzylbutyrolactones [1].

All of the groups of lignans have been isolated from *Taxus* species except for the dibenzocyclooctadiene, arylnaphthalene and cyclobutane groups. Most lignans belong to the dibenzylbutyrolactones, the furofurans and the dibenzylbutanes. Reports of the isolation of neolignans from *Taxus* species are rare in the literature. A benzofuran-type neolignan, 3-O-demethyl-dehydrodiconiferyl alcohol, was isolated from a *Taxus* species *T. mairei* [17]. In addition, three other neolignans have been reported, two of which (named

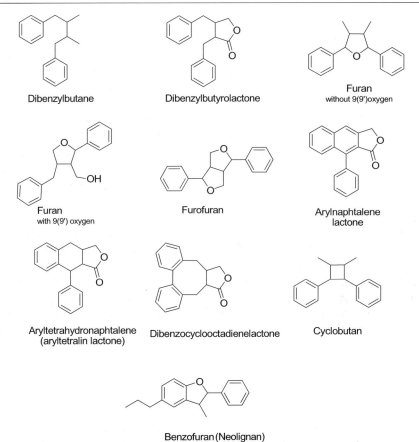

Dibenzylbutane

Dibenzylbutyrolactone

Furan
without 9(9')oxygen

Furan
with 9(9') oxygen

Furofuran

Arylnaphtalene
lactone

Aryltetrahydronaphtalene
(aryltetralin lactone)

Dibenzocyclooctadienelactone

Cyclobutan

Benzofuran (Neolignan)

Fig. 1 Representative skeletons of lignans

taxuyunin A and taxuyunin B) were oxyneolignans isolated from *T. yunna-nensis* [18], while the other was brevitaxin from *T. brevifolia* [19], which has a diterpene-lignan structure (Table 1).

3
Biosynthesis and Synthesis of Lignans

3.1
Biosynthesis of Lignans

In 1933, Erdtman suggested [20] that the basic lignan structure involved the coupling of two phenylpropanoid monomer units. Isotope tracer experiments on lignans were later described [2, 21, 22]. Significant advances have however been made in the chemistry and biosynthesis of lignans by Lewis

Table 1 Investigated *Taxus* species and Isolated Lignans

Taxus species and the plant part investigated	Crude extract	Lignans isolated	Refs.
***T. baccata* L.**			
a) Heartwood	H_2O	isotaxiresinol	[68]
b) Heartwood	CH_2Cl_2 : MeOH (1:1)	taxiresinol, isotaxiresinol, secoisolariciresinol,	[69]
c) Needles		(–)-secoisolariciresinol, 3-demethylsecoisolariciresinol, isotaxiresinol, (+)-isolariciresinol	[70]
d) Needles	CH_2Cl_2 : MeOH (1:1)	(–)-4-O-methyl-3′-O-demethylsecoisolariciresinol, (–)-secoisolariciresinol, 3-O-demethyl secoisolariciresinol	[71]
e) Twigs		suchilactone, 4′-O-demethylsuchilactone, lignan diol, (–)-secoisolariciresinol	[72]
f) Needles and Twigs	Not given	(–)-matairesinol, α-conidendrin, (–)-hibalactone, (–)-isohibalactone, (–)-2-(3,4-methylene dioxybenzyl)-3-(3,4-methylenedioxybenzylidene)butane-1,4-diol[a], (–)-2-(3,4-methylenedioxybenzyl)-3-(3,4-dimethoxy-benzylidene)-butane-1,4-diol[a]	[73]
g) Heartwood	EtOH	isolariciresinol, (–)-3′-demethylisolariciresinol-9′-hydroxyisopropylether, (–)-3-demethylisolariciresinol	[74]
h) Heartwood	EtOH	taxiresinol, lariciresinol (–)-3-demethylisolariciresinol-9-hydroxyisopropylether, (–)-3′demethylisolariciresinol	[75]
***T. brevifolia* Nutt** (Bark)	MeOH	isolariciresinol, secoisolariciresinol, isotaxiresinol, isotaxiresinol-6-methyl ether, brevitaxin	[19]

Table 1 (continued)

Taxus species and the plant part investigated	Crude extract	Lignans isolated	Refs.
T. cuspidata Sieb et Zucc.			
a) Wood	EtOH	isotaxiresinol, isolariciresinol	[77]
b) Roots	MeOH : CH$_2$Cl$_2$	(+)-taxiresinol, (+)-lariciresinol, (–)-secoisolariciresinol, (+)-pinoresinol	[78]
c) Roots	MeOH : CH$_2$Cl$_2$	isotaxiresinol, 7'-hydroxynortrachelogenin, nortrachelogenin, epi-nortrachelogenin, matairesinol, hydroxymatairesinol, allohydroxymatairesinol, oxomatairesinol	[79]
T. floridana Chapm.		(+)-isolariciresinol, (–)-secoisolariciresinol, isotaxiresinol, isotaxiresinol-6-methyl ether	[67]
T. mairei (Lemee and Levi) S.Y. Hu			
a) Roots	EtOH	(–)-α-conidendrin, (–)-secoisolariciresinol, isotaxiresinol, taxumairin	[80]
b) Bark	MeOH	Several lignans (their names were not given in the abstract)	[6]
c) Roots	EtOH	D-sesamin, 4-hydroxysesamin	[8]

Table 1 (continued)

Taxus species and the plant part investigated	Crude extract	Lignans isolated	Refs.
d) Twigs	Acetone	(−)-α-conidendrin, (+)-β-conidendrin, (−)-matairesinol, 7-(+)-hydroxymatairesinol, (S)-(−)-7′-hydroxymatairesinol, (R)-(+)-7′-hydroxymatairesinol, (−)-3′-O-demethylmatairesinol, (R)-(+)-7′-acetoxymatairesinol, (S)-(+)-7′-acetoxymatairesinol, (R)-(+)-7′-hydroxy-3′-O-demethylmatairesinol, 7′-oxomatairesinol, nortrachelogenin, (−)-epinortrachelogenin, 7′-(+)-hydroxynortrachelogenin, (+)-lariciresinol, (+)-isolariciresinol, (+)-isolariciresinol-9;9′-acetonide^b, (−)-secoisolariciresinol (as tetraacetate), Secoisolariciresinol-9,9′-acetonide^b, taxiresinol, 9-O-acetyltaxiresinol, (+)-isotaxiresinol, (+)-isotaxiresinol-9,9′-acetonide, (+)-pinoresinol, (+)-epipinoresinol, (+)-3′-O-demethylepipinoresinol, meso-neoolivil, (−)-dihydrocubebin, (+)-tanegool, (−)-8′-epitanegool, (+)-3′-O-demethyltanegool, (−)-dihydrodehydrodiconiferylalcohol, (−)-3-O-demethyldihydrodehydrodiconiferyl alcohol, (−)-3,3′-dimethoxy-4,4′,9-trihydroxy-7,9′-epoxylignan-7′-one, (−)-threo-3,3′-dimethoxy-4′,8-epoxyligna-4,7,9,9′-tetraol (as tetraacetate), (−)-sesquipinsapol B (as pentaacetate), (+)-sesquimarocanol B (as hexaacetate)	[17]

Table 1 (continued)

Taxus species and the plant part investigated	Crude extract	Lignans isolated	Refs.
T. media cv. **Hicskii** Roots	CHCl₃ : EtOH (1 : 1)	hydroxymatairesinol, *epi*-nortrachelogenin	[84]
T. wallichiana Zucc.			
a) Roots, stems, needles	EtOH	α-conidendrin, β-conidendrin, 7-hydroxymatairesinol, isoliovil	[85]
b) Heartwood	MeOH	taxiresinol, isotaxiresinol, (–)-secoisolariciresinol	[86]
T. yunnanensis **Cheng et L.K. Fu**			
a) Bark	EtOH	taxuyunin A, taxuyunin B, brevitaxin	[87]
		α-conidendrin, taxiresinol, isotaxiresinol, secoisolariciresinol,	
b) Wood	H₂O	(7'R)-7'-hydroxylariciresinol, 2-[2-hydroxy-5-(3-hydroxypropyl)-3-methoxyphenyl]-1-(4-hydroxy-3-methoxyphenyl)-propan-1,3-diol	[88]

a Lignan names are given as written in the literature, in general [8]. However, the two marked compounds[a] were also termed 9,9'-dihydroxy-3,4-dimethoxy-3',4'-methylenedioxylign-7-ene and 9,9'-dihydroxy-3,4:3',4'-bis-methylenedioxylign-7-ene which correspond to compounds **5** and **6** in Table 2, respectively

b Probably an artifact

and Davin [23] and Umezawa [1, 15], especially over the last decade. Lignan biosynthesis has been found to be closely related to, but fairly distinct from, the biosyntheses of other phenylpropanoids, such as norlignans, lignins and neolignans.

A report is about the in vitro formation of the first optically pure lignan (–)-secoisolariciresinol, from an achiral coniferyl alcohol in *Forsythia intermedia* which was used as an enzyme source [24]. From the same *Forsythia intermedia* source, (–)-matairesinol was also obtained by the selective oxidation of (–)-secoisolariciresinol in the presence of NADPH and H_2O_2 with an enzyme preparation [25]. Enzymatic reduction of (+)-pinoresinol to (–)-seco isolariciresinol via (+)-lariciresinol was also well characterized [26–28]. In 1997, by following approaches described in several reports, Davin et al. isolated [29] a unique protein (dirigent protein) from *Forsythia* spp., which led to the enantioselective formation of (+)-pinoresinol by the coupling of coniferyl alcohol in the presence of laccase/O_2 or a single-electron oxidant. Based on these studies, it was considered that lignan synthesis mediated by *Forsythia* dirigent protein and enzymes is well controlled in terms of stereochemistry. Homologous genes of the dirigent protein have been found in many plant species, and it seems to be present in almost all plant species.

Investigations of gymnosperms and angiosperms showed the existence of a common lignan biosynthetic pathway where it acts as enzymes, particularly pinoresinol/lariciresinol reductase [30] and secoisolariciresinol dehydrogenase [31].

Furthermore, experiments where various plant species are fed labeled coniferyl alcohol with either [^{2}H] or [^{14}C] have afforded pinoresinol, lariciresinol, secoisolariciresinol and/or matairesinol through what appears to be a common lignan biosynthetic pathway. As seen in Fig. 2, the coupling of coniferyl alcohol produces pinoresinol. This compound is reduced twice, leading to secoisolariciresinol. The lactone ring is then closed to give matairesinol, which may be the starting point for all lignans with a lactone ring. This first step in lignan biosynthesis seems to occur in many plants, and different lignans are formed from matairesinol in many distinct species—a clear example of divergent evolution [23]. Upstream lignans in the biosynthetic pathway consisting of pinoresinol, lariciresinol and secoisolariciresinol have been isolated from various plant species, and include both enantiomers. In contrast, all dibenzylbutyrolactone lignans, including matairesinol, have been identified as optically pure lignans using a chiral column on an HPLC [22]. A literature survey indicated that all of the dibenzylbutyrolactone lignans isolated so far from Thymelaceae plants are dextrorotatory [22] and have the same absolute configuration at C8 and C8′ with respect to the carbon skeleton, except for (–)-matairesinol. Sequences of conversion of coniferyl alcohol to lignans and interconversion of lignans in Thymelaceae plants are similar to the sequence catalyzed by *Forysthia* enzymes. However,

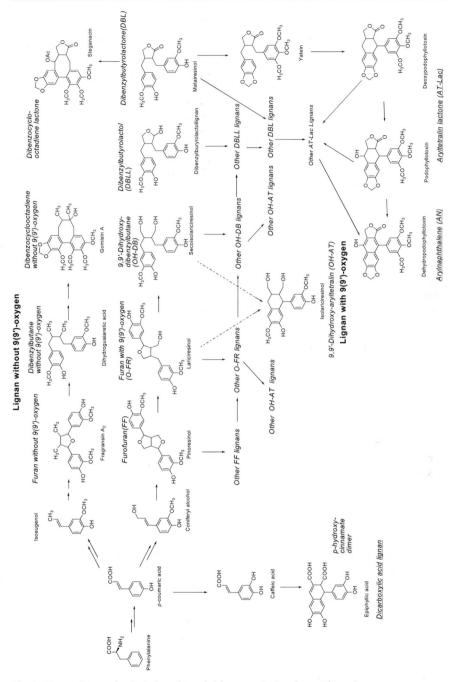

Fig. 2 Lignan biosynthesis (taken from [1] by permission from Kluwer)

the studies on these plants revealed that the dibenzylbutyrolactone lignans found in this family of plants, including *Wikstroemia sikokiana*, are the opposite enantiomers to those that occur in other plant species, such as *Forsythia* ssp. [32].

Broomhead and coworkers showed that matairesinol is the precursor of aryltetraline lactone podophyllotoxin in *Podophyllum hexandrum* (subclass Magnoliidae) [33], but this ^{14}C labeling experiment on *Diphylleia cymosa* indicated that there was no incorporation into the arylnaphthalene lignan diphyllin. Xia and coworkers showed that matairesinol also serves as a precursor in the biosynthesis of 6-methoxypodophyllotoxin in *Linum flavum* (subclass Rosidae) [34]. Noting the structural similarity between arylnaphthalenes and aryltetralins, arylnaphthalenes are considered to be formed by the dehydrogenation of the corresponding aryltetralins or by the hydroxylation of aryltetralins followed by dehydration (Fig. 2).

There are also many examples that highlight the stereochemical diversity of lignan biosynthesis, as observed in a cell-free extract of *Arctium lappa* petiole [35], which afforded secoisolariciresinol with the opposite antipode (+) to that formed by *Forsythia* spp.

A notable example of convergent evolution is the last part of the biosynthesis of 2,7′-cyclolignan [23]. For example, the cytotoxic 2,7′-cyclolignan podophyllotoxin (Fig. 2) can be found in at least ten plant families, such as Magnoliidae, Pinales, Dilleniidae, Rosidae and Asteridae. In some of these families' genera (for example *Juniperus*), most of the species produce 2,7′-cyclolignans, while synthesis seems limited to only one or two species in other families, such as *Anthriscus sylvestris* in the Apiaceae, although how extensively all of the species have been investigated is not clear. So far, podophyllotoxin has not been isolated from any *Taxus* species. As a result, the evolution of the last part of the biosynthesis of 2,7′-cyclolignan is an interesting subject for the convergent evolution of biosynthetic pathways. The biosynthesis of these lignans is only partially understood, but all lignan syntheses start with the stereospecific coupling of two monolignol units, usually coniferyl alcohol.

As interesting combination of flavonoid and lignan structures is found in a group of compounds called flavonolignans. They arise by oxidative coupling processes between a flavonoid and a phenylpropanoid, usually coniferyl alcohol. Silybin, silychristin and silydianin are well-known examples that are collectively termed "silymarins", and are isolated from *Silybum marianum* [36] as flavonolignans.

An optically active dicarboxylic acid lignan (caffeic acid dimer), named epiphyllic acid was isolated from liverwort as well as its derivatives, which do not belong to any typical lignan-producing subgroups. They also occur in some vascular plants, even as a triterpene ester of epiphyllic acid in *Rhoiptelea chiliantha* [37]. The conversion of caffeic acid to epiphyllic acid was demonstrated by Tazaki et al. [38]. However, little is known about the biosynthesis of this type of lignan in vascular plants yet.

3.2
Strategies for Lignan Synthesis

The lignans are an important class of natural products derived formally from the dimerization of substituted 3-phenylpropane precursors. Medicinal chemists have been especially intrigued by the lignans that display cytotoxicity, since such compounds could provide valuable leads in the search for novel antitumor agents [39, 40]. Complex lignan prototypes such as steganacin and podophyllotoxin, as well as simpler lignans, have attracted many synthetic organic chemists [39–41]. However, most of the work on lignan synthesis has focused on podophyllotoxin and derivatives rather than other types of lignans [39, 41].

3.2.1
General Approaches to Lignan Synthesis

There are various approaches to lignan synthesis. Although some of them are well known reactions that have been used for many years, there are also much newer approaches too. The approaches are listed and explained below.

3.2.1.1
Tandem Conjugate Addition Approach

The use of tandem conjugate addition reactions in lignan synthesis was introduced by Ziegler and Shwartz [42]. This methodology was subsequently extended to permit the synthesis of a wide variety of lignans [43–45] by Pelter et al. The advantage of this reaction is the generation of the required *trans*-2,3-dibenzylbutyrolactone framework in one step. For example, the treatment of the dithioacetal derivative obtained from piperonal with butyllithium followed by butenolide leads to an enolate anion that can be trapped with an aromatic aldehyde, such as 3,4,5-trimethoxybenzaldehyde, to give a dibenzyl-butyrolactone (Fig. 3). Thiophenylether substituents can be removed using Raney nickel. Acid-catalyzed cyclization then yields deoxyisopodophyllo-toxin [45].

3.2.1.2
Diels–Alder Approach

Diels–Alder reactions have been widely used to synthesize lignans, particularly the arylnaphthalene and aryltetralin types. This was one of the first examples of the use of an *inter*molecular (as opposed to an *intra*molecular) Diels–Alder reaction for lignan synthesis. The approach includes the use of an arylisobenzofuran (Fig. 4) generated in situ as diene and dimethylacetyl-enedicarboxylate (DMAD) as the dienophile (Fig. 4). The Diels–Alder prod-

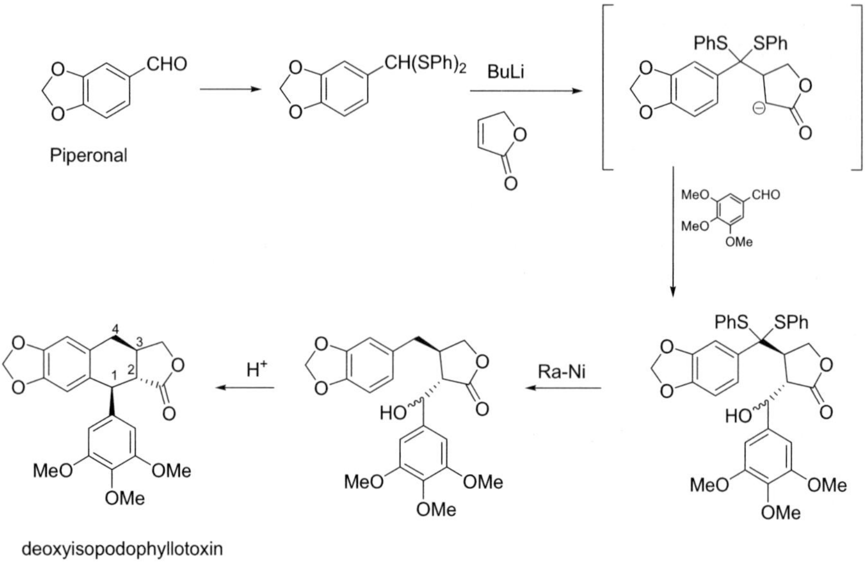

Fig. 3 Synthesis of deoxyisopodophyllotoxin by a tandem conjugate addition reaction (taken from [39] by permission from Kluwer)

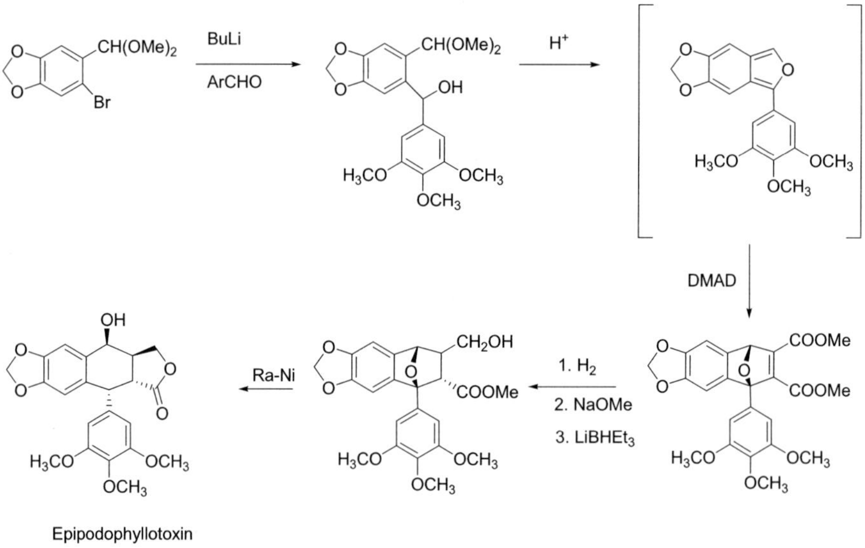

Fig. 4 Synthesis of epipodophyllotoxin through a Diels–Alder reaction (taken from [39] by permission from Kluwer)

uct (Fig. 4) was subsequently converted to a variety of different compounds, including epipodophyllotoxine. The advantage of this type of Diels–Alder reaction is the generation of the 1,2-*cis*-configuration. One of the first applications of this approach was the synthesis of podophyllotoxin derivatives, as carried out by Rodrigo et al. [46, 47].

3.2.1.3
Radical Carboxyarylation Approach

This is a new concept in lignan synthesis, based on a highly stereoselective domino radical sequence, which has only recently been developed. The eight lignans, seven of which are biologically important natural lignans [(–)-matairesinol, (–)-7(S)-hydroxymatairesinol, (–)-7(R)-hydroxymatairesinol, (+)-7-oxomatairesinol, (–)-arctigenin, 7(S)-hydroxyarctigenin, and 7(R)-hydroxyarctigenin], were synthesized in this way by Fischer et al. [48], who achieved the first asymmetric total synthesis of the four naturally occurring C7-oxygenated lignans. Syntheses of 7(S)-hydroxymatairesinol (isolated from many plants including *Taxus* species) and 7(S)-hydroxyarctigenin via this strategy are shown in Fig. 5. Thus, the synthesis of both isomers of 7-hydroxyarctigenin by Fischer et al. [48] also highlighted the correct assignment of natural 7-hydroxyarctigenin, which had previously been reported to be 7(S)-hydroxyarctigenin. This study indicated that natural 7-hydroxyarctigenin should be 7(R)-hydroxyarctigenin, which yields identical NMR data to those of synthetic 7(R)-hydroxyarctigenin [48].

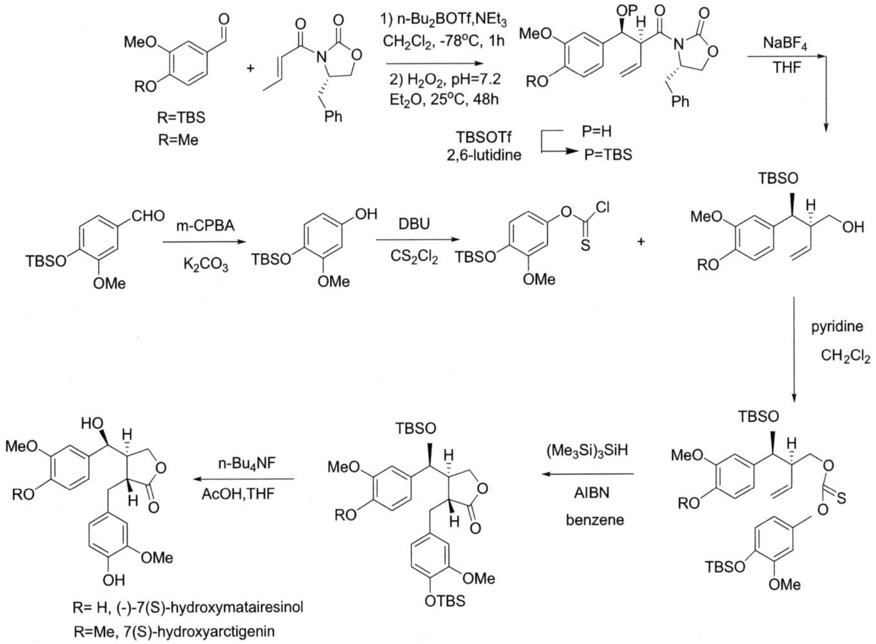

Fig. 5 Synthesis of 7-hydroxymatairesinol by a radical carboxyarylation reaction

3.2.1.4
Biomimetic Oxidative Coupling Approach

Biomimetic oxidative coupling reactions are of interest from a mechanistic standpoint and provide particularly efficient routes to stegane and isostegane derivatives.

Apart from one-electron oxidants based on iron(III) or copper(II), hypervalent iodine reagents such as PIDA (phenyliodonium diacetate) or PIFA [phenyliodinum bis(trifluoroacetate)] would seem to mimic the enzyme catalyzed steps that occur in nature most closely. Furthermore, reactions involving these reagents are interesting because they permit the isolation of the intermediate spirodienones in suitable cases. In fact, in all cases the reactions follow the expected reaction pathway, either leading directly to an eight-member product or giving an initial spirodienone, depending on the position of the phenolic OH group. The synthesis of isostegane is shown in Fig. 6. The use of PIFA in TFE (2,2,2-trifluoroethanol) afforded a spirodienone that underwent a dienone-phenol rearrangement with TFA (trifluoroacetic acid) to give isostegane as the major product [49, 50]

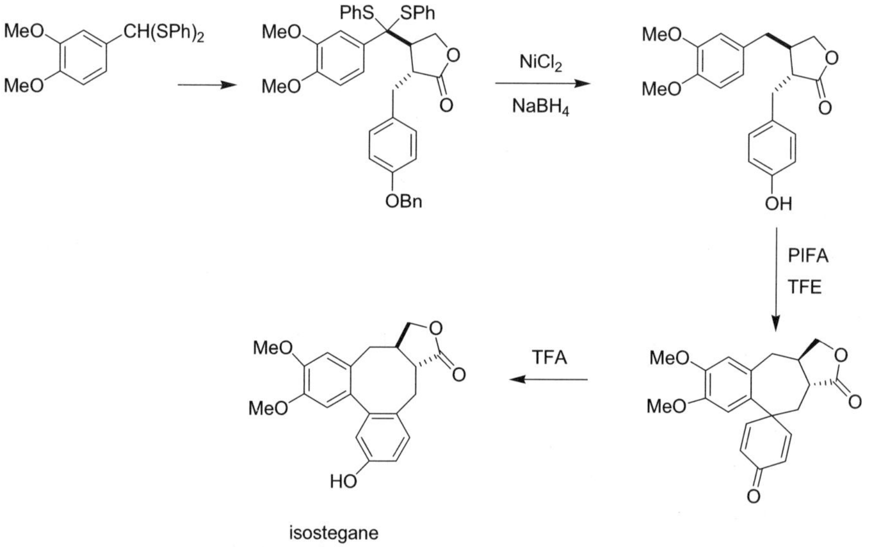

Fig. 6 Synthesis of isostegane using hypervalent iodine reagent PIFA (taken from [39] by permission from Kluwer)

3.2.1.5
Intramolecular CH Insertion

There are several approaches that can be used to achieve a 3-methyl-2,3-dihydrobenzofuran skeleton [51, 52], but all of the methods afford only

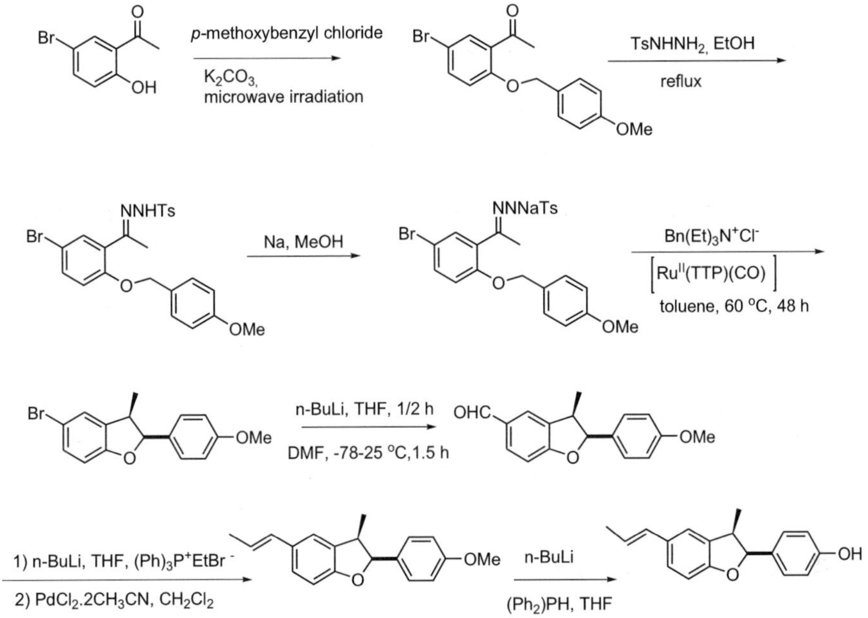

Fig. 7 Total synthesis of the neolignan (±)-*epi*-conocarpan

the *trans*-3-methyl-2,3-dihydrobenzofuran as the major product. Therefore, the synthesis of the intermediate *cis*-2,3-dihydrobenzofuran derivative
is the key step (Fig. 7). Intra- and intermolecular metal carbene C – H
insertion has become a general strategy for the formation of carbocy-
cles and heterocycles [53, 54]. Highly valent oxo- and imido-ruthenium
porphyrins are used to perform alkane hydroxylations and amidations
via C – H bond activation [55]. Recently, using aryl tosylhydrazone salts
as the carbene sources, the first ruthenium porphyrin-catalyzed intramolecular carbenoid C – H insertion was achieved, which selectively afforded
cis-2,3-disubstituted-2,3-dihydrobenzofurans. A total synthesis of (±)-*epi*-
conocarpan was carried out in eight steps with a 20% overall yield from
5-bromo-2-hydroxyacetophenone [56]. Ruthenium(II) porphyrin-catalyzed
intramolecular C – H insertion using aryl tosylhydrazone salt as the carbene
source provided a general route to the key intermediate 5-bromo-*cis*-2-(4-
methoxyphenyl)-3-methyl-2,3-dihydrobenzofuran with high stereoselectivity
(> 98% *cis*) for the synthesis of neolignan *epi*-conocarpan. This approach
seems to be practical for the synthesis of neolignans with a *cis*-2,3-dihydro-
benzofuran skeleton [56].

3.2.1.6
Intramolecular Radical Cyclization Reaction

The application of a radical cyclization reaction involving epoxides as a transition metal radical source is one of the most efficient methods for the total synthesis of various furan and furofuran lignans. Although the structures of furan and furofuran lignans look simple, obtaining good control over the stereochemistry in the cyclization step is a challenge to organic chemists. A short and stereoselective total synthesis of several furan and furofuran lignans was achived by Roy et al. [57]. Figure 8 shows syntheses of a furan lignan (lariciresinol) and a furofuran lignan (pinoresinol), both in racemic mixtures, as well as the formation of a dibenzylbutane lignan, secoisolariciresinol, through the reduction of pinoresinol. The three lignans were isolated from several *Taxus* species as well as from many different plant families.

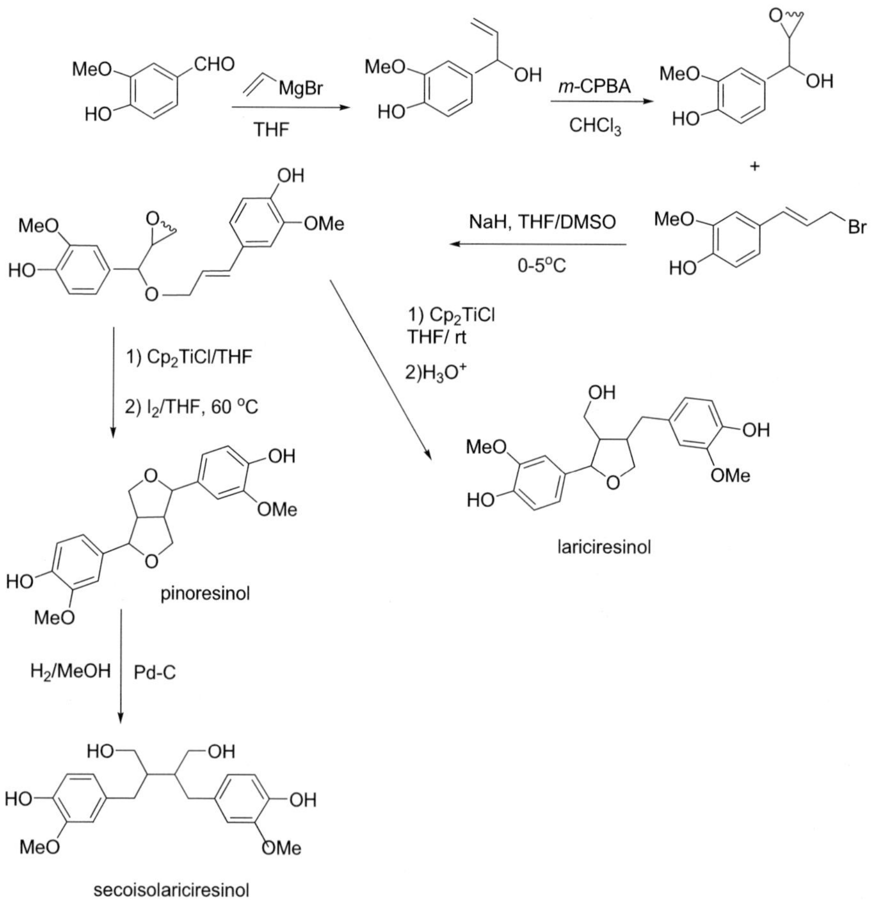

Fig. 8 Synthesis of some furan and furofuran lignans

3.2.2
Asymmetric Synthesis of Lignans

Asymmetric synthesis of lignans may involve the use of several approaches like a tandem conjugate addition reaction, a Diels–Alder reaction or a radical carboxyarylation reaction, as mentioned above.

A stereocontrolled synthesis of the biologically active neolignan (+)-dehydrodiconiferyl alcohol, which was isolated from several *Taxus* species, was achieved via Evans' asymmetric aldol condensation [58] using ferulic acid amide derived from D-phenylalanine. The reaction steps are shown in Fig. 9. This stereocontrolled reaction is also useful for preparing the enantiomer of (+)-dehydroconiferyl alcohol using chiral auxiliary oxazolidinone prepared from L-phenylalanine. This reaction also enables the syntheses of other natural products that possess the same phenylcoumaran framework.

Some approaches to the stereoselective synthesis of α-hydroxylated lactone lignans have been reported [59, 60]. As a short and efficient example, the synthesis of a dibenzylbutyrolactone lignan wikstromol from two diastereoselective alkylations of malic acid (+) has recently been reported [61]. In order to get high stereoselectivity, isoPr malate was chosen for the synthesis of (+)-wikstromol; its formation in six steps with a 20% overall yield is shown in Fig. 10.

3.3
Extraction and Isolation of Lignans

In general, to extract lignans from plants, the first step is to apply polar solvents to different parts of the plants involved. The resulting extracts are then dissolved in water and re-extracted with nonpolar solvents. Extracts from *Taxus* species are obtained in a similar manner to this [1, 3, 10].

During the last decade, increasing attention has focused on new and appropriate methods for the analysis of lignans from plant sources and in body fluids. Conventional chromatographic methods, including reversed-phase HPLC, still remain the most useful and most commonly applied techniques [62]. Isolation of lignans from the plant *Torreya jackii* (family Taxaceae) has been achieved by stopped-flow high-performance liquid chromatography–nuclear magnetic resonance (LC–NMR) spectroscopy [63], while GC–MS has been used to analyze trimethylsilyl derivatives of the lignans present in galls and shoots of *Picea glauca* [64]. Among the novel techniques employed, supercritical CO_2 has been demonstrated to provide an effective route to the extraction of lignans from leaves, seeds and fruits of *Schisandra chinensis* [65]. Furthermore, micellar electrokinetic chromatography (MEKC) was used to separate twelve lignans originating from *Phyllanthus* plants [66]. However, conventional methods, including HPLC, have been employed in general so far to isolate *Taxus* lignans.

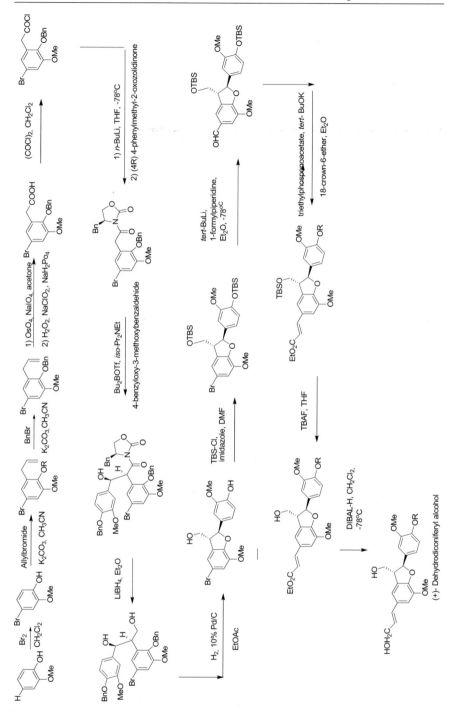

Fig. 9 Stereocontrolled synthesis of (+)-dehydrodiconiferyl alcohol

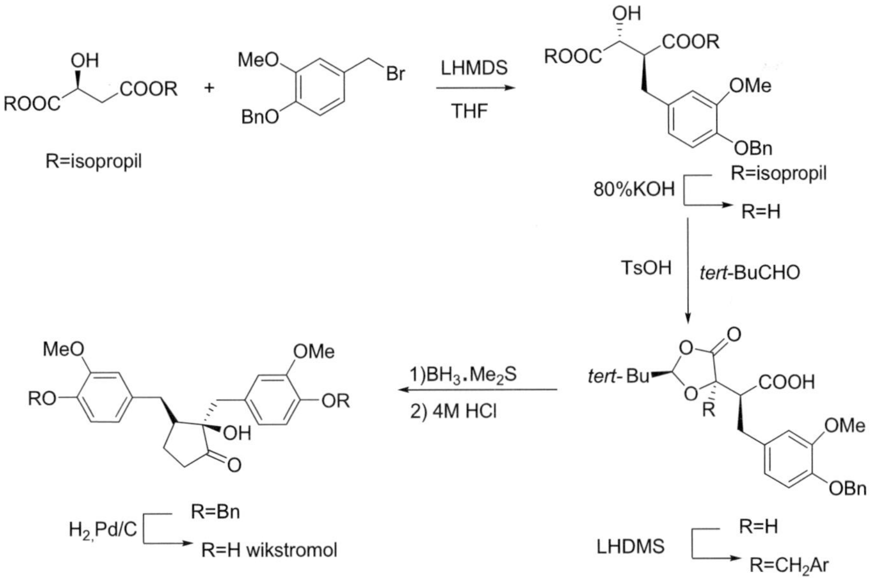

Fig. 10 Stereoselective synthesis of (+)-wikstromol

3.4
Isolation and Structural Elucidation of Lignans from *Taxus* Species

Over 40 lignans have been isolated from eight *Taxus* species to date (Table 1 and 2). α-Conidendrin isolariciresinol, secoisolariciresinol, isotaxiresinol and lariciresinol are found in several species of the genus *Taxus*. Indeed, iso-taxiresinol and secoisolariciresinol are found in all *Taxus* species.

Table 1 lists crude extracts (mother liquor) of the plants studied; however, the lignans were generally isolated after re-extractions rather than directly from the crude extracts. Extraction and isolation details are provided in the references [68–88] given in Table 1. As seen from the table, crude extracts were prepared with a polar solvent, mostly ethanol, sometimes water, but higher diversity was obtained by working with a direct acetone extract of *T. maireii* twigs, considering the number of the lignans isolated. An early study of *Taxus* lignans by Erdtman and Tsuno reported a brief comparative examination of several *Taxus* woods [67].

Lignans were found in the *Taxus* extracts prepared from bark, heart-wood, needles, roots, or other parts. In the studies on European yew *T. bac-cata* L. carried out by Das et al., the extracts were prepared by extracting from the needles rather than the heartwood of the plant using a solvent system $CH_2Cl_2 - MeOH$ (1 : 1) [70–73], although in the studies on Turk-ish *T. baccata*, the extracts were prepared by applying EtOH to the needles and twigs of the plant [74–76]. So far, after many studies of *T. baccata* extracts [68–76], over 15 distinct lignans have been isolated, but the lignan diversity is observed to be higher in *T. mairei*, which is the only species endemic to Taiwan [17]. From the latter species, over 35 lignans (two of which are sesquilignans) have been isolated, which is almost equal to the total number of lignans isolated from eight other *Taxus* species. The structures of the sesquilignans were determined using ^{1}H, ^{13}C (DEPT), COSY, NOESY, HMQC and HMBC NMR techniques, and FAB–MS spectra. One of them contains a linkage between dihydrodehydroconiferyl alcohol and another nine-carbon moiety, and the linkage was determined to be between C8′ and C8″. The second sesquilignan is constructed from one of the constituents of the plant, (–)-3,3′-dimethoxy-4′,8-epoxyligna-4,7,9,9′-tetraol, and a nine-carbon unit. Both lignans have previously been isolated from *Abies marocana* and were named (–)-sesquipinsapol B (+)-sesquimarocanol B; however, they were isolated as their pentaacetate and hexaacetate, respectively, after peracetylation of the obtained fractions during EtOAc elution of the EtOAc-soluble part of the crude *T. mairei* extract. Besides lignans and taxane diterpenoids, many other types of phenolic compounds, including flavonoids, biflavonoids, sugar-containing simple phenolics as well as steroids, have been isolated from *Taxus* species, as illustrated by the *T. mairei* study [17].

The formulae of all the lignans isolated from all of the *Taxus* species investigated are given in Table 2.

A new lignan, taxumairin, was isolated from the roots of the Formosan yew *T. mairei*, along with the known lignans (–)-α-conidendrin, (–)-secoisolariciresinol, isotaxiresinol. The structure of the new lignan was characterized as (+)-7,8-*trans*-8,8′-*trans*-7′,8′-*trans*-7-(3-ethoxymethyl tetrahydrofuran) based on spectral analyses [80].

The American tree *T. brevifolia* is famous for its well-known taxane diterpene alkaloid taxol, and its derivatives have been isolated [5,7,9]. In fact, brevitaxin, the first known terpenolignan, was also isolated from the twigs of the Himalayan yew *T. brevifolia* [19]. Its was identified using spectral methods, particularly 1-D and 2-D NMR techniques such HMBC and SINEPT experiments, which were used to establish regiochemistry of the terpene–lignan linkage.

Although structural elucidation of lignans is not a difficult task, the similarities between the structures can create problems. In particular, the determination of stereochemistry at the chiral center requires NOE/ NOESY NMR experiments and/or X-ray analyses. The enantiomeric excesses of the known lignans (+)-lariciresinol, (–)-secoisolariciresinol and (+)-taxiresinol, isolated from Japanese yew *T. cuspidata* roots, were determined by chiral high-performance liquid chromatographic analyses [78]; except for (+)-pinoresinol (77% enantiomeric excess), they were found to be optically pure by Kawamura et al. In an earlier study, the presence of taxiresinol in *Taxus* species was reported by Mujumdar et al. [69] after they had isolated it from the heartwood of *T. baccata*, although they did not study its stereochemistry.

Table 2 Lignans isolated from *Taxus* species

Lignan	Plant source	Refs.
(1) (–)-Secoisolariciresinol	*Taxus baccata*	[69–72]
	T. brevifolia	[19]
	T. cuspidata	[78]
	T. floridana	[67]
	T. mairei	[17, 80]
	T. wallichiana	[86]
	T. yunnanensis	[88, 105]
(2) (–)-3-*O*-Demethylsecoisolariciresinol	*T. baccata*	[70, 71, 74]
(3) 3,4,9,9′-Tetrahydroxy-3′,4′-dimethoxylignan	*T. baccata*	[71, 73]
(4) 9,9′-Dihydroxy-3′,4′-dimethoxy-3,4-methylenedioxylign-7-ene (Lignan diol)	*T. baccata*	[72, 73]
(5) 9,9′-Dihydroxy-3,4-dimethoxy-3′,4′-methylenedioxylign-7-ene	*T. baccata*	[73]

Table 2 (continued)

Lignan	Plant source	Refs.
(6) 9,9′-Dihydroxy-3,4:3′,4′-bis-(methylenedioxy)lign-7-ene	*Taxus baccata*	[73]

Lignan	Plant source	Refs.
(7) (–)-Matairesinol	*T. cuspidata* *T. mairei*	[79] [17, 116]

Lignan	Plant source	Refs.
(8) 7-Hydroxymatairesinol	*T. cuspidata* *T. mairei* *T. media* cv Hisckii *T. wallichiana*	[79] [17] [84] [85]

Lignan	Plant source	Refs.
(9) 7′-Hydroxymatairesinol	*T. mairei*	[17]

R_1 = OH R_2 = H (R)-(+)7′-Hydroxymatairesinol
R_1 = H R_2 = OH (S)-(–)7′-Hydroxymatairesinol

Table 2 (continued)

Lignan	Plant source	Refs.
(10) Allohydroxymatairesinol	*T. cuspidata*	[79]

Lignan	Plant source	Refs.
(11) 7′-(+)-Oxomatairesinol	*T. cuspidata**	[79]
	T. mairei	[17]

Lignan	Plant source	Refs.
(12) (a) (−)-Nortrachelogenin (8-α-OH)	*T. cuspidata*	[76, 79]
(b) (−)-Epinortrachelogenin (8-β-OH)	*T. mairei*	[17]
	T. media cv Hicskii	[84]

Lignan	Plant source	Refs.
(13) 7′-Hydroxynortrachelogenin	*T. cuspidata*	[79]
	T. mairei	[17]

Lignan	Plant source	Refs.
(14) Suchilactone	*T. baccata*	[72]

Table 2 (continued)

Lignan	Plant source	Refs.
(15) 4′-*O*-Demethylsuchilactone	*T. baccata*	[72]
(16) (–)-Isohibalactone	*T. baccata*	[73]
(17) (–)-Hibalactone	*T. baccata*	[73]
(18) *Meso*-neoolivil	*T. mairei*	[17]
(19) Isoliovil	*T. wallichiana*	[85]

Table 2 (continued)

Lignan	Plant source	Refs.
(20) Taxumairin	*T. mairei*	[80]

(21) Taxiresinol	*T. baccata*	[69, 75, 97]
	T. cuspidata	[78]
	T. mairei	[17]
	T. wallichiana	[86]
	T. yunnanensis	[88, 105]

(22) 9-*O*-Acetyltaxiresinol	*T. mairei*	[17]

(23) Lariciresinol	*T. baccata*	[75]
	T. cuspidata	[78]
	T. mairei	[17]

(24) (a) (+)-Tanegool (β-H)	*T. mairei*	[17]
(b) (−)-8′-*Epi*-tanegool (α-H)		

Table 2 (continued)

Lignan	Plant source	Refs.
(25) 3′,4,4′,7′,9-Pentahydroxy-3-methoxy-7,9′-epoxylignan. (3′-*O*-Demethyltanegool)	*T. mairei*	[17]

Lignan	Plant source	Refs.
(26) (−)-3,3′-Dimethoxy-4,4′,9-trihydroxy-7,9′-epoxylignan-7′-one	*T. mairei*	[17]

Lignan	Plant source	Refs.
(27) (a) (+)-Pinoresinol (7α-aryl) **(b)** (+)-*Epi*-pinoresinol (7β-aryl)	*T. cuspidata* (only pinoresinol) *T. mairei*	[78] [17]

Lignan	Plant source	Refs.
(28) 3′-*O*-Demethylepipinoresinol	*T. mairei*	[17]

Table 2 (continued)

Lignan	Plant source	Refs.
(29) D-Sesamin	*T. mairei*	[8]

(30) 4-Hydroxysesamin	*T. mairei*	[8]

(31) (−)-α-Conidendrin	*T. baccata*	[73]
	T. mairei	[17, 80]
	T. wallichiana	[85]
	T. yunnanensis	[88]

(32) β-Conidendrin	*T. mairei*	[17]
	T. wallichiana	[85]

(33) Isotaxiresinol	*T. baccata*	[19, 68–70]
	T. cuspidata	[77, 79]
	T. floridana	[67]
	T. mairei	[17, 80]
	T. wallichiana	[86]
	T. yunnanensis	[88]

Table 2 (continued)

Lignan	Plant source	Refs.
(34) (a) Isolariciresinol 34 **(b)** 3-Demethylisolariciresinol	*T. baccata*	[70, 74, 97]
	T. brevifolia	[19]
	T. cuspidata	[77]
	T. floridana	[67]
	T. mairei	[17]

(35) (–)-3′-Demethylisolariciresinol-9′-hydroxyisopropylether	*T. baccata*	[74, 97]

(36) (–)-3-*O*-Demethyldihydrodehydrodiconiferyl alcohol	*T. mairei*	[17]

(37) 2-[2-Hydroxy-5-(3-hydroxypropyl)-3-methoxyphenyl]-1-(4-hydroxy-3-methoxyphenyl)-propan-1,3-diol	*T. yunnanensis*	[88]

(38) Taxuyunin A	*T. yunnanensis*	[87]

Table 2 (continued)

Lignan	Plant source	Refs.
(39) Taxuyunin B	*T. yunnanensis*	[87]

| (40) (−)-Sesquipinsapol B (pentaacetate) | *T. mairei* | [17] |

| (41) (+)-Sesquimarocanol B (hexaacetate) | *T. mairei* | [17] |

| (42) Brevitaxin | *T. brevifolia* | [19] |

The stereochemistry of (+)-taxiresinol was clarified through NOE difference spectroscopic experiments as (7′S,8R,8′R) in the mentioned study [78].

Recently, an epimer of (+)-taxiresinol was isolated from *T. wallichiana*, the Himalayan yew, which is a small medium-sized evergreen tree growing in the temperate Himalayas, and its absolute stereochemistry was determined as (7′R,8R,8′R) by single-crystal X-ray spectroscopic analysis [86].

Investigations on the ethanolic extract from the bark of *T. yunnanensis* led to the isolation of two new neolignans. One of them was named taxuyunin A, and has a neolignan structure containing a sugar xylose, while the other neolignan, taxuyunin B, carries a rare C-3 side chain [87].

Neolignans were also isolated from *Taxus* species. A new neolignan, 2-[2-hydroxy-5-(3-hydroxypropyl)-3-methoxyphenyl]-1-(4-hydroxy-3-methoxyphenyl) propane-1,3-diol, was isolated from the wood of *Taxus yunnanensis* [88].

4
Biological Activities of Lignans

Lignans are found in a variety of foods, vegetables and fruits that have potentially protective effects on human health. Many lignans and neolignans have been adopted as lead compounds for the development of new drugs. The importance of lignans in cancer therapy and treatments for other diseases were well covered by Lee & Xiao in a recent review article [3]. Lignans have shown cytotoxic, antitumor/anticancer, antiviral, antifungal, antileishmanial, antiangiogenic, hypolipidemic and antirheumatic activities, and are selective inhibitors of some enzymes, such as 5-lipooxygenase and phosphodiesterases IV and V [89]. There is growing interest in this class of compounds because of their possible weakly estrogenous or estrogen- antagonistic effects [90]. Enterolactone and enterodiol, the two main lignans identified in human urine and plasma [91], are considered to derive from colonic bacterial metabolism of the plant-derived precursors matairesinol and secoisolariciresinol, respectively [92, 93]. Both experimental and epidemiological studies suggest that high plasma and urinary concentrations of phytoestrogens, including lignans, are associated with a decreased risk for hormone-dependent diseases, e.g., breast cancer and coronary heart disease [94–96]. The main activities of lignans, especially those isolated from *Taxus* species, are described below.

One of the most intensely studied *Taxus* species, due to its biological activities, is *T. baccata* L. The needles of this plant have shown hepatoprotective, tranquilizing and sedative properties, presumably related to the benzodiazepine-like activity of its biflavones. In contrast to the European yew *T. baccata*, the Himalayan yew *T. wallichiana* has a long history of use in medicines, as well as in coloring materials and as incense [86].

However, some isolated pure constituents, including lignans and their extracts, from *T. baccata* L growing in Turkey were evaluated for various biological activities [97–102]. Crude extracts prepared from the heartwood of *T. baccata* were also evaluated for pharmacological activity, which covers antiplatelet and vasorelaxing effects. All of the tested extracts exhibited a significant inhibitory effect on platelet aggregation induced by arachidonic acid, collagen and platelet-activating factor at a concentration of 400 μg/mL, but not thrombin. However, the extracts showed a weak inhibitory effect on high potassium depolarized smooth muscle [102].

4.1
Cytotoxic Activity

T. baccata L. growing in Turkey was also investigated for various other biological activities. Three lignans, (–)-taxiresinol, (–)-3′-demethylisolariciresinol-9′-hydroxy isopropylether and (–)-3-demethylisolariciresinol, isolated from the heartwood of *T. baccata* L., were investigated for cytotoxicity against a panel of cell lines (breast, colon, ovary, prostate, lung and a normal adult bovine aortic endothelial cell line) [76]. However, none of the lignans tested demonstrated much cytotoxic potency compared to the reference drug etoposide.

Cytotoxic activity assays were carried out on the lignans (–)-α-conidendrin, (–)-secoisolariciresinol, isotaxiresinol and taxiresinol isolated from Formosan *T. mairei*, which are also found in several *Taxus* species, and they exhibited potent cytotoxicity against KB-16, A-549 and HT-29 tumor cell lines [103].

Antiproliferative Activity

One of the most commonly investigated *Taxus* species for biological activity is *T. yunnannensis*, a tree commonly called "Hongdoushan" and found mainly in the Yunnan Province in the People's Republic of China. The wood of *T. yunnanensis* was reportedly used in Chinese traditional medicine according to several ethnic groups in Yunnan Province [104].

The three lignans (secoisolariciresinol, taxiresinol and isotaxiresinol) isolated as major constituents (along with six taxane diterpenes) from the wood of *T. yunnanensis* were evaluated for their antiproliferative activity against murine colon carcinoma and human fibrosarcoma cell lines [105], and among the compounds tested (including the taxane diterpenes), the highest activity was observed for secoisoisolariciresinol against the HT-1080 fibrosarcoma cell line, with an EC_{50} value of 5.9 μg/mL, while a taxane, hongdushan C, was found to be active against 26-L5 murine colon carcinoma.

4.2
Antitumor/Anticancer Activity

The best example of a lignan used as a lead compound is podophyllotoxin, an antimitotic compound that binds to tubulin, although podophyllotoxin has not been isolated from *Taxus* species,. Etoposide and teniposide are well-known compounds derived from podophyllotoxin, and their antitumor activity is due to the inhibition of topoisomerase II.

As reported by Apers et al. [89], podophyllotoxin derivatives can be divided into two groups in terms of their mechanism of action: 1 Inhibitors of tubulin polymerization (such as podophyllotoxin), 2 Inhibitors of DNA topoisomerase II (such as etoposide and teniposide). Combining both pharmacophores leads to compounds with a dual mechanism of action, such as azatoxin.

Podophyllotoxin derivatives (cyclolignans) have been used as anticancer agents as well as immunosuppressive drugs to prevent the rejection of transplanted organs [106]. In addition to podophyllotoxin, some other lignans and neolignans have also been found to act as inhibitors of tubulin polymerization. Among dihydrobenzofuran neolignans, 3′,4-di-O-methylcedrusin was obtained as one of the minor constituents of the red latex *Croton* species (Euphorbiaceae), called "dragon's blood" in South America, which exhibited potential antiproliferative and antitumoral activities [107]. In another study, a set of dihydrobenzofuran neolignans were screened against a panel tumor cell line by NCI, and the results showed that the leukemia cell lines and the breast cancer cell lines were more sensitive to these compounds than other cell lines [89].

Another lignan, taxiresinol, isolated from *Taxus wallichiana*, exhibited notable in vitro anticancer activity against colon, liver, ovarian and breast cancer [86].

Butyrolactone and bistetrahydrofuran lignans such as matairesinol, nortrachelogenin and pinoresinol are also known to possess antileukemia and cAMP-inhibitory activities [108].

4.3
Antiviral Activity

Besides antitumor/anticancer activity, lignans are among the most potent antiviral compounds, although most of the lignans evaluated for antiviral potency belong to families of plants other than Taxaceae [3].

Podophyllotoxin—an aryltetralin lignan—and some other type of lignans, such as dibenzylbutyrolactones related to arctigenin, dibenzocyclooctadiene lignans, and dibenzylbutanes have been found to act as antiviral agents, including against HIV. Natural and synthetic analogs of podophyllotoxin were also assayed against various viruses, including HIV, and four aryltetralin lignans showed significant inhibition of HIV replication, with EC_{50} values below 0.001 µM and therapeutic indices greater than 120 [109].

The dibenzylbutyrolactone lignan matairesinol was also found to be an anti-HIV agent [110], although the sample used in this bioassay was not isolated from a *Taxus* species. Another dibenzylbutyrolactone lignan, arctigenin, as well as its unsubstituted benzyl derivative exhibited anti-HIV replication activity, with EC_{50} values of 0.16 and 22 µg/mL and therapeutic index values of 5 and 9.1, respectively [111].

4.4
Antioxidant Activity

Recently, isotaxiresinol and secoisolariciresinol, two major lignans of the wood *T. yunnanensis*, have been shown to possess potent DPPH radical scavenging activity, with IC_{50} values of 21.7 and 28.9 µM, respectively, and significant inhibitory activity against nitric oxide (NO) production in lipopolysaccharide-activated murine macrophage-like J774.1 cells [112].

4.5
Hypoglycemic Activity

Hypoglycemic activity of the H_2O and MeOH extracts of the wood of *T. yunnanensis* was investigated in streptozotocin (STZ)-induced diabetic rats because of its use in traditional medicine as an antidiabetic agent. A 100 mg/kg dose of the H_2O extract significantly lowered the fasting blood glucose level by 33.7% upon intraperitoneal administration. Three lignans, isotaxiresinol, secoisolariciresinol and taxiresinol, isolated as major components from the active H_2O extract of the wood, were tested for their hypoglycemic effects on the same experimental model. At a dose of 100 mg/kg (i.p.), isotaxiresinol reduced the fasting blood glucose level of diabetic rats by 34.5%, while secoisolariciresinol and taxiresinol reduced it by 33.4% and 20.9%, respectively, showing stronger effects than a mixture of tolbutamide (200 mg/kg) and buformin [113].

4.6
Hepatoprotective Activity

The effects of secoisolariciresinol and isotaxiresinol on tumor necrosis factor-α (TNF-α)-dependent hepatic apoptosis, induced by D-galactosamine and lipopolysaccharide (DGalN/LPS) in mice, were investigated [96]. The results showed that both lignans prevent D-galactosamine/lipopolysaccharide-induced hepatic injury by inhibiting hepatocyte apoptosis through blocking TNF-α and IFN-γ production. Similar to secoisolariciresinol and isotaxiresinol, the hepatoprotective effects of taxiresinol and (7'R)-7'-hydroxylariciresinol (two tetrahydrofuran-type lignans isolated from the wood of *T. yunnanensis*) on

DGalN/LPS-induced hepatic liver injury in mice were investigated [114]. They significantly inhibited hepatocyte DNA fragmentation and apoptotic body formation at doses of 50 and 10 mg/kg (i.p.). Pretreatment with these two lignans further suppressed hepatic necrosis, which occurred at a later stage of DGalN/LPS intoxication, as demonstrated by a significant dose-dependent reduction in serum glutamic pyruvic transaminase (sGPT) and serum glutamic oxaloacetic transaminase (sGOT) at 8 h after intoxication. Both compounds significantly inhibited the elevation of the serum tumor necrosis factor-alpha (TNF-alpha) level when applied in doses of 50 and 10 mg/kg. Moreover, both of these lignans significantly protected hepatocytes from D-GalN/TNF-alpha-induced cell death in primary cultured mouse hepatocytes. These results suggest that both lignans protect the hepatocytes from apoptosis by inhibiting TNF-alpha production by activated macrophages and directly inhibiting apoptosis induced by TNF-alpha in D-GalN/LPS-treated mice [114].

4.7
Antiulcerogenic Activity

The in vivo antiulcerogenic potencies of four lignans, lariciresinol, taxiresinol, isolariciresinol and 3-demethylisolariciresinol, isolated from the heartwood of *T. baccata* growing in Turkey, were investigated on an ethanol-induced ulcerogenesis model in rats at two different doses, 50 and 100 mg/kg. All compounds were shown to possess significant antiulcerogenic activity at both doses. However, the effect of taxiresinol was the most prominent [99].

4.8
Anti-Inflammatory Activity

Five lignans isolated from the heartwood of *T. baccata* L. growing in Turkey were evaluated for their anti-inflammatory and antinociceptive activities in vivo. All of the compounds were shown to possess significant antinociceptive activities against *p*-benzoquinone-induced abdominal contractions and significantly inhibited carrageenan-induced hind paw edema in mice [100].

4.9
Antibacterial Activity

The ethanolic extract of Turkish *T. baccata* heartwood showed significant activity against selected gram-negative bacteria and against five out of nine tested fungi [101].

In another study, three lignans, (–)-taxiresinol, (–)-3′-demethylisolariciresinol-9′-hydroxyisopropyl ether and (–)-3-demethylisolariciresinol, all isolated from the heartwood of *T. baccata* L., were tested for antimicrobial activity. The two lignans (–)-taxiresinol and (–)-3-demethylisolariciresinol and the chloroform extract were found to possess antifungal activity, whereas only the chloroform extract exhibited antibacterial activity against *Pseudomonas aeruginosa* [76].

Crude extracts prepared from different parts of Turkish *T. baccata* have also been evaluated for their antimycobacterial activity. A $CHCl_3$ fraction of the heartwood and ethanol extract of the leaves exhibited a minimum inhibitory concentration value (MIC) against *Mycobacterium tuberculosis* $H_{37}Ra$ strain of 200 µg/mL [102].

4.10
Antiosteoporotic Activity

Isotaxiresinol, the main lignan isolated from the water extract of wood of *T. yunnanensis*, was investigated for its effect on bone loss, on serum biochemical markers for bone remodeling, and on uterine tissue, using ovariectomized rats as a model of postmenopausal osteoporosis. After oral administration of isotaxiresinol (50 and 100 mg/kg/d) for six weeks, bone mineral content and bone mineral density in total and cortical bones were higher than those of ovariectomized control rats, and decreases in three bone strength indices induced by ovariectomized surgery were prevented. The assay showed that isotaxiresinol slightly increased bone formation and significantly inhibited bone resorption without side effects on uterine tissue. These results suggest that isotaxiresinol may be useful for treating postmenopausal osteoporosis, especially for the prevention of bone fracture induced by estrogen deficiency [115].

4.11
Other Activities

The aqueous extracts and constituents of *T. yunnanensis* showed a remarkable inhibitory effect on induced histamine release from the human basophilic cell line KU812. The isolated lignans secoisolariciresinol and taxiresinol were also found to be antiallergic compounds, but this was not the case for the new neolignan 2-[2-hydroxy-5-(3-hydroxypropyl)-3-methoxyphenyl]-1-(4-hydroxy-3-methoxyphenyl)-propan-1,3-diol [88]. Some vasorelaxing and antihypertensive effects of lignan-containing plant extracts (including *Taxus* extracts [103] and pinoresinol diglucoside) were also reported.

5
Conclusions

Although lignans are widely distributed in vascular plants and are a very well-known class of natural compounds, *Taxus* lignans have not been investigated as much as their taxane diterpenoids in any respect. There is still some confusion over the identities and nomenclature of *Taxus* species. Lignans exhibit a remarkable diversity in terms of enantiomeric composition, biosynthesis, phylogenetic distribution and bioactivity. There is also increasing interest in the synthesis of lignans as valuable lead compounds for new medicines; this interest started with podophyllotoxin derivatives, which were synthesized using tandem conjugate addition and Diels–Alder reactions. Attention is now focused on the synthesis of neolignans by different approaches involving hypervalent iodine reagents, radical carboxyarylation, or C – H insertion. Based on a mechanistic approach, biomimetic oxidative coupling reactions seem particularly efficient routes to the synthesis of stegane and isostegane derivatives.

Lignans have been found in almost all parts of the plants, including bark, heartwood, needles and roots, unlike other secondary metabolites in the plants. Considering the number of lignans isolated from *Taxus* species, it is not easy to explain the high structural diversity exhibited by lignans in *T. maireii* compared to those of other *Taxus* species. The reason could be targeted isolation, but different type of compounds (such as taxane diterpenes, steroids and other phenolics) have been isolated from the same plant extract. On the other hand, there is also some uncertainty over the determinations of stereochemical centers of natural lignans, particularly in previous studies, due to the lack of high-resolution NMR measurements. The increase in the number of synthetic lignans would improve matters, as would the application of NOE/ NOESY NMR experiments along with single-crystal X-ray spectroscopic analysis. Unfortunately, ^{13}C NMR spectra are yet to be reported for most natural lignans.

Over the last decade, most studies of the lignans or lignan-containing plants, including *Taxus* species, have focused on their bioactivities. In addition to the well-known antitumor lignan podophyllotoxin, podophyllotoxin derivatives and other lignans have been investigated for various biological activities, including cytotoxic, antitumor, antimicrobial, antiviral, antiulcer, hepatoprotective, and antioxidant properties. In recent years, the extracts and isolated constituents of *Taxus* species, especially lignans from *T. baccata* growing in Turkey, have been assessed for various potential activities by a Turkish group, and some in vitro and in vivo bioassays have been carried out on extracts from *T. yunnanensis* and its constituents by a Japanese group, which may led to discover new lead medicinal drugs in the future. In one of these studies, the antiosteoporotic activity of taxiresinol was investigated in vivo; especially its ability to prevent bone fracture induced by estrogen deficiency, which may lead to advanced investigations into the use of taxiresinol as a candidate drug for the treatment of postmenopausal osteoporosis. In a second investigation, the antiallergic activities of wood extract and constituents from *T. yunnanensis* were studied using an in vitro histamine release test, and a higher inhibitory effect was observed for the extract than its constituents. The results indicated that this activity may be centered on the lignans isotaxiresinol, secoisolariciresinol and taxiresinol, which were found to be the most abundant compounds in the extract. In conclusion, we can state that lignans and neolignans possess great potential, based on the results from various pharmacological and biological studies, to provide lead drugs in the near future.

References

1. Umezawa T (2003) Phytochem Rev 2:371
2. Ayres DC, Loike JD (1990) Lignans. Chemical, biological and clinical properties. Cambridge University Press, Cambridge
3. Lee K-H, Xiao Z (2003) Phytochem Rev 2:341
4. Willis JC (1996) A dictionary of the flowering plants and ferns, 7th edn. Cambridge University Press, Cambridge, p 1132
5. Appendino G (1995) Nat Prod Rep 12:349
6. Shi Q-W, Oritani T, Sugiyama T, Kiyota H, Yamada T (1998) Tennen Yuki Kagobutsu Toronkai Koen Yoshishu 40:347
7. Kingston DGI, Molinero AA, Rimoldi JM (1993) In: Herz W, Kirby GW, Moore RE, Steglich W, Tamm C (eds) Progress in the chemistry of organic natural products. Springer, New York, p 1
8. Parmar VS, Jha A, Bisth KS, Taneja P, Singh SK, Kumar A, Poonam D, Jain R, Olsen CE (1999) Phytochem 50:1267
9. Baloglu E, Kingston DGI (1999) J Nat Prod 62:1448
10. Das B, Kasinath A (1996) J Sci Ind Res 55:246
11. Haworth RD (1936) Natural Resins Ann Rep Prog Chem 33:266
12. McCredie RS, Ritchie E, Taylor WC (1969) Aust J Chem 22:1011
13. Gottlieb OR (1972) Phytochemistry 11:1537
14. Gottlieb OR (1978) Fortschr Chem Org Naturst 35:1
15. Umezawa T (2001) Regul Plant Growth Develop 36:57
16. Moss GP (2000) Nomenclature of lignans and neolignans (IUPAC Recommendations). Pure Appl Chem 72:1493
17. Yang S-J, Fang J-M, Cheng Y-S (1999) J Chin Chem Soc 46:811
18. Li S-H, Zhang H-J, Yao P, Niu X-M, Xiang W, Sun H-D (2003) Chin J Chem 21:926
19. Arslanian RL, Bailey DT, Kent MC, Richheimer SL, Thornburg KR, Timmons DW, Zheng QY (1995) J Nat Prod 58:583
20. Erdtman Z (1933) Biochem Z 258:172
21. Dewick PM (1989) Biosynthesis of lignans. In: Atta-ur-Rahman (ed) Studies in natural product chemistry, vol 5: Structure elucidation (Part B). Elsevier, Amsterdam, p 459
22. Umezawa T (1997) Lignans. In: Higuchi T (ed) Biochemistry and molecular biology of wood (Springer Series in Wood Science). Springer Berlin p 181
23. Lewis NG, Davin LB (1999) Lignans: Biosynthesis and function. In: Sankawa U (ed) Comprehensive natural product chemistry, vol 1. Elsevier, Amsterdam, p 639
24. Umezawa T, Davin LB, Lewis NG (1990) Biochem Biophys Res Commun 171:1008
25. Umezawa T, Davin LB, Yamamoto E, Kigston DGI, Lewis NG (1990) Biochem Biophys Res Commun 20:1405
26. Katayama T, Davin LB, Lewis NG (1992) Phytochemistry 31:3875
27. Katayama T, Davin LB, Chu A, Lewis NG (1993) Phytochemistry 33:581

28. Umezawa T, Kuroda H, Isohata T, Higuchi T, Shimada M (1994) Biosci Biotech Biochem 58:230
29. Davin LB, Wang H-B, Crowell AL, Bedgar DL, Martin DM, Sarkanen S, Lewis NG (1997) Science 275:362
30. Katayama T, Masaoka T, Yamada H (1997) Mokuzai Gakkaishi 43:580
31. Xia Z-Q, Costa MA, Pelissier HC, Davin LB, Lewis NG (2001) J Biol Chem 276:12614
32. Okunishi T, Umezawa T, Shimada M (2000) J Wood Sci 46:234
33. Broomhead AJ, Rahman MMA, Dewick PM, Jackson DE, Lucas JA (1991) Phytochem 30:1489
34. Xia Z-Q, Costa MA, Proctor J, Davin LB, Lewis NG (2000) Phytochem 55:537
35. Suzuki S, Umezawa T, Shimada M (2002) Biosci Biotech Biochem 66:1262
36. Dewick PM (2002) Medicinal natural products; A biosynthetic approach, 2nd edn. Wiley, New York
37. Jiang Z-H, Tanaka T, Kouno I (1996) Chem Pharm Bull 44:1669
38. Tazaki H, Hayashida T, Ishikawa F, Taguchi D, Takasawa T, Nabeta K (1999) Tetrahedron Lett 40:101
39. Ward RS (2003) Phytochem Rev 2:391
40. Glinski MB, Freed JC, Drust T (1987) J Org Chem 52:2749
41. Ward RS (1982) Chem Soc Rev 75
42. Ziegler FE, Schwarts JA (1978) J Org Chem 43:985
43. Pelter A, Ward RS, Satyanarayana P, Collins P (1983) J Chem Soc Perkin Trans I:643
44. Pelter A, Ward RS, Pritchard MC, Kay IT (1988) J Chem Soc Perkin Trans I:1603
45. Pelter A, Ward RS, Pritchard MC, Kay IT (1988) J Chem Soc Perkin Trans I:1615
46. Rodrigo R (1980) J Org Chem 45:4538
47. Forsey SP, Rajapaska D, Taylor NJ, Rodrigo R (1989) J Org Chem 54:4280
48. Fischer J, Aaron JR, Sharp LA, Sherburn M (2004) J Org Lett 6:1345
49. Pelter A, Satchwell P, Ward RS, Blake K (1995) J Chem Soc Perkin Trans I:2201
50. Ward RS, Hughes DD (2001) Tetrahedron 57:4015
51. Snider BB, Han L, Xie C (1997) J Org Chem 62:6978
52. Engler TA, Letavic MA, Iyengar R, La Tessa KO, Reddy JP (1999) J Org Chem 64:2391
53. Sulikowski GA, Cha KL, Sulikowski MM (1998) Tetrahedron Asymm 9:3145
54. Davies HML, Antoulinakis EG (2000) J Organomet Chem 617
55. Che CM, Yu WY (1999) Pure Appl Chem 71:281
56. Zheng SL, Yu W-Y, Xu M-X, Che C-M (2003) Tetrahedron Lett 44:1445
57. Roy SC, Rana KK, Guin C (2002) J Org Chem 67:3242
58. Okazaki M, Shuto Y (2001) Biosci Biotechnol Biochem 65:1134
59. Khamlach MK, Dhal R, Brown E (1989) Tetrahedron Lett 30:2221
60. Khamlach MK, Dhal R, Brown E (1992) Tetrahedron Lett 46:10115
61. Sefkow M (2001) Tetrahedron Asymm 12:987
62. Wu M-D, Huang R-L, Kuo L-MY, Hung C-C, Ong C-W, Kuo Y-H (2003) Chem Pharm Bull 51:1233
63. Cavin A, Potterat O, Wolfender J-L, Hostettmann K, Dyatmyko W (1998) J Nat Prod 61:1497
64. Kraus C, Spiteller G (1997) Phytochem 44:59
65. Bartlova M, Opletal L, Chobot V, Sovova H (2002) J Chromatogr B 770:283
66. Kuo CH, Lee SS, Chang HY, Sun SW (2003) Electrophoresis 24:1047
67. Erdtman H, Tsuno K (1969) Phytochem 8:931
68. King FE, Jurd L, King TJ (1952) J Chem Soc 17
69. Mujumdar RB, Srinivasan R, Venkataramaran K (1972) Ind J Chem 10:677
70. Das B, Takhi M, Srinivas KVNS, Yadav JS (1993) Phytochem 33:1489

71. Das B, Takhi M, Srinivas KVNS, Yadav JS (1994) Phytochem 36:1031
72. Das B, Rao SP, Srinivas KVNS, Yadav JS (1995) Phytochem 38:715
73. Das B, Rao SP, Sirinivas KVNS, Yadav JS (1995) Fitoterapia 66:475
74. Erdemoğlu N, Şener B, Özcan Y, Ide S (2003) J Mol Str 655:459
75. Erdemoğlu N, Şahin E, Şener B, Ide S (2004) J Mol Str 692:57
76. Belletire JL, Douglas FF (1988) J Org Chem 53:4724
77. Youzuo Z, Chaomei Y, Yuanlong Z (1982) Zhongcaoyao 13:1
78. Kawamura F, Kikuchi Y, Ohira T, Yatagai M (2000) J Wood Sci 46:167
79. Kawamura F, Ohira T, Kikuchi Y (2004) J Wood Sci 50:548
80. Shen Y-C, Chen C-Y, Lin Y-M, Kuo Y-H (1997) Phytochem 46:1111
81. Liu CL, Lin YC, Lin YM, Chen FC (1984) Tai-wan K'O Hsueh 38:119
82. Liu CL, Lin YC, Lin YM, Chen FC (1984) CA 103:165998
83. Das B, Rao SP, Srinivas KVNS, Yadav JS (1994) Fitoterapia 65:189
84. Appendino G, Cravotto G, Enriu R, Gariboldi P, Barboni L, Torregiani E, Gabetta B, Zini G, Bombardelli E (1994) J Nat Prod 57:607
85. Miller RW, Mclaughlin JL, Powell RG, Platner RD, Weisleder D, Smith CR Jr (1982) J Nat Prod 45:78
86. Chattopadhyay SK, Kumar TRS, Maulik PR, Srivastava S, Garg A, Sharon A, Negi AS, Khanuja SPS (2003) Bioorg Med Chem 11:4945
87. Li S-H, Zhang H-J, Yao P, Niu X-M, Xiang W, Sun H-D (2003) Chin J Chem 21:926
88. Koyama J, Morita I, Kobayashi N, Hirai K, Simamura E, Nobukawa T, Kadota S (2006) Biol Pharm Bull 29:2310
89. Apers S, Vlietinck A, Pieters L (2003) Phytochem Rev 2:201
90. Nicolle C, Manach C, Morand C, Mazur W, Adlercreutz H, Remesy C, Scalbert A (2002) J Agric Food Chem 50:6222
91. Fuss E (2003) Phytochem Rev 2:307
92. Axelson M, Sjövall J, Gustafsson BE, Setchell KDR (1982) Nature 298:659
93. Borriello SP, Setchell KDR, Axelson M, Lawson AM (1985) J Appl Bacteriol 58:37
94. Vanharanta M, Voutilainen S, Lakka TA, van der Lee M, Adlercreutz H, Salonen JT (1999) Lancet 354:2112
95. Ingram D, Sanders K, Kolybaba M, Lopez D (1997) Lancet 350:990
96. Banskota AH, Nguyen NT, Tezuka Y, Tran QL, Nobukawa T, Kurashige Y, Sasahara M, Kadota S (2004) Life Sci 74:2781
97. Erdemoğlu N, Şener B, Choudhary MI (2004) Z Naturforsch C 59:494
98. Erdemoğlu N, Şener B, Teng CM (2004) Pharm Biol 42:135
99. Gurbuz I, Erdemoğlu N, Yeşilada E, Şener B (2004) Z Naturforsch C 59:233
100. Küpeli E, Erdemoğlu N, Yeşilada E, Şener B (2003) J Ethnopharma 89:265
101. Erdemoğlu N, Şener B (2001) Fitoterapia 72:59
102. Erdemoğlu N, Şener B, Palittapongarnpim P (2003) Pharm Biol 41:614
103. Shen Y-C, Chen C-Y, Chen Y-J, Kuo Y-H, Chien C-T, Lin Y-M (1997) Chin Pharm J (Taipei) 49:285
104. Chiang S (ed)(1977) Dictionary of Chinese crude drugs. Shanghai Scientific Technologic Publishing, Shanghai, p 2342
105. Banskota AH, Usia T, Tezuka Y, Kouda K, Nguyen NT, Kadota S (2002) J Nat Prod 65:170
106. Gordaliza M, Castro MA, del Corral JMM, Lopez-Vazques MI, San Feliciano A, Faricloth GT (1997) Bioorg Med Chem Lett 7:2781
107. Pieters L, Van Dyck S, Gao M, Bai R, Hamek E, Vlietinck A, Lemiere G (1999) J Med Chem 42:5475
108. Tsukamoto H, Hisada A, Nishibe S (1984) Chem Pharm Bull 32:2730

109. Lee CTL, Lin VCK, Zhang SX, Zhu XK, Van Vliet D, Hu H, Beers SA, Wang ZQ, Cosentino LM, Morris-Natschke SL, Lee KH (1997) Bioorg Med Chem Lett 7:2897
110. Ishida J, Wang HK, Oyama M, Cosentino ML, Hu CQ, Lee KH (2001) J Nat Prod 64:958
111. Yang LM, Lin SJ, Yang TH, Lee KH (1996) Bioorg Med Chem Lett 6:941
112. Banskota AH, Tezuka Y, Nguyen NT, Awale S, Nobukawa T, Kadota S (2003) Planta Med 69:500
113. Banskota AH, Nguyen NT, Tezuka Y, Nobukawa T, Kadota S (2006) Phytomedicine 13:109
114. Nguyen NT, Banskota AH, Tezuka Y, Le Tran Q, Nobukawa T, Kurashige Y, Sasahara M, Kadota S (2004) Planta Med 70:29
115. Yin J, Tezuka Y, Subehan SL, Nobukawa M, Nobukawa T, Kadota S (2006) Phytomedicine 13:37
116. Fonseca SF, Campello JP, Barata LES, Ruveda EA (1978) Phytochem 17:499

Top Heterocycl Chem (2007) 11: 145–178
DOI 10.1007/7081_2007_074
© Springer-Verlag Berlin Heidelberg
Published online: 5 July 2007

Antioxidant Activities of Synthetic Indole Derivatives and Possible Activity Mechanisms

Sibel Süzen

Faculty of Pharmacy, Department of Pharmaceutical Chemistry, Ankara University,
Tandogan, 06100 Ankara, Turkey
sibel@pharmacy.ankara.edu.tr

Abstract Indole derivatives constitute an important class of therapeutical agents in medicinal chemistry including anticancer, antioxidant, antirheumatoidal, aldose reductase inhibitor, and anti-HIV agents. Reactive oxygen species are constantly generated in the human body and are involved in various physiologically important biological reactions. However, high levels of free radicals can cause damage to biomolecules such as lipids, proteins, and DNA within cells. Oxidative stress has been implicated in the development of neurodegenerative diseases like Parkinson's disease, Alzheimer's disease, Huntington's disease, epileptic seizures, stroke, and as a contributor to aging and some types of cancer. Indolic compounds are very efficient antioxidants, protecting both lipids and proteins from peroxidation and it is known that the indole structure influences the antioxidant efficacy in biological systems. Due to its free radical scavenger and antioxidant properties, synthesis of indole derivative compounds are under investigation to determine which exhibit the highest activity with the lowest side effects. Epidemiological studies have been strongly suggesting that antioxidants can decrease the rate of many diseases. However, more clinical studies are required to determine the efficacy and safety of these compounds. This chapter gives another perspective on antioxidant activities of synthetic melatonin analogues.

Keywords Indole · Melatonin · Antioxidant activity · Oxidative stress · Mechanism

Abbreviations

ABTS	2,2′-azino-bis(3-ethylbenzthiazoline-6-sulfonic acid)
AFMK	$N(1)$-acetyl-$N(2)$-formyl-5-methoxykynuramine
AMK	N^1-acetyl-5-methoxykynuramine
CAT	Catalase
COX	Cyclooxygenase
DPPH	Diphenylpicrylhydrazyl
EPR	Electron paramagnetic resonance
ESR	Electron spin resonance
G6PD	Glucose-6-phosphate dehydrogenase
GPx	Glutathione peroxidase
GR	Glutathione reductase
HOCl	Hypochlorous acid
LP	Lipid peroxidation
LDL	Low-density lipoprotein
MDA	Malondialdehyde
MAO	Monoamine oxidase
MPO	Myeloperoxidase
NSAID	Non-steroidal anti-inflammatory drugs
RNS	Reactive nitrogen species
ROS	Reactive oxygen species
SAR	Structure activity relationship
SOD	Superoxide dismutase

1
Importance of Reactive Oxygen and Nitrogen Species for Human Body

Free radicals are atomic or molecular species with unpaired electrons that are highly reactive. They take part in chemical reactions and play an important role in many chemical processes, including human physiology. Reactive oxygen species (ROS) and reactive nitrogen species (RNS) have gained a lot of importance because of their active role in many diseases [1, 2]. The sources of ROS generation in different disease settings are of great interest. Despite the existence of these multiple sources, a large number of studies in the last decade indicate that a major ROS source involved in redox signaling is a family of complex enzymes, namely NADPH oxidases [3].

ROS include oxygen-based free radicals such as superoxide, hydroxyl, alkoxyl, peroxyl, and hydroperoxyl (Table 1). Other ROS, such as hydrogen peroxide and lipid peroxides, can be converted into free radicals by transition

Table 1 List of reactive oxygen and nitrogen species

| Reactive oxygen species (RNS) | | Reactive nitrogen species (RNS) | |
Radicals	Non-radicals	Radicals	Non-radicals
Hydroxyl OH	Peroxinitrite $ONOO^-$	Nitrous oxide $NO^{\cdot}$	Peroxynitrite $OONO^-$
Superoxide $O_2^{\cdot-}$	Hypochlorous acid HOCl	Nitrogen dioxide $NO_2^{\cdot}$	Peroxynitrous acid ONOOH
Nitric oxide NO	Hydrogen peroxide H_2O_2		Nitroxyl anion NO^-
Peroxyl $RO_2^{\cdot}$	Singlet oxygen $^{-1}O_2$		Nitryl chloride NO_2Cl
Lipid peroxyl LOO	Ozone O_3		Nitrosyl cation NO^+
Alkoxyl $RO^{\cdot}$	Lipid peroxide LOOH		Dinitrogen trioxide N_2O_3
Hydroperoxyl $ROOH^{\cdot}$			Nitrous acid HNO_2

metals. Reactive nitrogen species (RNS) include mainly nitric oxide, nitrogen dioxide and the potent oxidant peroxynitrite. ROS and RNS are well recognized for playing a dual role as both harmful and beneficial species [4].

The damage to animal or plant cells and tissues caused by ROS is called oxidative stress, which is caused by an imbalance between the production of reactive oxygen and a biological system's ability to detoxify the reactive intermediates or repair the resulting damage. A particularly negative side of oxidative stress is the production of ROS, which includes free radicals.

Nitrosative stress occurs when the generation of RNS in a system exceeds the system's ability to eliminate them. Since ROS and RNS are generally highly reactive, they react with key organic substances such as lipids, proteins, and DNA [5]. Oxidation of these biomolecules can damage them and may be responsible to a variety of diseases.

Free radicals are typically formed as a result of normal cellular metabolism, however, studies have shown that their numbers in the organism increase when its cells are exposed to harmful environmental influences [6] such as pollutants [7], sunlight, radiation [8, 9], emotional stress, smoking [10], excessive alcohol [11], infection [12], and some drugs [13].

1.1
Diseases Associated with Increased Oxidative Stress

The free radicals are often referred to as ROS because the most biologically important free radicals are oxygen-centered. ROS not only exist in living organisms, they also exist in our environment. Combustion sources and photochemical reactions are the major sources [14]. ROS are continuously generated in the human body and take place in various physiologically important biological reactions. However, high levels of free radicals can cause damage to biomolecules such as lipids, proteins, and DNA within cells, resulting in mutations that can lead to malignancy [15]. Damage to DNA by ROS has been widely accepted as a major cause of cancer [16]. Oxidative stress induces a cellular redox imbalance that has been found to be present in various cancer cells compared with normal cells. DNA mutation is a critical step in carcinogenesis and elevated levels of oxidative DNA lesions have been noted in various tumors [17].

Increased levels of ROS due to oxidative stress have been consistently found in cardiovascular diseases as atherosclerosis or hypertension [18]. There is certain evidence that the free radicals involved in Parkinson's disease are mainly due to the production of increased levels of free radicals during oxidative metabolism of dopamine [19]. Oxidative stress, manifested by protein oxidation and lipid peroxidation (LP), among other alterations, is a characteristic of Alzheimer's disease [20] and in the pathogenesis of diabetes related complications. Treatment with antioxidants seemed to be a promising therapeutic option for these diseases [21]. The inflammatory nature of rheumatoid arthritis implies that a state of oxidative stress may also exist in this disease [22, 23]. Also, free radicals have a certain role in Huntington's disease [24, 25], age related degeneration [26], and some autoimmune disorders [27].

Singlet molecular oxygen is one of the major agents responsible for oxidative damage in biological systems, including human skin and eyes [28]. Nitrogen oxide is a relatively stable and highly reactive radical that develops in the organism by oxidation of the guanidine nitrogen of the amino acid L-arginine by the action of NO-synthase. It participates in a number of physiological and pathological processes [29].

Peroxynitrite has been suggested to be formed from nitric oxide and superoxide in vivo. It is a highly reactive oxidant, and causes nitration on the aromatic ring of free tyrosine and protein tyrosine residues. It was reported that peroxynitrite induced various oxidative damage in vitro, for example LDL oxidation, lipid peroxidation, and DNA strand breakage [30].

2
Enzymatic and Non-enzymatic Antioxidants

Because free radicals are necessary for life, the human body has a number of mechanisms to minimize free-radical-induced damage and to repair the damage that does occur, such as the primary enzymes superoxide dismutase (SOD), catalase (CAT), glutathione peroxidase (GPx), and glutathione reductase (GR), glucose-6-phosphate dehydrogenase (G6PD), thioredoxin reductase, heme oxygenase and biliverdin reductase, are the most important antioxidant enzymes. The glutathione system (glutathione, GPx, and GR) is a key defense against hydrogen peroxide and other peroxides. GPx reduces hydrogen peroxide by transferring the energy of the reactive peroxides to a very small sulfur-containing protein called glutathione. Also, when SOD comes in contact with superoxide, it reacts with it and forms hydrogen peroxide that is dangerous in the cell because it can easily transform into a hydroxyl radical. CAT reacts with the hydrogen peroxide and forms water and oxygen [31].

Other enzymes that have antioxidant properties include paraoxonase [32], glutathione-S transferases [33], aldehyde dehydrogenases [34], peroxiredoxins [35] and the recently discovered sulfiredoxin [36]. The existence of antioxidant enzymes in biological systems proves the importance of oxidative damage as a real threat to cellular and organismal survival.

Some non-enzymatic antioxidants play a key role in these defense mechanisms. These are often vitamins (A, C, E, K), minerals (zinc, selenium), caretenoids, organosulfur compounds, allyl sulfide, indoles, antioxidant cofactors (coenzyme Q_{10}), and polyphenols (flavonoids and phenolic acids) [1, 37]. Further, there is good evidence that bilirubin and uric acid can act as antioxidants to help neutralize certain free radicals [38]. Alpha-carotene, lycopene, lutein, and zeaxanthine [39] can be considered subgroups of carotenoids [40] that are effective antioxidant compounds.

In summary, the human body is constantly exposed to ROS generated from both endogenous and exogenous sources. Antioxidants, both enzymatic and non-enzymatic, prevent oxidative damage to biological molecules by various mechanisms [41].

2.1
Antioxidant Properties of Melatonin

Melatonin, N-acetyl-5-methoxytryptamine, is the main secretory product of the pineal gland and is released in the circulation in a circadian manner, with highest concentrations at night. Synthesis of melatonin also occurs in other areas of the body, including the retina, the gastrointestinal tract, the skin, bone marrow, as well as in lymphocytes [42]. It is produced to assist our bodies to regulate our sleep-wake cycles. Natural production of melatonin is increased by darkness and suppressed by sunlight. Its synthesis and secretion decrease significantly by middle age and declines even further during old age. Studies have shown that melatonin modulates biological rhythms [43, 44]. In addition to sleep, melatonin has many other functions. It is a highly conserved molecule that acts as a receptor-independent free-radical scavenger and a broad-spectrum antioxidant. The ability of melatonin to react with free radicals was first shown in 1993 when Tan et al. [45] identified its interaction with hydroxyl radicals. Melatonin and its metabolites successively scavenge ROS/RNS, which is referred to as the free-radical scavenging cascade. Rapid melatonin consumption during stress may provide a protective mechanism of organisms against oxidative damage [46]. O-methyl and N-acetyl residues of melatonin assist its amphilicity, enabling the molecule to enter all organs and all subcellular compartments, and crucial for its antioxidant properties [47]. Antioxidative protection exceeds direct interactions with free radicals and is effective in detoxifying ROS in particular, the hydroxyl radical and the generation of primary and secondary products in order to eliminate radicals [48, 49].

Recent evidence indicates that the original melatonin metabolite may be N(1)-acetyl-N(2)-formyl-5-methoxykynuramine (AFMK), which also has antioxidant properties. There is a high correlation between the concentration at which melatonin and closely related indoles exert a direct antioxidant effect in vitro and a neuroprotective effect [50]. Radical scavenging by indolic compounds is strongly modulated by their functional residues. All indolamines have a heteroaromatic ring system of high electroreactivity and they only differ in carrying functional groups in their side chains. These substituents determine to a great extent the reactivity, potency, and efficiency of radical scavenging activity [47]. Like other indole derivatives and tryptophan metabolites, melatonin has redox properties because of the presence of an electron-rich aromatic ring system, which allows the indoleamine to easily function as an electron donor. A number of oxygen-centered radicals and other reactive species have been shown to be capable of oxidizing melatonin in various experimental systems [51]. Light exposure, either natural or artificial, causes extremely rapid destruction of melatonin.

According to Tan et al. [45], the methoxy group in the 5th position of the indole ring keeps melatonin from exhibiting prooxidative activity. If the methoxy group is replaced by a hydroxyl group, the antioxidant capacity of this molecule may be enhanced. The changes could be decreased lipophility and increased prooxidative activity.

SAR studies on melatonin analogues showed that methoxy and amido moieties are important for binding to melatonin receptors; various substituents on the 2nd position of the indole ring enhance the binding affinity [52–54]. The conformational flexibility of the C-3 ethylamido side chain is probably responsible for the broad spectrum of biological activities. As a powerful antioxidant melatonin was shown to have significantly broader actions including oncostatic effects [55], immune system stimulation [56], and anti-inflammatory functions [57, 58].

Melatonin has side effects, but much less so than pharmaceutical sleeping pills. Long-term safety is not known. Prolonged use may have an influence on sex organs and reduce libido. It may slightly lower blood pressure. People with the symptoms of severe mental illness, severe allergies, auto-immune diseases, or immune system cancers such as leukemia should not take melatonin. Despite its possible contribution in regulating many physiological processes, two main problems limit its therapeutic use at present. The first is a very short biological half life, due to its fast metabolism to 6-hydroxymelatonin and AFMK, and the second is the lack of selectivity of melatonin at target sites [59, 60].

Melatonin oxidation seems quite important for the production of other biologically active metabolites such as AFMK and N^1-acetyl-5-methoxykynuramine (AMK). AMK interacts with reactive oxygen and nitrogen species, conveys protection to mitochondria, inhibits and downregulates COX-2. Thus, melatonin may be considered as a prodrug [61].

The antioxidant properties of melatonin include scavenging free radicals and the regulation of the activity of antioxidant and pro-oxidant enzymes. It seems the potential application of melatonin in an antioxidant therapy could be complicated by the pharmacokinetic behavior of the molecule, which is subjected to a limited bioavailability and plasma half-life after oral administration [62].

Plasma levels of melatonin could not be delegate of its concentration in tissues [63], and its half-life in the target tissues could be much lower than expected. The availability of new compounds having the beneficial properties of melatonin, but revealing different pharmacokinetic properties could therefore be useful in the search for good antioxidant efficiency.

3
Synthetic Indole Derivatives having in Vitro Antioxidant Activity

Indole derivatives are biologically important chemicals present in microorganisms, plants, and animals, and represent an important class of therapeutical agents in medicinal chemistry. Anticancer [64–67], antioxidant [2, 68, 69], anti-rheumatoidal [70–73], aldose reductase inhibitory [74–76], antibacterial [77–79], antifungal [80, 81], antiviral [82–84], antimalarial [85, 86] and anti-HIV activities [87–90] can be suggested as the most significant activities of synthetic indole derivatives. Indolic compounds are very efficient antioxidants, protecting both lipids and proteins from peroxidation, and it is known that the indole structure influences the antioxidant efficacy in biological systems [91–93].

In recent years, many physiological properties of melatonin have been described resulting in much attention in the development of synthetic compounds possessing indole ring [94]. These compounds have structural similarity to melatonin. However, the therapy of oxidative stress-related diseases has not found satisfactory application in clinical practice. This may be due to the insufficient efficacy of drugs available, their unsuitable pharmacokinetics, side effects, and toxicity.

3.1
Anti-inflammatory Indole Derivatives

Cyclooxygenase (COX) is a rate-limiting enzyme in the synthesis of prosta-noids from arachidonic acid. Two isoforms of COX have been identified: COX-1 and COX-2. COX reaction products have long been hypothesized to participate in the regulation of the cerebral circulation. An indole deriva-tive indomethacin is an agent that inhibits both COX-1 and COX-2 [95]. Indomethacin pretreatment effectively inhibited free radical production [96]. COX-2 plays a critical role in the inflammatory response and its over-expression has been associated with several pathologies, including neurode-generative diseases and cancer. Melatonin has well-documented antioxidant and immuno-modulatory effects [97, 98]. Thus, scavenging activity against ROS by anti-inflammatory drugs may be of great therapeutical value. Recent studies have suggested that free radicals may be involved in the pathogenesis of brain injury and brain edema, and compounds that have free-radical scavenger activity can reduce brain edema with trauma or ischemia. Stimulation of the COX activity shows peroxides in biological fluids. Hydrogen peroxide has been shown to be formed during inflammatory processes and is implicated in its pathophysiology. These findings imply that non-steroidal anti-inflammatory drugs (NSAIDs) may be able to scavenge many forms of free radicals.

Electron spin resonance (ESR) study has demonstrated that etodolac (1,8-diethyl-1,3,4,9-tetrahydropyrano-[3,4-b]indole-1-acetic acid) and in-domethacin had direct superoxide scavenging activity [99].

In a study NSAIDs indomethacin (2-[1-(4-chlorobenzoyl)-5-methoxy-2-methyl-indol-3-yl]acetic acid), acemetacin (1-[p-chlorobenzoyl]-5-methoxy-2-methylindole-3-acetic acid carboxymethyl ester), and etodolac ((±)-1,8-diethyl-1,3,4,9-tetrahydropyrano-(3,4-b)indole-1-acetic acid) were tested on different ROS generating systems (Fig. 1). There are some satisfactory results that confirm that the anti-inflammatory activity of these compounds may be also partly due to their ability to scavenge ROS and RNS. The observed effect-ive scavenging activity may contribute to the anti-inflammatory therapeutical effects [92, 100].

The indolic nitrogen is the active redox center of indoles, due to its lone pair of electrons [101, 102]. Indeed, if the oxygen replaces the nitrogen in the indolic ring, the antioxidant activity of the resulting benzofurane is much lower [47]. Delocalization of this electron pair over the aromatic system seems to be of great importance for antioxidant activity of indole derivatives.

Evaluation of the scavenging activity for H_2O_2 by NSAIDs, namely indole derivatives (indomethacin, acemetacin, etodolac), pyrrole derivatives (tol-metin, ketorolac), oxazole derivative (oxaprozin), indene derivative (sulin-dac) and its metabolites (sulindac sulfide and sulindac sulfone) was per-formed by Costa et al. [103]. The obtained results against endogenous antiox-idants melatonin and GSH demonstrated that all the studied NSAIDs display

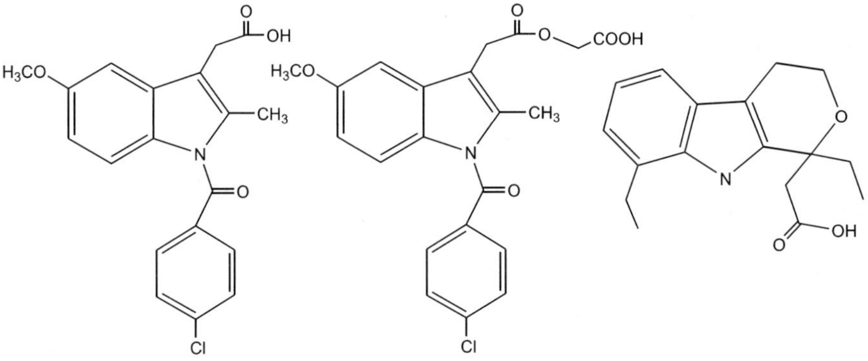

O
OH
H3CO
CH3
N
O
Cl
O
O
COOH
H3CO
CH3
N
O
Cl
O
N
OH
O

1 2 3

Fig. 1 Chemical structures of indomethacin (**1**), acemetacin (**2**) and etodolac (**3**)

H_2O_2 scavenging activity, although in different degrees. The ranking order of potency found was sulindac sulfone > sulindac sulfide > GSH > sulindac > indomethacin > acemetacin > etodolac > oxaprozin > ketorolac approximately melatonin > tolmetin. The inhibition of prostaglandin synthesis constitutes the primary mechanism of the anti-inflammatory action of these drugs. According to the results, the anti-inflammatory activity of NSAIDs may be partly due to NSAIDs ability to scavenge ROS and RNS [100, 101].

Biochemical responses against free radicals were investigated on *N*-substituted indole-2 and 3-carboxamides (Fig. 2) [104–107], which are proposed to be selective COX-2 inhibitors since they show good binding capability to an enzyme active site [108]. Also, some indole-3-acetamides have significant antioxidant activity [91, 105]. Especially the compounds that have chlorophenyl and chloropiperidine as side groups showed the best activity by inhibiting superoxide radical. The antioxidant profiles of congeners at the 2nd or 3rd positions of the indole ring were found to be similar, but it can be con-

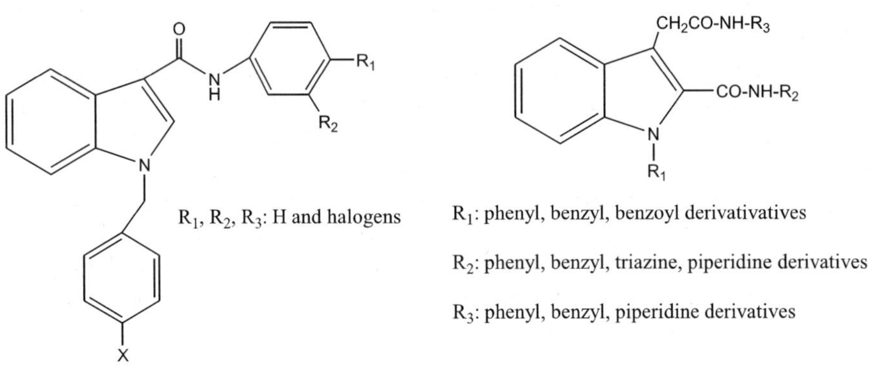

O
N
H
R1
R2
N
X
R1, R2, R3: H and halogens
CH2CO-NH-R3
CO-NH-R2
N
R1
R1: phenyl, benzyl, benzoyl derivativatives
R2: phenyl, benzyl, triazine, piperidine derivatives
R3: phenyl, benzyl, piperidine derivatives

Fig. 2 *N*-substituted indole-2-carboxamides and indole-3-acetamides [105–107]

cluded that indole-2-carboxamide derivatives are stronger inhibitors for LP than indole-3-carboxamides. The special effect of the substitution feature at the 2nd position might be due to the lipophilic groups being involved in the interaction with the aromatic ring located at position 1.

Also, N-H and N-substituted indole esters (Fig. 3), which were reported to be COX-2 enzyme inhibitors, have inhibitory activity on LP and superoxide anion production [91]. Especially pyrrolidine and o-methylphenyl groups were found to be the most important side chains (R_1 group) that elevated both activities for indole esters.

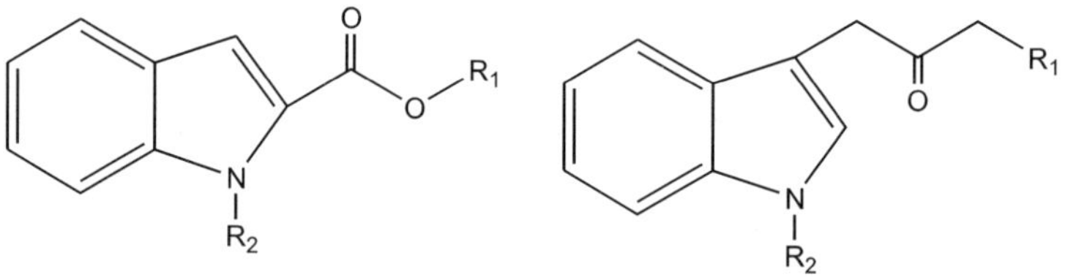

R_1: phenyl, pyrrolidine piperidine and cyclopropyl derivatives

R_2: H, benzoyl derivatives

Fig. 3 N-H and N-substituted indole esters [91]

3.2
Fused-Ring Indole Derivatives

One of the most important condensed ring systems is indole. Whether the indole nitrogen is substituted or not, the favored site of attack is C-3 of the heterocyclic ring. Bonding of the electrophile at that position permits stabilization of the intermediate by the nitrogen without disruption of the benzene aromaticity. Indole can exist in two tautomeric forms, the more stable enamine and the 3-H-indole or imine forms. C-2 to C-3 pi-bond of indole is more capable of cycloaddition reactions then the other pi bonds of the molecule. Intermolecular cycloadditions are not favorable, whereas intramolecular variants are often high-yielding.

Stobadine ((−)-cis-2,8 dimethyl-2,3,4,4a,5,9b-hexahydro-1-H-pyrido (4,3b) indole) is a model drug with pyridoindole structure with cardioprotective and antioxidant properties [109]. Based on stobadine, (−)-cis-2,8-dimethyl-2,3,4,4a,5,9b-hexahydro-1H-pyrido[4,3-b]indole (Fig. 4), a well-known antioxidant, free-radical scavenger and neuroprotectant, stobadine derivatives with improved pharmacodynamic and toxicity profiles were developed [110]. A stobadine molecule was modified mostly by electron donating substitution on the benzene ring and by alkoxycarbonyl substitution at N-2 position. Significant antioxidant activity was observed in the new compounds. A link

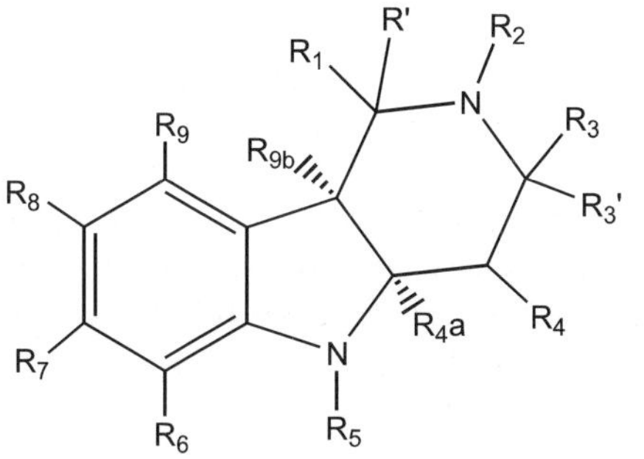

Fig. 4 General formula of stobadine derivative antioxidant compounds [110]

between the neuroprotective and antioxidant/scavenger properties in the compounds was suggested. Acute toxicity of some of the new pyridoindoles was diminished compared to stobadine. Molecular modelling pointed that indole nitrogen, responsible in stobadine and derivatives, was found capable to react with free radicals, creating a more stable and less-reactive nitrogen-free centered radical. The pyridoindole molecules extend the range of available neuroprotectants interfering with oxidative stress in neuronal tissue.

The synthesis [111] and examination of 1-p-toluenesulfonyl-6,7,8,9-tetrahydro-*N,N*-di-*n*-propyl-1*H*-benz [g]indol-7-amine (TPBIA in Fig. 5) for behavioral effects in rats related to interactions with central dopamine receptors showed remarkable antioxidant activity. Because TPBIA has increased lipophilicity, penetrating the blood-brain barrier in a considerable degree was expected and it was found that it completely inhibits the peroxidation of rat liver microsome preparations [112].

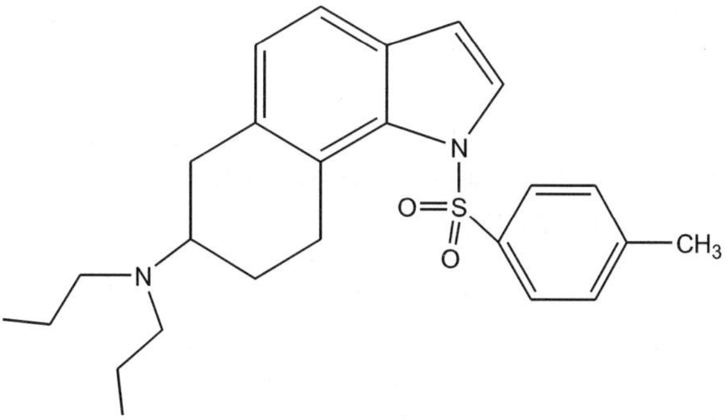

Fig. 5 1-p-toluenesulfonyl-6,7,8,9-tetrahydro-*N,N*-di-*n*-propyl-1*H*-benz [g]indol-7-amine (TPBIA) [112]

Studies of influence of the acetamidoethyl side chain on the capacities of melatonin-related compounds to inhibit low-density lipoprotein

(LDL) oxidation showed that nonanoyl and phenyl derivatives were particularly active [113]. The capacity of the two melatonin-related compounds DTBHB (*N*-[2-(5-methoxy-1*H*-indol-3-yl)ethyl]-3,5-di-*tert*-butyl-4-hydroxybenzamide) and GWC20 [(R,S)-1-(3-methoxyphenyl)-2-propyl-1,2,3,4-tetrahydro-β-carboline] (Fig. 6) inhibit Cu^{++} and free-radical-induced LDL oxidation was compared to the antioxidant effect of melatonin. The compounds were found considerably more active than melatonin in inhibiting this oxidation. It can be seen that there is a strong relationship between the consequence of modifications in the acetamidoethyl side chain of melatonin and the capacity of the resulting molecules to inhibit LP. The replacement of the acetamido function by a benzoyl gives highly antioxidant molecules [114].

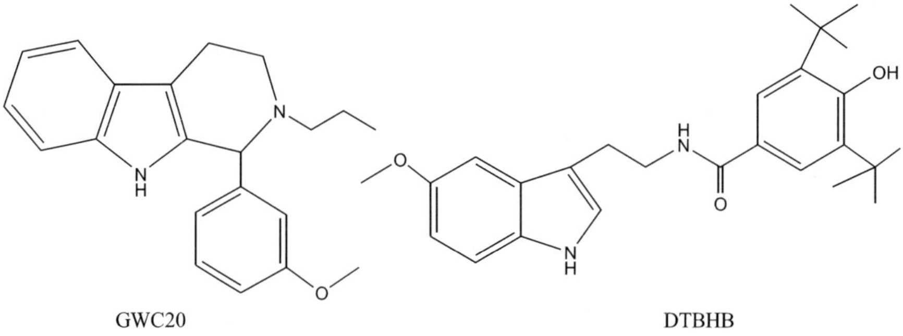

Fig. 6 Molecular formula of GWC20 and DTBHB [114]

Oxidative stress appears to have a central role in the induction of apoptosis following the exposure of cells to a range of cytotoxic insults. Anti-apoptotic properties of the antioxidant, 4b,5,9b,10-tetrahydroindeno[1,2-b]indole, in Jurkat T cells subjected to a number of cytotoxic insults. Peroxide and superoxide anion production following UV treatment showed that indole derivative was found to only partially inhibit superoxide anion production and exhibited strong inhibition of caspase-3 activation in UV [115].

3.3
Indole Derivatives Connected to a Known Antioxidant Molecule

Antioxidants of a different chemical nature have been investigated for use as therapeutic agents, either alone or in combination. Combination of a known antioxidant compound with indole ring or melatonin analogue compounds is a new trend in antioxidant chemistry.

It is known that retinoid-related compounds (Fig. 7) represent classes of promising antioxidative potential. A series of melatonin retinoids was synthesized using the condensation reaction sequence involving tetrahydrote-

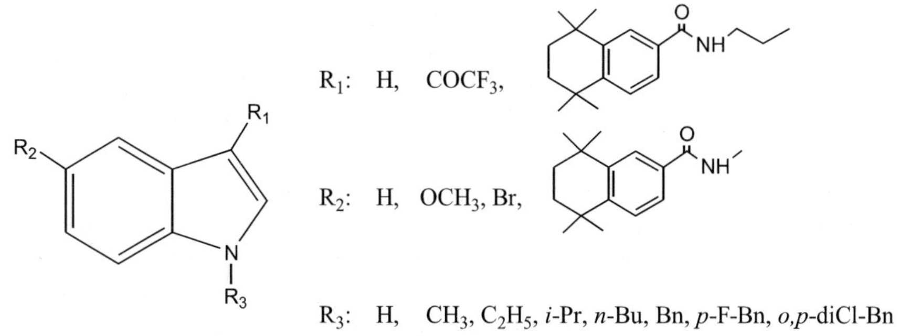

Fig. 7 Chemical structure of tetrahydronaphthalene-indole derivatives [116]

tramethylnaphthalene carboxylic acid and appropriate melatonin-type moieties [116]. These derivatives were reported to be very effective compounds on LP.

Also, antioxidant properties of conjugates based on indole and lipoic acid moieties (Fig. 8) were studied. The target compounds showed reasonable antioxidant properties using rat liver microsomal, NADPH-dependent LP in-

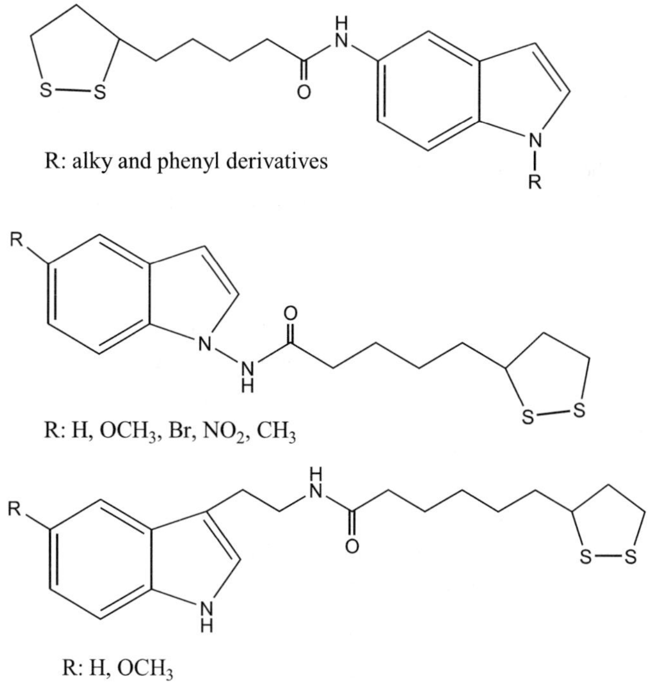

Fig. 8 Chemical structure of indole α-lipoic acid derivatives [117]

hibition. Some of the target compounds, especially those containing amidFe
linker at position 5 of the indole ring, proved to be highly effective in inhibiting LP compared to α-lipoic acid. Removal of alkyl substituents from
the indole nitrogen resulted drop in the inhibitory activity. Non-substituted
derivatives at position 5 also showed better activity than those with an
electron-donating substituent [117].

Benzimidazole and indole rings and some 6-fluoro-5-substituted-benz-
imidazole derivatives in which indole and 1,1,4,4-tetramethyl-1,2,3,4-tetra-
hydro-naphthalene groups were attached to the 2-position of the benzimidazole ring (Fig. 9) were synthesized and tested for antioxidant properties [118]. A connection of both indoles to the benzimidazole ring using
appropriate o-phenylendiamine-related indole derivative with Na$_2$S$_2$O$_5$ gave
5-substituted-6-fluoro-2-(5-substituted-1H-indol-3-yl)-1H-benzimidazole
derivatives. These compounds showed very good free-radical scavenging
properties in vitro by determining their capacity to scavenge superoxide anion formation. Compounds that are bear p-phenyl piperazine on the 5th
position of indole ring showed significant activity.

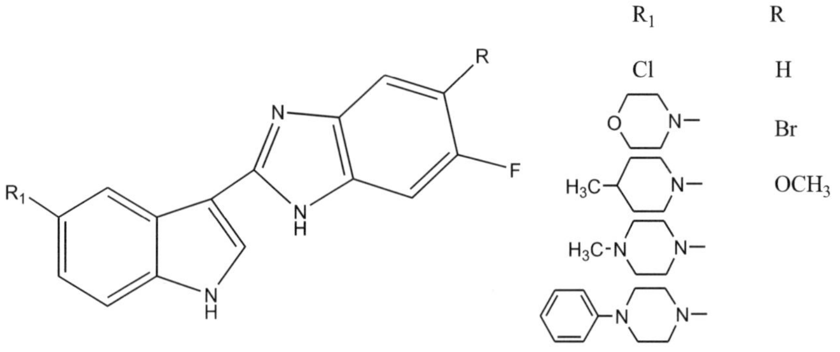

Fig. 9 Chemical structure of indole-benzimidazole derivatives [118]

Asakai et al. [119] reported that bisindolylmaleimide inhibited necrotic cell
death induced by oxidative stress in a variety of primary-cultured cells. Also
structure–activity relationship of bisindolylmaleimide derivatives (Fig. 10)
as inhibitors of H$_2$O$_2$-induced necrotic cell death was studied by Katoh
et al. [120]. Based on the SAR of the bisindolylmaleimides it was hypothesized
that the coplanarity of one indole ring with the maleimide ring is important for the activity, and it was expected that removal of the second indole
ring would favor such a coplanar conformation. Analogs of indolylmaleimide
derivatives [121] were synthesized and tested for cell death-inhibitory activity
of necrotic cell death induced by H$_2$O$_2$. 2-(1H-Indol-3-yl)-3-pentylamino-
maleimide was the most effective cell-death inhibitor among the compounds.
Heteroatom substitution at the maleimide C3 position was found import-

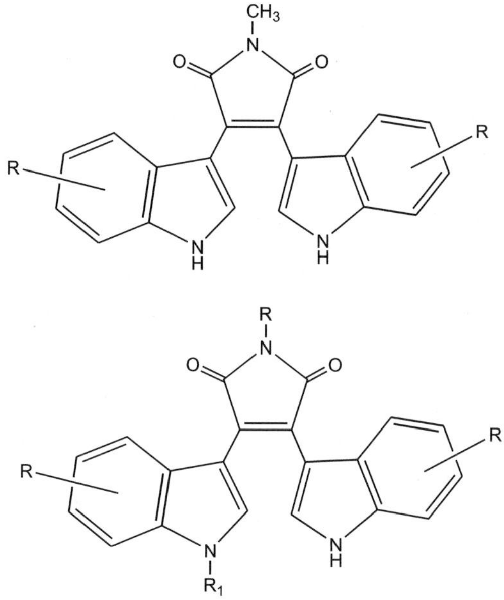

R: H, CH$_3$, OCH$_3$, Br, Cl, F

R: H, CH$_3$

R$_1$: H, (CH$_2$)$_3$-NH$_2$, (CH$_2$)$_3$N(CH$_3$)$_2$

Fig. 10 Analogs of bisindolylmaleimide derivatives [120, 121]

ant, so that the lone-pair electrons could be delocalized on the maleimide ring.

3.4
Alkyl and/or Aryl Substituted Indole Derivatives

In the process of screening indole compounds as antioxidant agents, new properties were discovered for an endogenous species. Indole-3-propionic acid (IPA) has been identified in the plasma and cerebrospinal fluid of humans [122]. It was shown that IPA prevented oxidative stress. The radical-scavenging efficiency of IPA has a better activity of several previously reported antioxidants, including melatonin. Because hydroxyl radicals cannot be enzymatically detoxified, radical-scavenging compounds have on-site protection against these reactive radicals [123]. IPA (OXIGON) is a potent antioxidant devoid of pro-oxidant activity. IPA has been demonstrated to be an inhibitor of beta-amyloid fibril formation and to be a potent neuroprotectant against a variety of oxidotoxins [124].

Based on IPA, *N*-H, and *N*-substituted indole-3-propanamide derivatives (Fig. 11) have been prepared and their efficiencies were investigated towards SOD and LP. Compounds have very significant activity on inhibiting SOD as well as LP [125].

The anti-LP activity and anti-superoxide formation of *N*-H and *N*-substituted indole derivatives (Fig. 12) were evaluated to determine their anti-

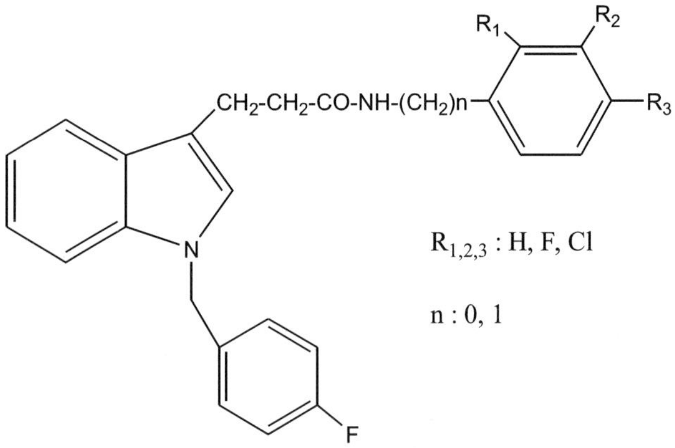
CH2-CH2-CO-NH-(CH2)n
R1
R2
R3
N
F
R1,2,3 : H, F, Cl
n : 0, 1

Fig. 11 Chemical structure of *N*-H and *N*-substituted indole-3-propanamide derivatives [125]

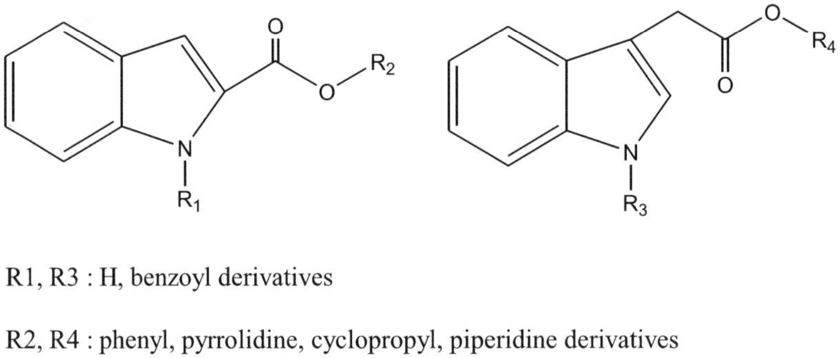
O
O
R2
N
R1
O
R4
O
N
R3
R1, R3 : H, benzoyl derivatives
R2, R4 : phenyl, pyrrolidine, cyclopropyl, piperidine derivatives

Fig. 12 Chemical structure of *N*-H and *N*-substituted indole derivatives [126]

oxidant activity [126]. Most of the compounds exhibit important antioxidant activity showing considerable inhibition of LP of mouse liver homogenate.

Many studies have established beneficial effects of antioxidants that reduce the myocardial infarct size by interventions, which either attenuate the generation or reduce the effects of reactive oxygen species [127, 128]. Indole derivatives containing a triazole moiety (Fig. 13) were synthesized [129] and their antioxidant activity was investigated using methods for interactions of these derivatives with reactive oxygen species in vitro [130–132]. All compounds showed significant activity that possibly depends on the attachment position of the triazole moiety on the indole nucleus. Some derivatives that were substituted on the nitrogen of the indolic nucleus showed better antioxidant properties than melatonin. Also, cardioprotective efficacy of 3-[(1*H*-1-indolyl)methyl]-4-amino 4,5-dihydro-1*H*,1,2,4 triazole-5-thione that has been proven to show significant antioxidant properties by inhibiting in vitro non-enzymatic rat hepatic microsomal LP re-

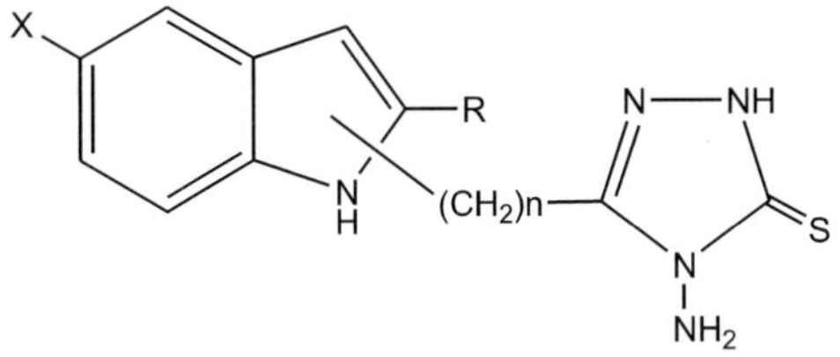

X: H, Br, Cl, NO_2

R: H, CH_3

Fig. 13 Chemical structure of indole derivatives containing a triazole moiety [130, 131]

duced significantly the level of malondialdehyde in rabbits under ischemia–reperfusion [133].

Indole derivatives, with changes in the 5-methoxy and acylamino groups (Fig. 14), the side chain position and the lipophilic/hydrophilic balance were tested for their in vitro antioxidant potency and for their cytoprotective activity by Mor et al. [134]. Substitutions on the indole ring, which were highly unfavorable for melatonin receptor affinity, were generally tolerated in terms of the antioxidant properties of the compounds. Structural modifications, such as the shift of the methoxy group from the 5- to the 6-position, the introduction of a bromine in the 2-position and the insertion of the acetylaminoethyl chain in the 2- instead of 3-position proved favorable for the antioxidant performance of these compounds. The melatonin analogue *N*-[2-(5-methoxy-1*H*-indol-2-yl)ethyl]acetamide, which proved to be among the most potent derivative showed, a low-affinity antagonist on melatonin membrane receptors.

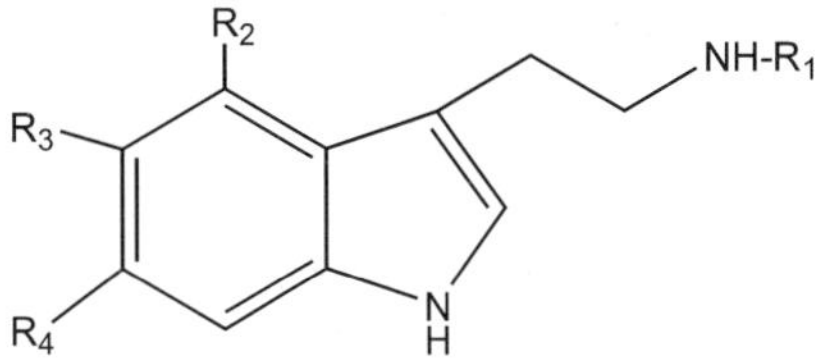

R_1: H, $COCH_3$,COC_2H_5

R_2: H, Br

R_3: H, OH, OCH_3, OCH_2CH_2OH

R_4: H, OCH_3

Fig. 14 Indole derivatives, with changes in the 5-methoxy and acylamino groups [134]

Investigation of replacement of the 5-methoxy group by substituents with different electronic and lipophilic properties and methylation of the indole nitrogen or its replacement by a sulfur atom was evidence for the shift of the 5-methoxy group to the 4-position of the indole nucleus led to the most active radical scavenger but much less effective as a cytoprotectant [135]. 5-alkoxy-2-(*N*-acylaminoethyl)indole (Fig. 15) appeared as the key feature to confer both antioxidant and cytoprotective activity to the structure. Antioxidant activity seems essential for cytoprotection, but it is not sufficient, and there is no statistically significant correlation between the two types of activity.

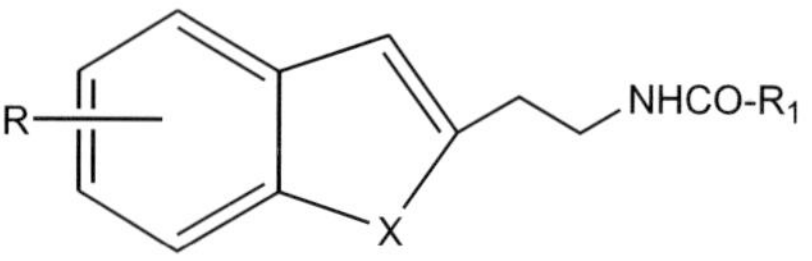

X: NH, S, N-CH₃

R: H, F, CH₃, OCH₃

R₁: CH₃, CH₂COOH

Fig. 15 2-*N*-acylaminoethyl derivatives [135]

Some of the tryptamine and *N*-alkyl-substituted melatonin analogues (Fig. 16) constituted significant activity against LP [60]. Alkyl substitution of the first position affects the antioxidant activity changeable due to the substitution pattern at the 5th position of the indole ring. Electron-donor groups of position 5 of the indole ring might be involved in the interaction with those located at the 1st position.

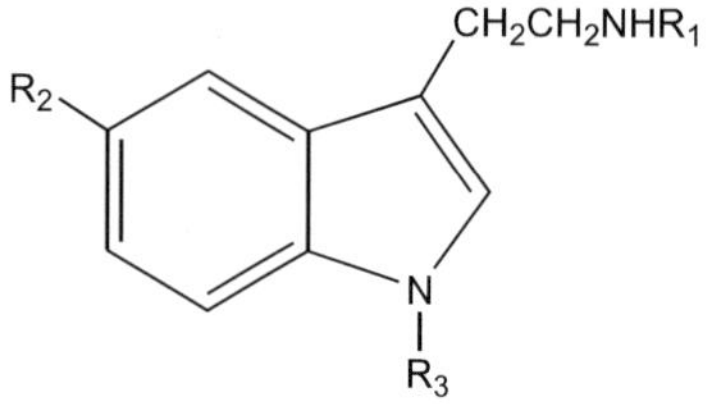

R₁:CH3, C₂H₅
R₂: H, OCH₃
R₃: alkyl and phenyl derivatives

Fig. 16 Tryptamine and *N*-alkyl substituted melatonin analogues [60]

N-(2-propynyl)2-(5-benzyloxy-indol)methylamine (PF 9601N in Fig. 17), a novel MAO B inhibitor has shown a neuroprotective effect antioxidant activity [136]. This neuroprotective effect could be explained in terms of the antioxidant capacity of PF 9601N. According to structure-activity relationship studies, the presence of a benzyloxy group, or a hydroxy or methoxy group, at position 5 of the indole ring enhanced these antioxidant characteristics, presenting a decreasing order of antioxidant activity of the primary > secondary > tertiary amines.

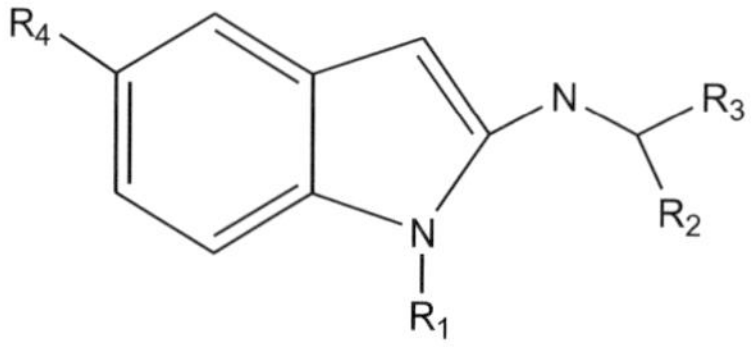

R₁, R₂, R₃ : H, alkyl derivatives
R₄ : benzyloxy, OCH₃, OH

Fig. 17 *N*-alkyl 2-substituted melatonin analogues [136]

HOCl released by activated leukocytes has been implicated in the tissue damage that characterizes chronic inflammatory diseases. The release of HOCl can be measured by the production of taurine-chloramine. A group of substituted indole derivatives including indole, 2-methylindole,

3-methylindole, 2,3-dimethylindole, 2,5-dimethylindole, 2-phenylindole, 5-methoxyindole, 6-methoxyindole, 5-methoxy-2-methylindole, melatonin, tryptophan, indole-3-acetic acid, 5-methoxy-2-methyl-3-indole-acetic acid, and indomethacin was found effective inhibitors of the generation of HOCl. The neutrophil enzyme myeloperoxidase (MPO) plays an essential part in the innate immune system by catalyzing the production of HOCl. The results show that the indole moiety should be considered to be a promising candidate in the search for new, reversible and selective MPO inhibitors [137].

A series of 2-phenyl indole derivatives (Fig. 18) were prepared using Fischer indole synthesis and their in vitro effects on rat liver LP levels, superoxide formation and DPPH stable radical scavenging activities were determined against melatonin, BHT and α-tocopherol [68]. The compounds significantly inhibited LP. Compounds bearing electron-withdrawing groups, such as F, Cl, NO_2, had the highest reduction in LP values, since these groups are inductive electron withdrawing due to their electronegativity and they help the indole ring to easily interact with free radicals. All of the examined compounds (and melatonin) had no significant inhibitory effect on superoxide anion formation, which clearly shows that melatonin and 2-phenyl indole derivatives act as potent hydroxyl radical scavengers in vitro. Interestingly, compounds that have a carboxaldehyde group on the 3rd position of the indole showed no antioxidant activity.

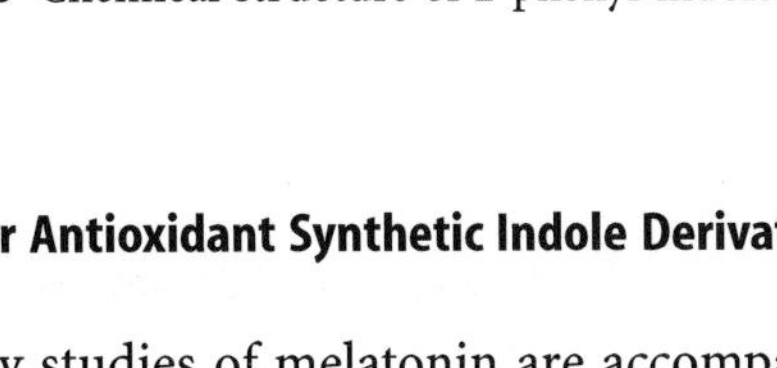

$R_{1,2}$: H, NO_2

$R_{3,4,5,6}$: H, OH Cl, NO_2, NH_2, CH_3

R_7: H, CHO

Fig. 18 Chemical structure of 2-phenyl indole derivatives [68]

3.5
Other Antioxidant Synthetic Indole Derivatives

Many studies of melatonin are accompanied on an idea that many of the effects credited to melatonin may be mediated by some melatonin analogues or metabolites [138], such as by AFMK, a product of the melatonin antioxidant cascade, which is also a potent scavenger of reactive species [139]. Also, the melatonin precursor, *N*-acetylserotonin, or the melatonin metabolite 6-hydroxymelatonin, which was found a better antioxidant than melatonin itself by authors Zhang and coworkers [140], is considered Melatonin-related indoles that protect and defend against both autooxidation and iron-induced peroxidation of lipids [141].

According to Tan et al. [142], melatonin, as an electron-rich molecule, may interact with free radicals via an additive reaction to form several stable end-products. Changes in the aromatic ring of the indolamine could reduce the antioxidant properties. Although both the 5-methoxy group and the acetamidoethyl side chain are important in the antioxidant properties of melatonin, the lack of both side chains implies an antioxidant ability higher than the lack of each of them separately. Zolpidem (Fig. 19) is a non-benzodiazepine-related hypnotic with a imidazopyridine structure [*N,N*,6-trimethyl-2-*p*-tolyl-imidazo (1,2-a) pyridine-3-acetamide L-(+)]. Comparison with the protection by melatonin and Zolpidem showed that Zolpidem tartrate was as effective as melatonin in preventing LP in liver and brain tissue, although it did not show in vitro protein protection against oxidation, contrary to melatonin [143].

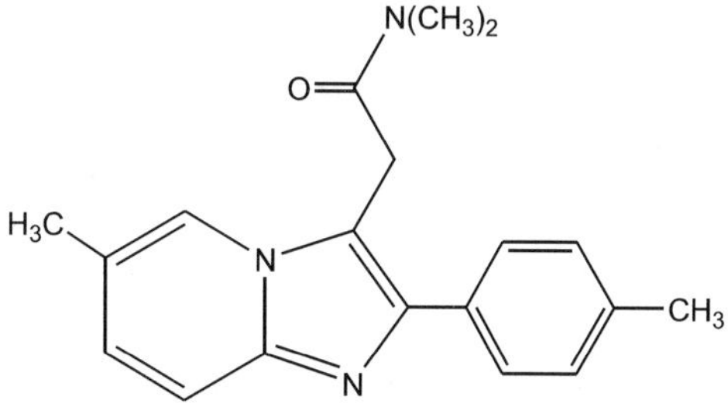

Fig. 19 Zolpidem [*N,N*,6-trimethyl-2-*p*-tolyl-imidazo (1,2-a) pyridine-3-acetamide L-(+)] [143]

Selective reduction of a double bond in the indole nucleus of melatonin using $NaBH_3CN/CF_3COOH$ system resulted as 2,3-dihydromelatonin (Fig. 20) with improved antioxidant activity in the model of quenching of DPPH radical as well as in lipoperoxidation induced by the system Fe/ascorbate in rat brain homogenates [144].

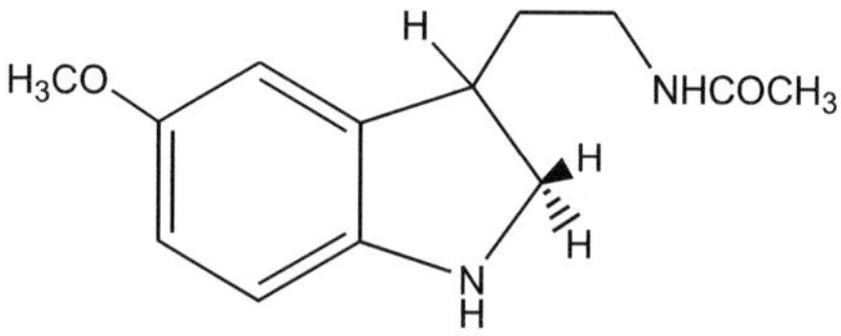

Fig. 20 Chemical structure of 2,3-dihydromelatonin [144]

Fluvastatin (Fig. 21) is a member of the drug class of statins used to treat hypercholesterolemia and to prevent cardiovascular disease. It is able to decrease ROS, such as hydroxyl radicals and superoxide anions generated by the Fenton reaction, and by the xanthine–xanthine oxidase system. The antioxidative effect of fluvastatin was thought to have caused not only the scav-

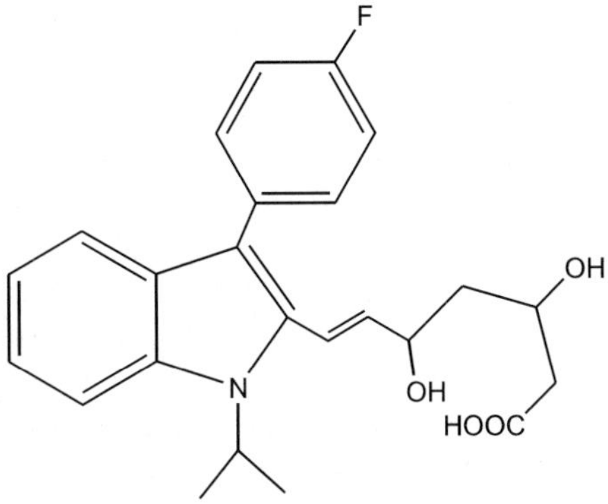

Fig. 21 Fluvastatin (7-[3-(4-fluorophenyl)-1-(1-methylethyl)-1H-indol-2-yl]-3,5-dihydroxy-hept-6-enoic acid) [145]

enging action of the radicals but also to have inhibited ROS generation by inhibiting the NADPH oxidase activity. This antioxidative potential of fluvastatin may be beneficial in preventing atherosclerosis [145].

The antioxidant behavior of a series of substituted indoline-2-ones and indolin-2-thiones (Fig. 22) was investigated using an oxygen radical absorbance capacity assay and 2,2′-azobis(2-amidino-propane) dihydrochloride as the radical generator. The results indicated that the examined indoline derivatives had effective activities as radical scavengers and may be considered as an effective source for combating oxidative damage. The compounds showed preventive antioxidative action similar to SOD and protective action against deoxyribose degradation by hydroxyl radical [146].

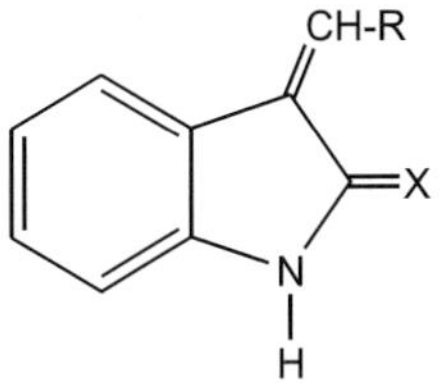

X: oxo and thio

R: flouro, chloro, nitro, methoxy and amino phenyl derivatives, imidazole, phenylpropenyl

Fig. 22 Chemical structure of indoline-2-ones and indolin-2-thiones [146]

4
Indole *N*-oxides and Analysis
with Electron Spin Resonance (ESR) Spectroscopy

Studying with free radicals is difficult in order to measure steady-state concentrations because of the extremely short half-life of these chemical species. Molecules were discovered that could "trap" free radicals, and these molecules were called spin traps. The spin-trapping technique [147] facilitates the detection of reactive free radicals by electron spin resonance (ESR) spectroscopy. This method involves the addition of a diamagnetic molecule (the spin trap), typically an organic nitrone or nitroso compound, to a system containing or producing a reactive radical. The trap reacts with the free radical to give a stable paramagnetic spin adduct that accumulates until it becomes observable by EPR. Spin trapping is considered one of the best techniques available for characterizing the presence and the nature of reactive radicals and, in some cases, it is particularly useful for the study of a reaction mechanism involving radical intermediates [148]. Examination of the antioxidant effects of indole compounds such as melatonin, tryptophan, and serotonin, on ROS generation by ESR showed that hydroxy radical scavenging activity of melatonin was higher than that of serotonin [149].

The spin-trapping technique was used to detect the free radical nitrogen dioxide in solution using 5,7-di-*tert*-butyl-3,3-dimethyl-3*H*-indole *N*-oxide and 6-*tert*-butyl-3,3-dimethyl-3*H*-indole *N*-oxide (Fig. 23) [150]. Formation of an acyl nitroxide radical was observed when the spin-trap solution was mixed with nitrogen dioxide. The first step of the reaction is likely represented by the nucleophilic attack of the oxygen atom of an NO_2 molecule to the C-2 atom of the indole derivatives with formation of a C-O bond. The spin adduct thus formed is unstable and rapidly rearranges to acyl nitroxide with loss of HNO, which decomposes to N_2O and water. It was shown that nitrogen dioxide, a biologically relevant species, is trapped by electrophilic neutral compounds such as nitrones by indole derivatives.

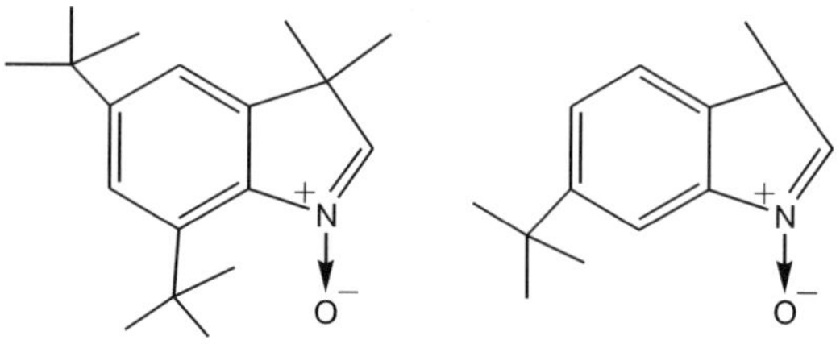

Fig. 23 Chemical structure of 5,7-di-*tert*-butyl-3,3-dimethyl-3*H*-indole *N*-oxide (1) and 6-*tert*-butyl-3,3-dimethyl-3*H*-indole *N*-oxide (2) [150]

Spin trapping/EPR spectroscopy is singular in its ability to characterize specific free radicals, generated in situ, and identified in animal models in real time. 2-Alkyl and 2-aryl substituted-3*H*-indol-3-one-1-oxides (Fig. 24) was prepared [151, 152] and evaluated for its radical trapping properties [153]. Spin trapping and electron paramagnetic resonance experiments demonstrate the ability of these indolone-1-oxides to trap hetero- and carbon-centered radicals. Indolone-1-oxide series lacking a β-hydrogen atom gave

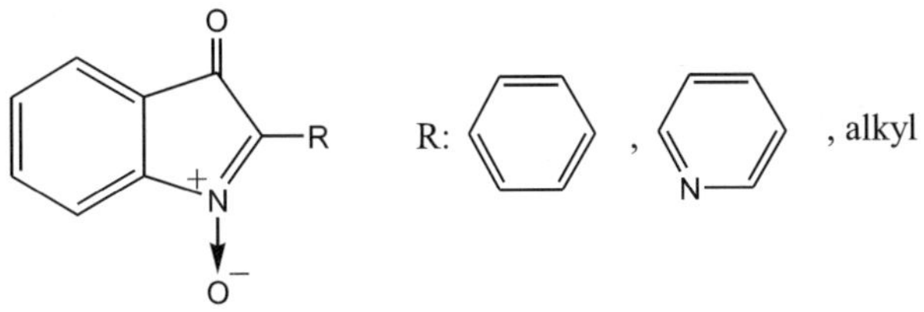

Fig. 24 Chemical structures of 2-alkyl and 2-aryl substituted-3H-indol-3-one-1-oxides [153]

rise highly stable adducts with free radicals. The high stability of spin adducts of indolone-1-oxide series associated with their strong lipophilicity and low working concentration make them possible candidates to trap free radicals in biological systems.

1,1,3-Trimethylisoindole N-oxide (TMINO in Fig. 25) was found as a scavenger for several Fenton-derived carbon- and oxygen-centered radicals including hydroxyl, formyl and alkyl radicals [154, 155]. The adduct display good stability and allowing the detection of the expected radicals. Trapping experiments were also undertaken with nitric oxide, which gave strong EPR signals attributed to the action of higher oxides of nitrogen. The selectivity of TMINO towards HO˙ with respect to superoxide radicals demonstrates its potential as a useful spin-trap.

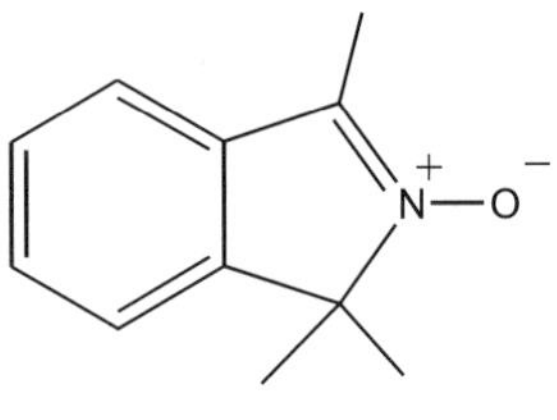

Fig. 25 Structure of 1,1,3-trimethylisoindole N-oxide [154, 155]

Aromatic aminoxyls having a conjugated benzene ring can react with virtually all kinds of oxygen-centered radicals. Protein peroxyl radicals are well-known intermediates of free-radical-dependent protein oxidation. Indolinic aminoxyls were synthesized using appropriate Grignard reagents [156]. 1,2-dihydro-2-ethyl-2-phenyl-3H-indole-3-phenylimino-1-oxyl and 1,2-dihydro-2-octadecyl-2-phenyl-3H-indole-3-phenylimino-1-oxyl (Fig. 26) were found to be a good candidate for substituting vitamin E in biological systems as it is a good antioxidant protecting membrane phospholipid. Due to high reactivity of aromatic aminoxyls with peroxyl species, very low concentrations are sufficient to defend cells. The protective effect of these molecules on membrane proteins points to the possibility of applying a new strategy for rational design of an antioxidant molecule, which would efficiently protect the pro-

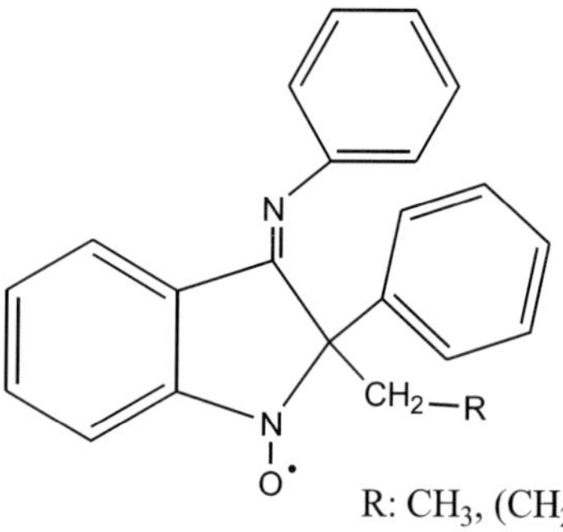

Fig. 26 Structure of aromatic indolinic aminoxyls [156]

teins against oxidative stress. It was proved that inhibition of peroxidation of certain membrane components depends on the length of acyl chain. Short-side chained aminoxyl derivative inhibits the lipid peroxidation process while the long chained is an efficient protector against protein oxidation.

5
Electrochemical Investigations of Indoles

The use of electrochemical techniques for the determination of compounds of pharmaceutical interest is continually gaining in importance [157]. Practical application of electrochemistry includes the determination of electrode oxidation mechanisms. Owing to the existing resemblance between electrochemical and biological reactions, it can be assumed that the oxidation mechanisms taking place at the electrode and in the body share similar principles [158]. Voltammetric techniques are most suitable to investigate the redox properties of a new drug. This can give insight into its metabolic fate. Some metabolites can be differentiated from the parent drug since metabolization often proceeds through the addition or the modification of a substituent. This will give rise to additional waves or to a shift of the main wave. Such a situation may allow the simultaneous determination of the initial compound and the metabolized drug [159]. This technique in general, and differential pulse voltammetry in particular, are considered to be useful tools for the determination of indole derivatives [160, 161].

Electrochemical behavior of indole-3-propionamide derivatives (Fig. 27) was investigated in order to establish experimental conditions for the electrochemical oxidation and determination of these derivatives using a glassy carbon electrode [158]. Cyclic voltammetry has been used in studying the redox mechanism that is related to antioxidant activity of the derivatives. The results showed that the compounds might have profound effects on the understanding of their in vivo redox processes and pharmaceutical activity. Indole-

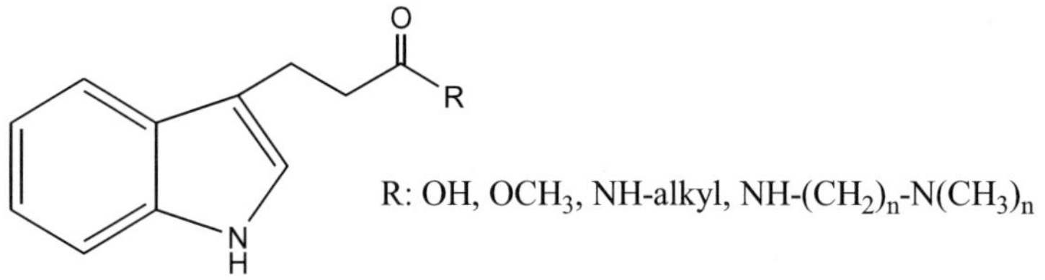

R: OH, OCH$_3$, NH-alkyl, NH-(CH$_2$)$_n$-N(CH$_3$)$_n$

Fig. 27 Chemical structure of indole-3-propionamide derivatives [158]

3-propionamide derivatives like melatonin have a heterocyclic aromatic ring structure with high resonance stability, which led to suspect antioxidant activity in the compounds. Oxidation step of indolic compounds is located on the nitrogen atom in the indole ring of the molecule, which is electro active in both acidic and basic media leading finally to hydroxylation of benzene ring.

Some indolylthiohydantoin derivatives that have aldose reductase inhibitory activity [75] were investigated electroanalytically by voltammetric determination. Based on this study, a simple, rapid, sensitive and validated voltammetric method was developed for the determination of the compounds that are readily oxidized at carbon-based electrodes. Oxidation of the indolic compounds occurs on the nitrogen atom in the indole ring of the molecule [162].

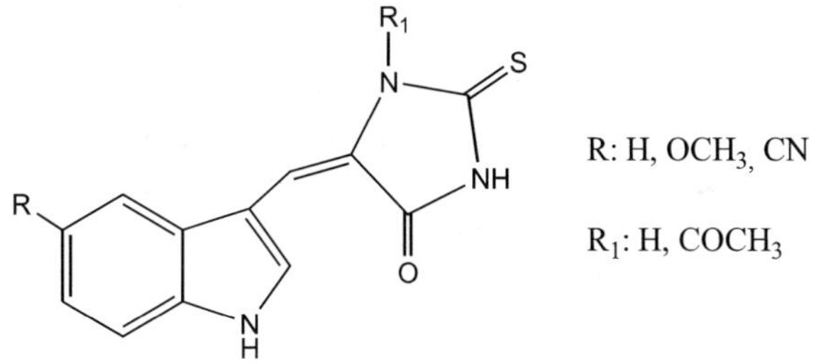

R: H, OCH$_3$, CN

R$_1$: H, COCH$_3$

Fig. 28 Chemical structure of 5-(3′-indolyl)-2-thiohydantoin derivatives [162]

Electroanalytically, investigation of some 2-phenyl indole by voltammetric determination also showed that all derivatives were oxidized in a broad pH range [163]. All the derivatives were electrochemically reduced from the indole ring as well. Indole reduction is irreversible and occurs in single step. It has more negative potentials than aromatic nitro group reduction. Interestingly, electrochemical oxidation of 2-phenyl indole derivatives showed that the indole ring is most likely from dimmers that are oxidized further to polymers in some cases.

Cyclic voltammetry is perhaps the most effective and versatile electroanalytical techniques available for the mechanistic study of redox system. The obtained results from the redox properties of drugs and biomolecules might have profound effects on our understanding of their in-vivo redox behavior or pharmaceutical activity.

Considering all the studies performed, it seems that the oxidation step of indolic compounds is located on the nitrogen atom in the indole ring of the molecule which is electroactive in both acidic and basic media leading finally to hydroxylation of benzene ring.

6
Discussion on Possible Antioxidant Mechanisms of Melatonin and Related Indole Derivatives

Mechanisms of the antioxidant effects of indole derivatives might involve dealing with melatonin receptors, which is less likely, and has not been proven yet or nonreceptor mechanisms such as stimulation of GPx, inhibition of LP. Moreover, suppression of phospholipase A2 activation, attenuation of tumor necrosis factor-α production, prevention of pathological opening of the mitochondrial permeability transition pores, and inhibition of sepiapterin reductase, the key enzyme of biosynthesis of tetrahydrobiopterin, the essential cofactor of nitric oxide synthase, might be involved in the antioxidant mechanism of indole derivatives [164]. Recently it was proposed that melatonin modulates antioxidant enzyme activities via its interaction with calmodulin, which in turn inhibits downstream processes that lead to the inactivation of nuclear RORα melatonin receptor [165]. In addition to its free-radical scavenging activities, melatonin has important actions in oxidative defense by stimulating enzymes that metabolize free radicals and radical products to innocuous metabolites. Furthermore, several metabolites that are formed when melatonin neutralizes damaging reactants are themselves scavengers suggesting that there is a cascade of reactions that greatly increase the efficacy of melatonin [166].

Melatonin and classic antioxidants possess the capacity to scavenge ABTS radical cation [167]. In terms of scavenging ABTS cation radical, melatonin exhibits a different profile than that of the classic antioxidants. Classic antioxidants scavenge one or less ABTS cation radical while each melatonin molecule can scavenge more than one. Intermediates, including the melatoninyl cation radical, the melatoninyl neutral radical, cyclic 3-hydroxymelatonin and AFMK seem to participate in these reactions. However, when melatonin is added to the reaction system in much lower quantities than ABTS cation radical, the number of radicals scavenged per melatonin molecule is considerably higher and can attain a value of ten [168]. Under conditions allowing for such a stoichiometry, novel products have been detected that are derived from AFMK. The identified substances are formed by re-cyclization and represent 3-indolinones carrying the side chain at C2; the N-formyl group can be maintained, but deformylated analogs seem to be also generated.

The mechanism of melatonin's interaction with reactive species probably involves donation of an electron to form the melatoninyl cation radical or through a radical addition at the site C3. Other possibilities include hydrogen donation from the nitrogen atom or substitution at position C2, C4, and C7 and nitrosation [169]. The mechanisms by which melatonin protects against LP most likely involve direct or indirect antioxidant and free-radical scavenging activities of this indoleamine [169, 171]. 2-Phenyl indole derivatives have redox properties because of the presence of an electron-rich aromatic ring system that allows the indoleamine to easily function as an electron donor. For these derivatives, the possible antioxidant mechanism might be most probably toward carbon-centered radicals described by Antosiewicz et al. [172].

Poeggler et al. showed that indole-3-propionate was a potent hydroxyl radical scavenger [15] and also found that indole-3-acetate, indole-3-lactate and indole-3-pyruvate derivatives, which have very polar and bulkier side chains, were poor inhibitors of hydrogen peroxide-stimulated and endogenous basal MDA formation in vitro. These results confirmed that for antioxidant activity, not only the indole type aromatic ring is important but also the side chain containing the amide group. In 2-phenyl indole derivatives, the amide chain is longer and bulkier than that of melatonin, and this increases the percent inhibition effect on MDA [173]. Also, Iakovou et al. [174] found that some β-substituted indol-3-yl ethylamido derivatives, which have similar structures to the compounds tested, were potent inhibitors of LP. On the other hand, the lack of methoxy group on the C5 of the indole ring made no significant difference on in vitro activity [173]. While the lack of a methoxy group at C5 did not significantly affect activity in a recent study [134], LP assay results showed that structural modifications of the methoxy group or replacing the methoxy group with a hydroxyl group increased antioxidant activity. There is still the question of the role of the methoxy group at C5. According to Tan et al. [142], the antioxidative mechanisms of melatonin seem different from classical antioxidants such as vitamin C, vitamin E, and glutathione. As electron donors, classical antioxidants undergo redox cycling; thus, they have the potential to promote oxidation as well as prevent it. Melatonin, as an electron-rich molecule, may interact with free radicals via an additive reaction to form several stable end-products that are excreted in the urine. The substitution of a hydroxy for the methoxy group led to phenolic compounds endowed with very high antioxidant activity. Replacing the amide with a ketone function did not affect the activity, while replacement with an amine group in some cases resulted in prooxidant compounds.

Examination of 3-indolyl compounds for relationships between antioxidation potential (using in vitro LP assays) and electronic, polar, and steric parameters, including bond dissociation energies, bond lengths, dipole moments, electronic charge densities, and molecular size parameters showed that antioxidant efficacy of 3-indolyl compounds was most strongly predicted by molecular size parameters and by the energy of electron abstraction as calculated from the difference in heat of formation between the parent compound and its cation radical [175].

Nevertheless, all the indole derivatives related to melatonin show some degree of antioxidant activity, revealing differences related to their electronic distribution and lipophilicity.

7
Conclusions

Drugs possessing antioxidant and free-radical scavenging properties are considered for prevention and/or treatment of such diseases that are directly related to the lack of the antioxidant capacity of the organism [2]. The scavenging ability of melatonin is related to its structure. This indolamine consists of an electron-rich indole heterocycle and additionally methoxy and aminoacetyl side chains, which seems to be essential for the radical scavenging ability of melatonin [176]. Various studies describe melatonin as a potent hydroxyl scavenger in relation to its structural analogues and also when compared to other antioxidants [47, 177].

Depending on the analyses of structure–activity relationships and electrochemical studies [158, 162], the indole nucleus is the reactive center of interaction with oxidants due to its high resonance stability and very low activation energy wall towards the free-radical reactions. However, the methoxy and amide side chains are also important for indole's antioxidant capacity [169].

Due to its free-radical scavenger and antioxidant properties, multiple melatonin-related compounds such as melatonin metabolites and synthetic analogues are under investigation to determine which exhibit the highest activity with the lowest side effects.

These findings could point to new strategies needed in designing new antioxidant indole derivatives with superior capacity that could have an indirect beneficial effect on the human antioxidant defense system.

Most of the examined compounds, including melatonin, have no significant inhibitory effect on superoxide anion formation but do have significant effects on LP. This shows that melatonin and the melatonin analogue indole derivatives act more like potent hydroxyl radical scavengers in vitro.

Epidemiological studies have strongly suggested that antioxidants can decrease the rate of many diseases [178]. However, more clinical studies are required to create the efficacy and safety of these compounds. This chapter gives another perspective of antioxidant activities of synthetic melatonin analogues.

References

1. Venkat RD, Ankola DD, Bhardwaj V, Sahana DK, Ravi Kumar MNV (2006) J Control Release 113:189
2. Suzen S (2006) Comb Chem High Throughput Screen 9:409
3. Shah AM, Channon KM (2004) Heart 90:486
4. Cooper CE, Vollaard NBJ, Choueiri T, Wilson MT (2002) Biochem Soc Trans 30:280
5. Vaiko M, Rhodes CJ, Moncol J, Izakovic M, Mazura M (2006) Chem Biol Inter 160:1
6. Drew B, Leeuwenburgh C (2002) Ann NY Acad Sci 959:66
7. Kelly FJ, Sandstrom T (2004) The Lancet 363:95
8. Wan XS, Ware JH, Zhou Z, Donahue JJ, Guan J, Kennedy AR (2006) Int J Radiat Oncol Biol Phys 64:1475
9. Ozturk D, Suzen S, Sabuncuoglu B, Akkus B (2005) Int J Rad Oncol Biol Phys 63:S437
10. Agarwal R (2005) BMC Nephrol 6:13
11. Wu D, Cederbaum AI (2003) Alcohol Res Health 27:277
12. Kumar Das S, Vasudevan DM (2005) Ind J Clin Biochem 20:24
13. Steven MT (2004) Exp Biol Med 229:607
14. Huang MF, Lin WL, Ma YC (2005) Indoor Air 15:135
15. Poeggeler B, Pappolla MA, Hardeland R, Rassoulpour A, Hodgkins S, Guidetti P, Schwarcz R (1999) Brain Res 815:382
16. Waris G, Ahsan H (2006) J Carcinog 5:14
17. Valko M, Rhodes CJ, Moncol J, Izakovic M, Mazur M (2006) Chem Biol Interact 160:1
18. Szasz T, Thakali K, Fink GD, Watts SW (2007) Exp Biol Med (Maywood) 232:27
19. Kedar NP, William CC, Bipin K (1999) J Am Coll Nutr 18:413
20. Varadarajan S, Yatin S, Aksenova M, Butterfield DA (2000) J Struct Biol 130:184
21. Yim S, Malhotra A, Veves A (2007) Curr Diab Rep 7:8
22. Remans PH, van Oosterhout M, Smeets TJ, Sanders M, Frederiks WM, Reedquist KA, Tak PP, Breedveld FC, van Laar JM (2005) Arthritis Rheum 52:2003
23. Surapaneni KM, Venkataramana G (2007) Indian J Med Sci 61:9
24. Tabrizi SJ, Workman J, Hart PE, Mangiarini L, Mahal A, Bates G, Cooper JM, Schapira AH (2000) Ann Neurol 47:80
25. Schapira AHV (1999) Biochim Biophys Acta (BBA)/Bioenergetics 1410:159
26. Miquel J (2002) Ann NY Acad Sci 959:508
27. Fernandez D, Bonilla E, Phillips P, Perl A (2006) Endocr Metab Immune Disord Drug Targets 6:305
28. Matuszak Z, Bilska MA, Reszka KJ, Chignell CF, Bilski P (2003) Photochem Photobiol 78:449
29. Kupkova Z, Cerna H, Lojek A, Ciz M, Benes L (2004) Ceska Slov Farm 53:310
30. Nakagawa H, Takusagawa M, Arima H, Furukawa K, Kinoshita T, Ozawa T, Ikota N (2004) Chem Pharm Bull (Tokyo) 52:146
31. Wozniak B, Wozniak A, Kasprzak HA, Drewa G, Mila-Kierzenkowska C, Drewa T, Planutis G (2007) J Neurooncol 81:21
32. Selek S, Aslan M, Horoz M, Gur M, Erel O (2007) Clin Biochem 40:287
33. Vontas JG, Small GJ, Hemingway J (2001) Biochem J 357:65
34. Berger A, Roberts MA, Hoff B (2006) Lipids Health Dis 5:10
35. Rhee SG, Chae HZ, Kim K (2005) Free Radic Biol Med 38:1543
36. Rey P, Becuwe N, Barrault MB, Rumeau D, Havaux M, Biteau B, Toledano MB (2007) Plant J 49:505
37. Mukai K, Tokunaga A, Itoh S, Kanesaki Y, Ohara K, Nagaoka S, Abe K (2007) J Phys Chem B Condens Matter Mater Surf Interfaces Biophys 111:652

38. Voss P, Siems W (2006) Free Radic Res 40:1339
39. Pulido R, Bravo L, Saura-Calixto F (2000) J Agric Food Chem 48:3396
40. Ishida BK, Chapman MH (2004) J Agric Food Chem 52:8017
41. Dominici S, Pieri L, Paolicchi A, De Tata V, Zunino F, Pompella A (2004) Ann NY Acad Sci 1030:62
42. Pandi-Perumal SR, Srinivasan V, Maestroni GJM, Cardinali DP, Poeggeler B, Hardeland R (2006) FEBS J 123:2813
43. Arendt J, Middleton B, Stone BM, Skene D (1999) Sleep 22:625
44. Kocher L, Brun J, Borson-Chazot F, Gonnaud PM, Claustrat B (2006) Chronobiol Int 23:889
45. Tan DX, Chen LD, Poeggeler B, Manchester. LC, Reiter RJ (1993) Endocr J 1:57
46. Tan DX, Manchester LC, Terron MP, Flores LJ, Reiter RJ (2007) J Pineal Res 42:28
47. Poeggeler B, Thuermann S, Dose A, Schoenke M, Burkhardt S, Hardeland R (2002) J Pineal Res 33:20
48. Hardeland R (1997) Melatonin: multiple functions in signaling and protection. In: Altmeyer P, Hoffmann K, Stücker M (eds) Skin Cancer, UV Radiation. Springer, Berlin Heidelberg New York, pp 186–198
49. Reiter RJ (1998) Prog Neurobiol 56:359
50. Herrera F, Martin V, Garcia-Santos G, Odriguez-Blanco J, Antolin I, Rodriguez C (2007) J Neurochem 100:736
51. Allegra M, Reiter RJ, Tan DX, Gentile C, Tesoriere L, Livrea MA (2003) J Pineal Res 34:1
52. Mathe-Allainmat M, Andrieux J, Langlois M (1997) Exp Opin Ther Patents 7:1447
53. Ates-Alagoz Z, Suzen S (2001) J Fac Pharm Ankara Univ 30:41
54. Bedini A, Di Giacomo B, Gatti G, Spadoni G (2005) Bioorg Med Chem 13:4651
55. Blask DE, Sauer LA, Dauchy RT (2002) Curr Top Med Chem 2:113
56. Guerrero JM, Reiter RJ (2002) Curr Top Med Chem 2:167
57. Cuzzocrea S, Reiter RJ (2002) Curr Top Med Chem 2:153
58. Reiter RJ, Tan DX, Mayo JC, Sainz RM, Leon J, Czarnock Z (2003) Acta Biochim Pol 50:1129
59. Depreux P, Lesieur D, Mansour HA, Morgan P, Howell HE, Renard P, Caignard DH, Pfeiffer B, Delagrange P, Guardiola B, Yous S, Demarque A, Adam G, Andrieux J (1994) J Med Chem 37:3231
60. Ates-Alagoz Z, Buyukbingol Z, Buyukbingol E (2005) Pharmazie 60:643
61. Hardeland R, Pandi-Perumal SR (2005) Nutr Metab Lond 2:22
62. De Muro RL, Nafziger AN, Blask DE, Menhinick AM, Bertino JS (2000) J Clin Pharmacol 40:781
63. Reiter RJ, Tan DX (2003) J Pineal Res 34:79
64. Suzen S, Buyukbingol E (2000) Il Farmaco 55:246
65. Chintharlapalli S, Papineni S, Safe S (2006) Mol Cancer Ther 5:1362
66. Pappa G, Strathmann J, Lowinger M, Bartsch H, Gerhauser C (2007) Carcinogenesis (in press)
67. Defant A, Guella G, Mancini I (2007) Arch Pharm (Weinheim) 340:147
68. Suzen S, Bozkaya P, Coban T, Nebioglu D (2006) J Enzyme Inh Med Chem 21:405
69. Oliveira DL, Pugine SM, Ferreira MS, Lins PG, Costa EJ, de Melo MP (2007) Cell Biochem Funct 25:195
70. Mavunkel BJ, Chakravarty S, Perumattam JJ, Luedtke GR, Liang X, Lim D, Xu YJ, Laney M, Liu DY, Schreiner GF, Lewicki JA, Dugar S (2003) Bioorg Med Chem Lett 13:3087
71. Cruz-Lopez O, Diaz-Mochon JJ, Campos JM, Entrena A, Nunez MT, Labeaga L, Orjales A, Gallo MA, Espinosa A (2007) ChemMedChem 2:88

72. Draheim R, Egerland U, Rundfeldt C (2004) J Pharmacol Exp Ther 308:555
73. Prasanna S, Manivannan E, Chaturvedi SC (2005) J Enzyme Inhib Med Chem 20:455
74. Suzen S, Buyukbingol E (2003) Curr Med Chem 10:1329
75. Buyukbingol E, Suzen S, Klopman G (1994) Il Farmaco 49:443
76. Djoubissie PO, Snirc V, Sotnikova R, Zurova J, Kyselova Z, Skalska S, Gajdosik A, Javorkova V, Vlkovicova J, Vrbjar N, Stefek M (2006) Gen Physiol Biophys 25:415
77. Tiwari RK, Singh D, Singh J, Yadav V, Pathak AK, Dabur R, Chhillar AK, Singh R, Sharma GL, Chandra R, Verma AK (2006) Bioorg Med Chem Lett 16:413
78. Yamamoto Y, Kurazono M (2007) Bioorg Med Chem Let 17:1626
79. Mahboobi S, Eichhorn E, Popp A, Sellmer A, Elz S, Mollmann U (2006) Eur J Med Chem 41:176
80. Ryu CK, Lee JY, Park RE, Ma MY, Nho JH (2007) Bioorg Med Chem Lett 17:127
81. Tiwari RK, Verma AK, Chhillar AK, Singh D, Singh J, Kasi Sankar V, Yadav V, Sharma GL, Chandra R (2006) Bioorg Med Chem 14:2747
82. Williams JD, Drach JC, Townsend LB (2005) Nucleos Nucleot Nucl Acids 24:1613
83. Chen JJ, Wei Y, Williams JD, Drach JC, Townsend LB (2005) Nucleos Nucleot Nucl Acids 24:1417
84. Chai H, Zhao Y, Zhao C, Gong P (2006) Bioorg Med Chem 14:911
85. Agarwal A, Srivastava K, Puri SK, Chauhan PM (2005) Bioorg Med Chem Lett 15:3133
86. Kgokong JL, Smith PP, Matsabisa GM (2005) Bioorg Med Chem 13:2935
87. Suzen S, Buyukbingol E (1998) Il Farmaco 53:525
88. De Martino G, La Regina G, Ragno R, Coluccia A, Bergamini A, Ciaprini C, Sinistro A, Maga G, Crespan E, Artico M, Silvestri R (2006) Antivir Chem Chemother 17:59
89. Jochmans D, Deval J, Kesteleyn B, Van Marck H, Bettens E, De Baere I, Dehertogh P, Ivens T, Van Ginderen M, Van Schoubroeck B, Ehteshami M, Wigerinck P, Gotte M, Hertogs K (2006) J Virol 80:12283
90. Silvestri R, Artico M (2005) Curr Pharm Des 11:3779
91. Olgen S, Coban T (2002) Arch Pharm Pharm Med Chem 7:331
92. Dannhardt G, Kiefer W (2001) Eur J Med Chem 36:109
93. Brown DW, Graupner PR, Sainsbury M, Shertzer HG (1991) Tetrahedron 25:4383
94. Suzen S, Ates-Alagoz Z, Puskullu O (2000) FABAD J Pharm Sci 25:113
95. Niwa K, Haensel C, Ross ME, Iadecola C (2001) Circ Res 88:600
96. Torres L, Anderson C, Marro P, Mishra OP, Delivoria-Papadopoulos M (2004) Biomed Life Sci 29:1825
97. Mayo JC, Sainz RM, Tan DX, Hardeland R, Leon J, Rodriguez C, Reiter RJ (2005) J Neuroimmunol 165:139
98. Reiter RJ, Calvo JR, Karbownik M, Qi W, Tan DX (2000) Ann NY Acad Sci 917:376
99. Ikeda Y, Matsumoto K, Dohi K, Jimbo H, Sasaki K, Satoh K (2001) Headache 41:138
100. Mouithys-Mickalad AM, Zheng SX, Deby-Dupont GP, Deby CMT, Lamy M, Reginster JYY, Henrotin YE (2000) Free Radic Res 33:607
101. Fernandes E, Costa D, Toste SA, Lima JLFC, Reis S (2004) Free Radic Biol Med 37:1895
102. Stolc S (1999) Life Sci 65:1943
103. Costa D, Gomes A, Reis S, Lima JLFC, Fernandes E (2005) Life Sci 76:2841
104. Aboul-Enein HY, Kruk I, Lichszteld K, Michalska T, Kladna A, Marczynski S, Olgen S (2004) Luminescence 19:1
105. Olgen S, Coban T (2004) J Fac Pharm Ankara 33:109
106. Olgen S, Kilic Z, Ada O, Coban T (2007) J Enz Inh Med Chem (in press)

107. Bozkaya P, Olgen S, Coban T, Nebioglu D (2007) J Enz Inh Med Chem (in press)
108. Olgen S, Guner E, Fabregat MA, Crespo MI, Nebioglu D (2002) Pharmazie 57:238
109. Majekova M, Koprda V, Bohacik L, Bohov P, Hadgraft J, Bezakova Z, Majek P (2006) Pharmacol Toxicol Pharm Pharm 13:51
110. Stolc S, Snirc V, Majekova M, Gasparova Z, Gajdosikova A, Stvrtina S (2006) Cell Mol Neurobiol 26:1493
111. Demopoulos VJ, Gavalas A, Rekatas G, Tani EK (1995) J Heterocyclic Chem 32:1145
112. Zika CA, Nicolaou I, Gavalas A, Rekatas GV, Tani E, Demopoulos VJ (2004) Ann General Hospital Psychiatry 3:1
113. Gozzo A, Lesieur D, Duriez P, Fruchart JC, Teissier E (1999) Free Radic Biol Med 26:1538
114. Bonnefont-Rousselot D, Cheve G, Gozzo A, Tailleux A, Guilloz V, Caisey S, Teissier E, Fruchart JC, Delattre J, Jore D, Lesieur D, Duriez P, Gardes-Albert M (2002) J Pineal Res 33:109
115. Devitt GP, Creagh EM, Cotter TG (1999) Biochem Biophys Res Commun 264:622
116. Ates-Alagoz Z, Coban T, Buyukbingol E (2006) Arch Pharm (Weinheim) 339:193
117. Gurkan AS, Karabay A, Buyukbingol Z, Adejare A, Buyukbingol E (2005) Arch Pharm (Weinheim) 338:67
118. Ates-Alagoz Z, Kus C, Coban T (2005) J Enzyme Inhib Med Chem 20:325
119. Asakai R, Aoyama Y, Fujimoto T (2002) Neurosci Res 44:297
120. Katoh M, Dodo K, Fujita M, Sodeoka M (2005) Bioorg Med Chem Lett 15:3109
121. Dodo K, Katoh M, Shimizu T, Takahashi M, Sodeoka M (2005) Bioorg Med Chem Lett 15:3114
122. Morita I, Kawamoto M, Yoshiba H (1992) J Chromatogr 576:334
123. Chyan YJ, Poeggeler B, Omar RA, Chain DG, Frangione B, Ghiso J, Pappolla MA (2002) J Biol Chem 274:21937
124. Bendheim PE, Poeggeler B, Neria E, Ziv V, Pappolla MA, Chain DG (2002) J Mol Neurosci 19:213
125. Olgen S, Kilic Z, Ada AO, Coban T (2007) Arch Pharm Chem Life Sci 340:140
126. Olgen S, Coban T (2003) Biol Pharm Bull 26:736
127. Das S, Falchi M, Bertelli A, Maulik N, Das DK (2006) Arzneimittelforschung 56:700
128. Riccioni G, Bucciarelli T, Mancini B, Di Ilio C, Capra V, D'Orazio N (2007) Expert Opin Investig Drugs 16:25
129. Varvaresou A, Tsantili-Kakoulidou A, Siatra-Papastaikoudi T, Tiligada E (2000) Arzneimittelforschung 50:48
130. Andreadou I, Tsantili-Kakoulidou A, Siatra-Papastaikoudi Th (2000) Res Commun Biochem Cell Mol Biol 4:269
131. Andreadou I, Tasouli A, Bofilis E, Chrysselis M, Rekka E, Tsantili-Kakoulidou A, Iliodromitis E, Siatra T, Kremastinos D (2002) Chem Pharm Bull 50:165
132. Andreadou I, Tsantili-Kakoulidou A, Spyropoulou E, Siatra T (2003) Chem Pharm Bull 51:1128
133. Andreadou I, Tasouli A, Iliodromitis E, Tsantili-Kakoulidou A, Papalois A, Siatra T, Kremastinos D (2002) Eur J Pharm 453:271
134. Mor M, Silva C, Vacondio F, Plazzi PV, Bertoni S, Spadoni G, Diamantini G, Bedini A, Tarzia G, Zusso M, Franceschini D, Giusti P (2004) J Pineal Res 36:95
135. Spadoni G, Diamantini G, Bedini A, Tarzia G, Vacondio F, Silva C, Rivara M, Mor M, Plazzi PV, Zusso M, Franceschini D, Giusti P (2006) J Pineal Res 40:259
136. Sanz E, Romera M, Bellik L, Marco JI, Unzeta M (2004) Med Sci Monit 10:BR477
137. Ximenes VF, Paino IMM, Faria-Oliveira OMM, Fonseca LM, Brunetti IL (2005) Braz J Med Biol Res 38:1575

138. Ressmeyer AR, Mayo JC, Zelosko V, Sainz RM, Tan DX, Poeggeler B, Antolin I, Reiter RJ, Hardeland R (2003) Redox Rep 8:205
139. Reiter RJ, Tan DX, Allegra M (2002) Neuroendocrinol Lett 23(Suppl 1):3
140. Zhang H, Squadrito GL, Uppu R, Pryor W (1999) Chem Res Toxicol 12:526
141. Karbownik M, Gitto E, Lewinski A, Reiter RJ (2001) J Cell Biochem 81:693
142. Tan DX, Manchester LC, Reiter RJ, Qi WB, Karbownik M, Calvo JR (2000) Biol Signals Recept 9:137
143. Garcia-Santos G, Herrera F, Martin V, Rodriguez-Blanco J, Antolin I, Fernandez-Mari F, Rodriguez C (2004) Free Rad Res 3:1289
144. Gasparova Z, Stolc S, Snirc V (2006) Pharmacol Res 53:22
145. Bandoh T, Sato EF, Mitani H, Nakashima A, Hoshi K, Inoue M (2003) Biol Pharm Bull 26:818
146. Aboul-Enein HY, Kladna A, Kruk I, Lichszteld K, Michalska T, Olgen S (2005) Biopolimers 78:171
147. Janzen EG, Haire DL (1990) Adv Free Radic Chem 1:253
148. Zubarev VE, Belevskii VN, Bugaenko LT (1979) Russian Chem Rev 48:729
149. Fukutomi J, Fukuda A, Fukuda S, Hara M, Terada A, Yoshida M (2006) Life Sci 80:254
150. Astolfi P, Greci L, Panagiotaki M (2005) Free Rad Res 39:137
151. Bernardes-Genisson V, Bouniol AV, Nepveu F (2001) Synlett 5:700
152. Boyer J, Bernardes-Genisson V, Nepveu F (2003) J Chem Res 8:507
153. Boyer J, Bernardes-Genisson V, Farines V, Souchard JP, Nepveu F (2004) Free Rad Res 38:459
154. Bottle SE, Hanson GR, Micallef AS (2003) Org Biomol Chem 1:2585
155. Bottle SE, Micallef AS (2003) Org Biomol Chem 1:2581
156. Kulawiak-Galaska D, Wozniak M, Greci L (2002) Acta Biochim Pol 49:43
157. Ozkan SA, Uslu B, Aboul-Enein HY (2003) Crit Rev Anal Chem 33:155
158. Suzen S, Ates-Alagoz Z, Demircigil BT, Ozkan SA (2001) Il Farmaco 56:835
159. Dogan B, Uslu B, Suzen S, Ozkan SA (2005) Electroanalysis 17:1886
160. Yïlmaz S, Uslu B, Ozkan SA (2002) Talanta 54:351
161. Ozkan SA, Uslu B (2002) Anal Bioanal Chem 372:582
162. Suzen S, Demircigil T, Buyukbingol E, Ozkan SA (2003) New J Chem 27:1007
163. Bozkaya P, Dogan B, Suzen S, Nebioglu S, Ozkan SA (2006) Can J Anal Sci Spec 51:125
164. Oxenkrug G (2005) Ann NY Acad Sci 1053:334
165. Tomas-Zapico C, Coto-Montes A (2005) J Pineal Res 39:99
166. Reiter RJ, Tan DX, Gitto E, Sainz RM, Mayo JC, Leon J, Manchester C, Vijayalaxmi Kilic E, Kilic U (2004) Pol J Pharmacol 56:159
167. Tan DX, Hardeland R, Manchester LC, Poeggeler B, Lopez-Burillo S, Mayo JC, Sainz RM, Reiter RJ (2003) J Pineal Res 34:249
168. Rosen J, Than NN, Koch D, Poeggeler B, Laatsch H, Hardeland R (2006) J Pineal Res 41:374
169. Tan DX, Reiter RJ, Manchester LC, Yan MT, El-Sawi M, Sainz RS, Mayo JC, Kohen R, Allegra M, Hardeland R (2002) Curr Top Med Chem 2:181
170. Reiter RJ, Tan D-X, Manchester LC, Karbownik MB, Calvo JR (2000) Biol Signal Recept 9:160
171. Tan DX, Manchester LC, Hardeland R, Lopez-Burillo S, Mayo JC, Sainz RM, Reiter RJ (2003) J Pineal Res 34:75
172. Antosiewicz J, Damiani E, Jassem W, Wozniak M, Orena M, Greci L (1997) Free Rad Biol Med 22:249
173. Ateş-Alagoz Z, Coban T, Suzen S (2005) Med Chem Res 14:169

174. Iakovou K, Varvaresou A, Kourounakis AP, Steed K, Sugden D, Tsotinis A (2002)
     J Pharm Pharmacol 54:147
175. Shertzer HG, Tabor MW, Hogan ITD, Brown SJ, Sainsbury M (1996) Biomed Life Sci
     70:830
176. Gulcin I, Buyukkokuroglu ME, Oktay M, Kufrevioglu OI (2002) J Pineal Res 33:167
177. Bromme HJ, Morke W, Peschke D, Ebelt H, Peschke D (2000) J Pineal Res 29:201
178. Geronikaki AA, Gavalas AM (2006) Comb Chem High Throughput Screen 9:425

Top Heterocycl Chem (2007) 11: 179–211
DOI 10.1007/7081_2007_066
© Springer-Verlag Berlin Heidelberg
Published online: 21 July 2007

Quinoxaline 1,4-Dioxide and Phenazine 5,10-Dioxide. Chemistry and Biology

Mercedes González[1] (✉) · Hugo Cerecetto[1] · Antonio Monge[2]

[1]Laboratorio de Química Orgánica, Facultad de Ciencias/Facultad de Química, Universidad de la República, Iguá 4225, Montevideo, 11400, Uruguay
megonzal@fq.edu.uy

[2]Centro de Investigaciones en Farmacobiología Aplicada (CIFA), Universidad de Navarra, Irunlarrea s/n, 31080 Pamplona, Spain

Abstract Since the beginning of the past century, quinoxaline 1,4-dioxide and phenazine 5,10-dioxide derivatives have been known to be potent bioactive compounds. Maybe the most relevant reported biological property has been their ability to act as bioreductive agents, selective cytotoxins against hypoxic cells; moreover, other significant activities have been identified, such as promotion-growth ability, antibacterial, antifungal, antiparasite, cytotoxic, as well as herbicide properties. Structure-activity relationship studies have been performed and important findings have been described. On the other hand, specific side effects for these compounds have been studied and identified being the most relevant mutagenic and clastogenic properties. These kinds of compounds have been also reported as components in the material sciences. This chapter includes old and more recent methods of synthesis of quinoxaline 1,4-dioxide and phenazine 5,10-dioxide derivatives, chemical and biological reactivity, biological properties and mode of action, structure-activity studies and other relevant chemical and biological properties.

Keywords Antibacterial properties · Bioreductive agents · Mutagenic effects · Phenazine 5,10-dioxide · Quinoxaline 1,4-dioxide

Abbreviations

Bfx	Benzofuroxan
E_{pc}	First cathodic peak
ESR	Electron spin resonance
HCR	Hypoxia cytotoxicity selectivity relationship
HIF-1 α	Hypoxia-inducible factor 1 α
MDR-TB	Multidrug-resistant tuberculosis
MIC	Minimum inhibitory concentration
P	Potency under hypoxia
PDO	Phenazine 5,10-dioxide
PMO	Phenazine mono-oxide
Pz	Phenazine
QDO	Quinoxaline 1,4-dioxide
QMO	Quinoxaline mono-oxide
Qx	Quinoxaline
ROS	Reactive oxygen species
sce	Saturated calomel electrode
SI	Selectivity index
TB	Tuberculosis
WHO	World Health Organization

1
Introduction

Since the beginning of the past century, quinoxaline 1,4-dioxide (QDO) and phenazine 5,10-dioxide (PDO) derivatives have been described as interesting biological active compounds reminding its interest in medicinal chemistry until now. Formally, these compounds are [1,4]diazine 1,4-dioxide derivatives, and specifically QDO should be named as benzo[b][1,4]diazine 1,4-dioxide and PDO as dibenzo[b,e][1,4]diazine 5,10-dioxide, however, the common trivial name of the base heterocycle, quinoxaline (Qx) and phenazine

(Pz), respectively, is accepted by IUPAC. The position of the N-oxide moiety is indicated as 1,4-, N,N'- or N^1,N^4-dioxide in the case of Qx or as 5,10-, N,N'- or N^5,N^{10}-dioxide for Pz. The heterocycles are systematically numbered according to Scheme 1.

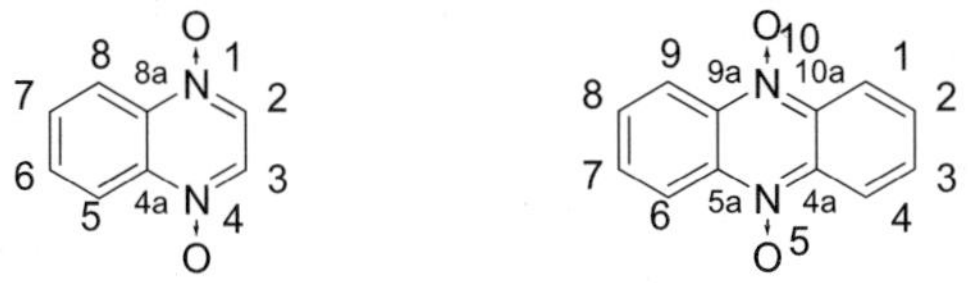

quinoxaline 1,4-dioxide (QDO)   **phenazine 5,10-dioxide (QDO)**

Scheme 1 Systematic numeration of QDO and PDO

The N-oxide functional group is characterized by the presence of a donative (or coordinate-covalent) bond between nitrogen and oxygen as a result of the overlap of the nonbonding electron pair on the nitrogen with an empty orbital on the oxygen atom. Formally this group is neutral, however, nitrogen and oxygen possess positive and negative formal charges, respectively. Consequently, the correct representation should be $N^+ - O^-$, however in the formulas of this chapter, the N-oxide moiety will be represented as $N \rightarrow O$. This function, occurring in three important structures such as N-oxide of tertiary aliphatic or aromatic amines, N-oxide of imines or nitrones, and N-oxide of nitriles [1, 2], could be considered as a masking moiety that changes physicochemical properties of the parent aza-heterocycle, i.e., protonation capability [3] and dipole moment (Table 1), between others.

Table 1 Dipole moments of pyridine, quinoxaline, and phenazine and its N-oxide and N,N'-dioxide

System		Dipole moments (D)	
	Heterocycle	N-oxide [a]	N,N'-dioxide [a]
Pyridine	2.02[b]/2.19[c]	4.24	–
Quinoxaline	0.61[c]	2.53	2.27
Phenazine	0.00[d]/0.00[c]	1.76	2.20

[a] From [4]
[b] From [2]
[c] Calculated using density functional theory (B3LYP/6-31G*//B3LYP/6-31G*)
[d] From [5]

The occurrence of QDO's and PDO's substructures in natural products has been barely reported, especially from a bacterial source (Fig. 1) [6], being the described derivatives mainly from a chemical synthetic origin.

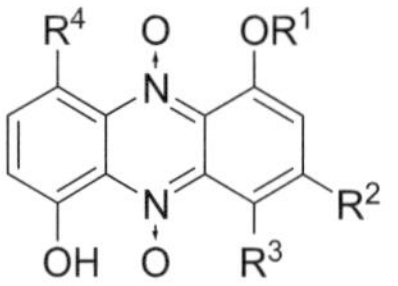

1, iodinin, R^1-R^4= -H
2, myxin, R^1= -CH$_3$, R^2-R^4= -H
3, lomondomycin, R^1= -H, R^2= -OH, R^3= -CHO, R^4= -CO$_2$CH$_3$

Fig. 1 Chemical structures of natural PDO derivatives

2
Chemical Syntheses

The main synthetic approaches for the preparation of QDO and PDO until the middle of the 20th century [7] had been associated with the oxidation of the parent heterocycles, Qx and Pz, respectively. However, since the description of the Beirut reaction, by Haddadin and Issidorides [8], the most important preparation procedure of both heterocycle systems is the expansion of benzofuroxans (Bfxs) with adequate synthons that introduces carbons 2 and 3 for QDO or carbons 1-4a and 10a in the case of PDO. Some other synthetic procedures have also been depicted, which are described in the next sections.

2.1
Oxidation from the Parent Heterocycles

N-Oxidation is one of the oldest processes in the synthetic approaches of *N*-oxide-containing heterocycles. In the preparation of QDO and PDO from Qx and Pz, respectively, a great number of invention patents have described the use of different oxidant agents that directly transfer oxygen to nitrogen atoms. In general, these procedures involved the use of a peracid commercially available or generated in situ from the corresponding acid and hydrogen peroxide (Fig. 2) [7, 9, 10]. In order to improve yields, increase the ratio dioxide/monoxide, and decrease the reaction times, some catalyzers, such as sodium tungstate, molybdic acid, sodium molybdates or phosphomolybdic acid, have been combined with hydrogen peroxide (Fig. 2) [11–14]. However, when the first *N*-oxidation takes place, in the most reactive Qx or Pz nitrogen, the second nitrogen atom is turned deactivates almost completely toward any further electrophilic attack, such as the oxygen transfers processes, producing absence of di-*N*-oxide in some of the cases. Not only the second nitrogen of the ring system makes the diazines less reactive than pyridine toward electrophilic substitutions in the *N*-oxidized form but also electron-withdrawing substituents, such as halogens, reduce the reactivity of the ring nitrogens even further. For example, neither AcOH-30% H$_2$O$_2$ dioxidizes 2,3-dichloroquinoxaline nor the oxidizer H$_2$SO$_4$/K$_2$S$_2$O$_8$, affording only 2,3-dichloroquinoxaline mono *N*-oxide [15]. Recently, it has successfully been studied different alternative reagents in the oxygen transfer process, for

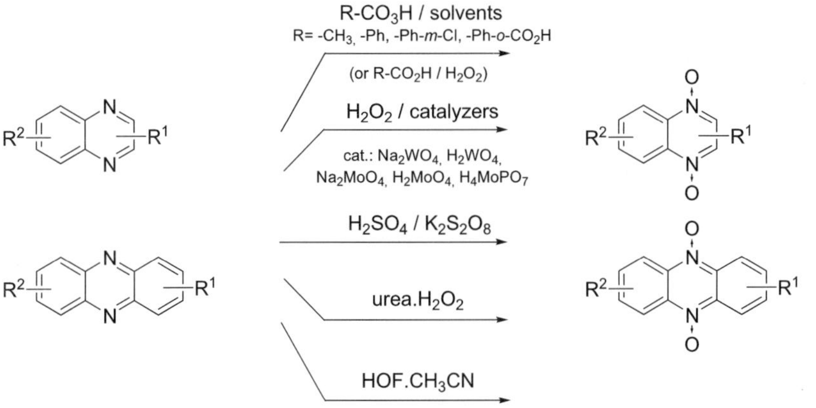

Fig. 2 Syntheses of QDO and PDO from oxidation of Qx and Pz

example the use of the complexes urea $\cdot$ H_2O_2, as anhydrous H_2O_2 [16], and HOF $\cdot$ CH_3CN, prepared passing a mixture of $F_2(g) : N_2(g)$ through a mixture of $CH_3CN : H_2O$ at $-15\,°C$ (Fig. 2) [15]. This last complex allows producing the desired products in very short reaction times, from seconds to minutes, and according to isotopic studies the oxidized species was originated from the water.

2.2
From Benzofuroxans

In the mid-1960s, Issidorides and Haddadin [8] described an elegant one-step synthesis of QDO and PDO, and for this first time, tetrahydro-PDO was described. This procedure was called the Beirut reaction, named after its origin from the capital of Lebanon. The Beirut reaction involves a condensation between adequate substituted Bfx and alkene-type substructure synthons, particularly enamine and enolate nucleophiles (Fig. 3). The enamine synthons were the first studied reagents in the QDO and PDO synthesis from Bfx [17–21] preparing the enamines previously or in situ from the corresponding carbonyl compound and amines or ammonia gas. On the other hand, a wide variety of enolates has been described as synthons in the Beirut reaction [22–31]. For example, β-dicarbonyl synthons yield QDO with 2-acyl, 2-alkyloxycarbonyl, or 2-carbamoyl substitutions while malononitrile produces 3-amino-2-quinoxalinecarbonitrile 1,4-dioxide derivatives. When phenolates are used as enolates, the final product corresponds to PDO. The reaction proceeds under mild conditions, in general at room temperature, and inorganic bases, like K_2CO_3 and NaOH, or organic bases like Et_3N and RONa, have been employed in the enolate's preparations. Other basic and also neutral conditions, such as $KF - Al_2O_3$, silica gel, molecular sieves,

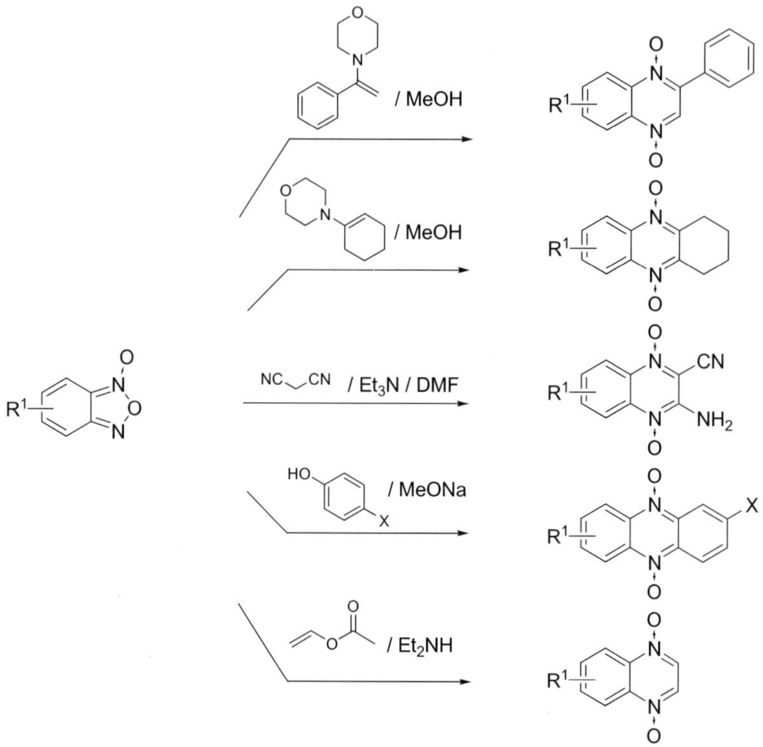

Fig. 3 Uses of the Beirut reaction in the syntheses of QDO and PDO

and zeolite, have recently been studied as conditions for the reaction between these methylene-active reagents and Bfx [32–36]. Alkenes, in basic and neutral medium, have been described as synthons for the Beirut reaction, i.e., styrene derivatives or vinyl acetate, and furthermore acetylenes, reacting in the presence of amines, have been employed in this process (Fig. 3) [37–40].

Mechanistically, the Beirut reaction is a heterocycle expansion-process where the nucleophile attack conducts, after atomic rearrangements, to a new diazine system. In general, a further elimination—i.e., of H_2O, amines, acetate—aromatizes the new heterocycle.

The Beirut process as a method for preparing QDO and PDO is sometimes inconvenient. One of the main problems has been evidenced when substituted Bfxs were used as a reagent for the Beirut reaction. In this case, a mixture of positional isomers, in general non-separable by ordinary chromatographic techniques, of QDO or PDO have been obtained. This finding is the result of the well-known tautomerism that affects the Bfx reactants at room temperature (see example in Scheme 2) [41, 42]. In general, the tautomeric equilibrium energy barrier is very low at room temperature and therefore

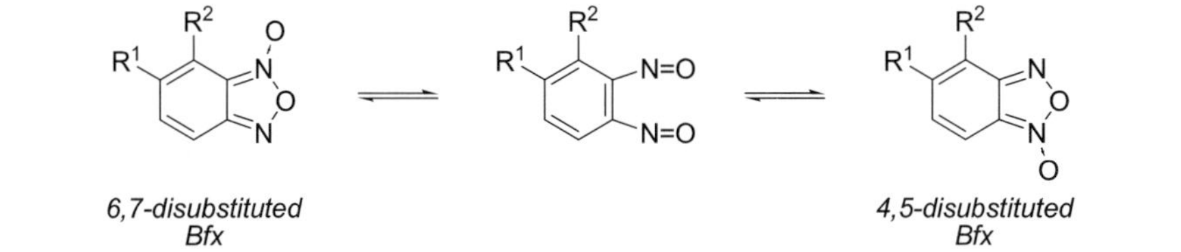

Scheme 2 Benzofuroxans ring-chain tautomerism

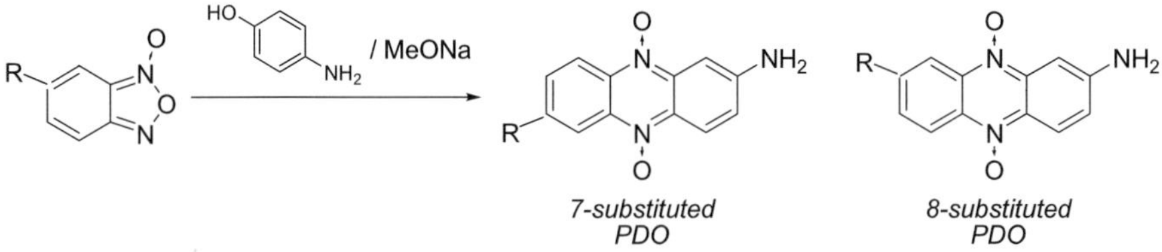

R	proportion 7:8	σ_p^+, R
OCH$_3$	95 : 5	- 0.78
Cl	63 : 37	0.11
NO$_2$	40 : 60	0.79

Fig. 4 Different isomeric 2-amino-7/8-substituted PDO obtained through the Beirut reaction, relationship with electronic substituent characteristics

both *N*-oxide tautomers co-exist in the reaction medium, producing a mixture of inseparable positional isomers (see example in Fig. 4). However, each Bfx tautomeric form has different stability that depend on substituent electronic characteristics. Consequently, a different proportion of isomers could be experimentally obtained. Recently, in the reaction between 5-substituted Bfx and *p*-aminophenol a mixture of the corresponding 7- and 8-substituted-2-aminophenazine 5,10-dioxide derivatives have been obtained in a isomeric proportion depending on the Bfx substituent electronic characteristics, σ_p^+ (Fig. 4) [43].

Another recognized inconvenience of the Beirut reaction involves the 4/7-mono and disubstituted Bfx that reacts significantly slower or does not react at all in producing the corresponding 5/8-mono or disubstituted QDO and 1/4-mono or disubstituted PDO. This is explained by the repulsive effect of the substituent and the *N*-oxide in the forming product [44].

Finally, in the PDO preparation through the Beirut reaction, the reactivity of the phenol is very important to the success of the process. For example, less nucleophilic phenol, like *p*-formylphenol dioxalane, does not conduct to the desired product [45].

The Beirut reaction has also been employed to prepare 1-hydroxybenz-imidazole 3-oxides or benzimidazole 1,3-dioxides, when nitroalkanes have been used as enolate-producer reagent [46, 47], and benzo[e][1,2,4]triazine 1,4-dioxides when Bfx reacts with sodium cyanamide [48–50].

2.3
Other Synthetic Approaches

In 1970, Abushanab described the synthesis of QDO through the con-densation between o-quinone dioxime and α-dielectrophile derivatives—i.e., α-dicarbonyl, α-hydroxycarbonyl, α-halocarbonyl, α-epoxycarbonyl, or α-epoxy halogenide—(Fig. 5) [51, 52]. After this, no more approaches using this methodology were described.

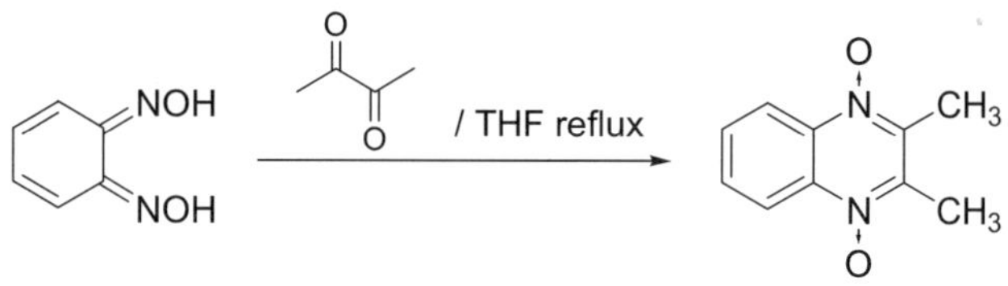

Fig. 5 Preparation of 2,3-dimethylquinoxaline 1,4-dioxide from o-quinone dioxime

3
Reactivity

These heterocycles are especially deactivated toward electrophilic and nucleo-philic substitution under standard conditions and most of the reactions of QDO and PDO involve N-oxide modifications or ring-substituent deriva-tizations. The present section outlines the well-known reactions described particularly in the last decade.

3.1
Deoxygenation

The reduction of the N-oxide moiety, in order to produce the heterocycle par-ent compounds Qx and Pz, maybe has been one of the most studied processes for these systems. In general, the procedures describe mild conditions and either mono- or di-reduction products are generated. In addition, some ex-amples of simultaneous deoxygenation and C-reactions have been reported. In Table 2 are listed substrates, conditions, products, and comments for the most recent reported deoxygenations.

Table 2 Described processes to reduce, deoxygenate, QDO and PDO

Substrates	Conditions	Products (yield)	Comments	Refs.
QDO PDO	$Na_2S_2O_4/$ MeOH-$H_2O/$ 65 °C	Qx (80%) Pz (70–90%)	in the case of PDO reduction PMO was also isolated	[53, 54]
PDO	$Na_2S_2O_3/$ NaOH/H_2O	Pz (78%)	described for benzo[a]phenazine dioxide	[55]
PDO	$H_2N(HN=)CSO_2H/$ NaOH/EtOH	Pz (76%)	–	[56]
QDO	Mg/$NH_4OAc/$ MeOH	Qx (24%)	–	[57]
PDO	Zn/$NH_4OCOH/$ MeOH	Pz (62%)	–	[58]
PDO	$TiCl_4$	Pz (superior to 90%)	Different conditions were studied	[59–61]
QDO	L-Ascorbic acid/ H_2O	QMO (48%), Qx (9%)	–	[62]
QDO	Ac_2O	1-hydroxyquinoxalin-2(1H)-one (30%), 1-acetylhydroxy-quinoxalin-2(1H)-one (70%)	other reduction-products were obtained in the assayed conditions	[63]

3.2
Lateral Chain Modifications

Lateral chain modifications on QDO and PDO have been well documented in the literature. The described reactions have shown modifications in specific functional group—i.e., acylations, alkylations, decarboxylations, hydrolysis, rearrangements, substitutions—or the production of a new heterocycle, fusioned or not. Examples of the most recent descriptions are depicted in Fig. 6 [64–66].

3.3
Complexation with Metals

The metal complexes of QDO and PDO have been the subject of some studies since the middle of the 20th century [67]. In these cases, the heterocycle N-oxide has acted either as a unidentate N-oxide oxygen ligand or as a chelating or bridging bidentate ligand when heterocycle substituents were involved in the coordination processes. A wide variety of metals have been used in the coordination reactions, among Cr(III), Mn(II), Fe(III), Co(II), Ni(II), Cu(II), and Zn(II).

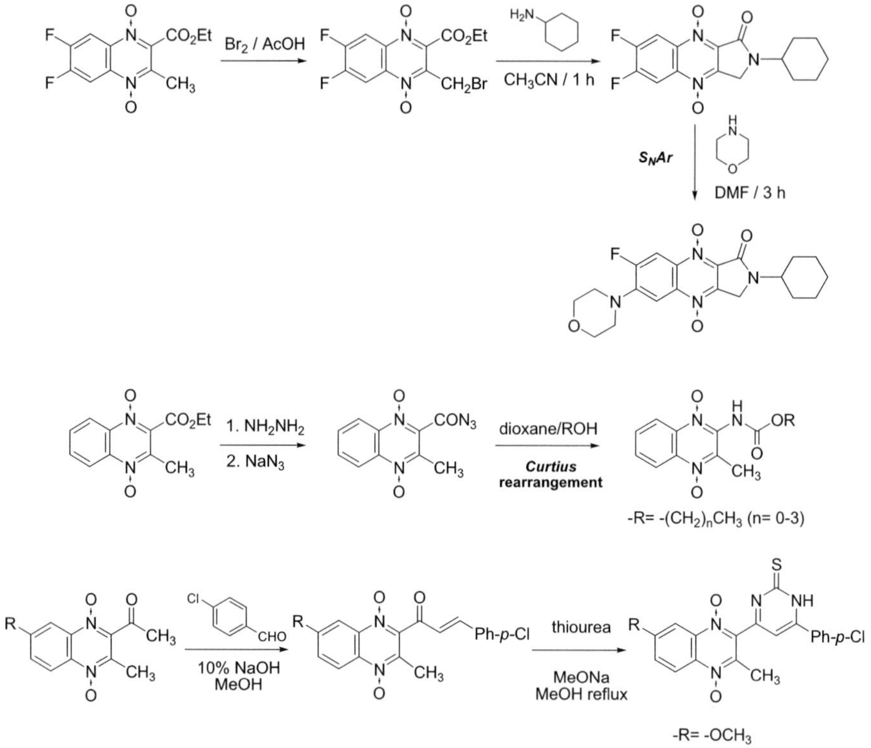

Fig. 6 Examples of lateral chain modifications of QDO

Recently, other metals have been studied as coordination agents. For example, Ce(III) and Nd(III) have been coordinated with 2,3-*bis*(diphenyl-phosphino)quinoxaline dioxide (L) yielding complexes with the formula CeL_2 $(NO_3)_3 \cdot H_2O$, $NdL_2(NO_3)_3 \cdot 3H_2O$, and $[NdL_2(NO_3)_2(H_2O)]NO_3 \cdot 1.5H_2O$ which was characterized by X-ray analyses [68]. Other recent studies have involved the use of 3-amino-2-quinoxalinecarbonitrile 1,4-dioxide derivatives as bidentate ligands coordinating with Cu(II) and with V(IV) as oxovanadium entities [69, 70]. In both cases, the QDO ligand is deprotonated (Fig. 7).

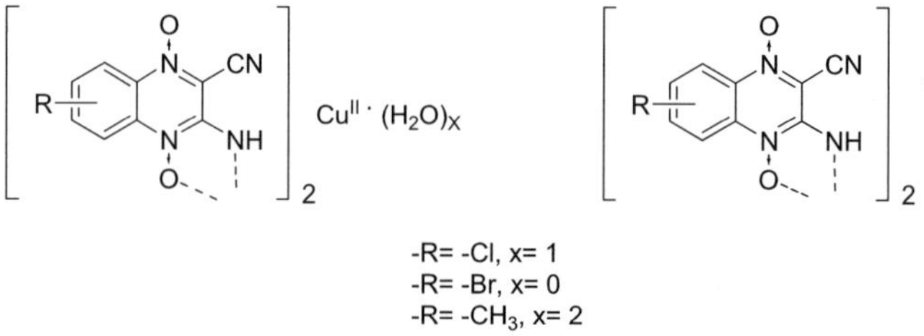

-R= -Cl, x= 1
-R= -Br, x= 0
-R= -CH₃, x= 2

Fig. 7 Cu(II) and V(IV) complexes of 3-amino-2-quinoxalinecarbonitrile 1,4-dioxide

3.4
Other Relevant Reactions

The 2-amino-substituted QDO has been converted to [1,2,4]oxadiazolo[2,3-a]quinoxaline derivatives when it reacts with adequate isocyanate (Fig. 8) [71]. Besides, the oxadiazoloquinoxaline 5-oxide has been employed as intermediate in the synthesis of QDO 2-carbamate derivatives.

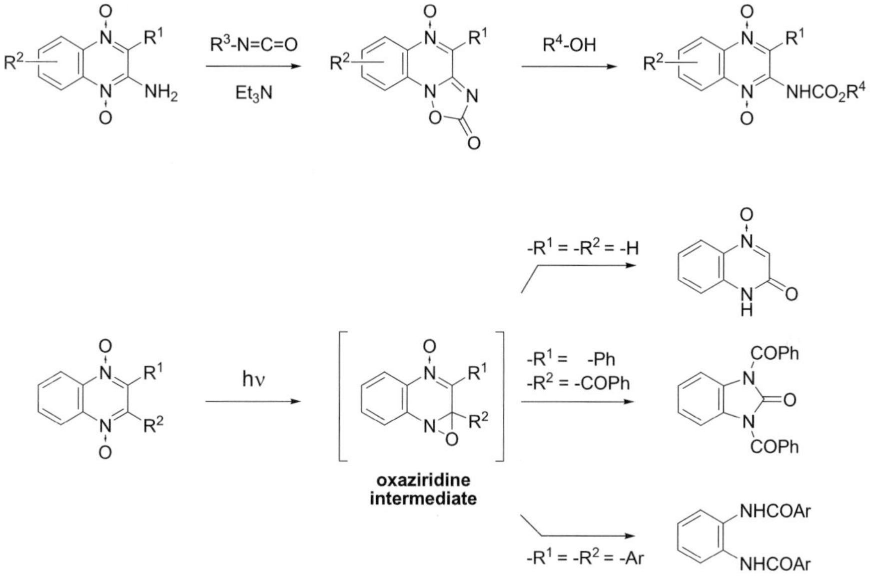

Fig. 8 Synthesis of QDO 2-carbamate derivatives and photochemical reactions of QDO

Other relevant described reactions, which affect the stability [72] and are related to some adverse effects of these compounds (see below) [73], have been the photochemical transformations. In fact, QDO undergoes rearrangements in the presence of UV-irradiation, passing through an oxaziridine intermediate, like an acyclic nitrone (Fig. 8). The final products depend on the QDO 2,3-substitutions and reaction conditions [72, 74, 75].

4
Structural Studies

4.1
Solid State

Structural studies in solid state have been performed for QDO and PDO derivatives in terms of X-ray diffraction experiments. According to these

Table 3 Characteristic crystallographic data of QDO, 2,3-dimethylquinoxaline 1,4-dioxide, 3-ethylsulfonyl-2-phenylquinoxaline 1,4-dioxide, and PDO

| Derivative | Crystallographic findings | | |
	Crystal characteristics	N – O bond length	N – O bond order[a]
Quinoxaline 1,4-dioxide	monoclinic	1.29	1.35
2,3-dimethylquinoxaline 1,4-dioxide	orthorhombic	1.30	1.31
3-ethylsulfonyl-2-phenylquinoxaline 1,4-dioxide	monoclinic	1.28	–
Phenazine 5,10-dioxide	monoclinic	1.33	1.30

[a] The N – O bond order in pyridine N-oxide hydrochloride is 1.17 [76]

studies, the systems are planar and the N – O bond orders, for both heterocycles, indicate some double-bond character for N-oxide moiety being the bond orders a little higher than the corresponding value for pyridine N-oxide [76–78]. Table 3 collects relevant values for some of these derivatives. On the other hand, it has been evidenced that these systems are able to form complex, through donor-acceptor interactions, with 1,1,2,2-tetracyanoethylene in methylenchloride solution. In solid state, complexes between this acceptor and the donors 2,3-dimethylquinoxaline 1,4-dioxide or phenazine 5,10-dioxide have been obtained and studied showing the presence of weakly donor-acceptor interactions. The nature of these weak donor-acceptor interactions is postulated to be both electrostatic and covalent [79].

4.2
Solution

Since the 1970s, general works on the QDO and PDO properties in solution have been reported in terms of diverse spectroscopic techniques. For example, electron spectroscopy has allowed studying π-electron charge densities and bond orders of PMO and PDO [80] or QDO intramolecular charge transfer process through N-oxide moiety [81]. Nuclear magnetic resonance spectroscopies involving ^{1}H-, ^{13}C-, and ^{14}N-nucleus, have been performed and N-oxide effects have been studied. Chemical shift changes on position peri atoms, protons, and carbons, have been the most analyzed aspect [82–85].

Perhaps the most extensively studied fact in solutions is the QDO and PDO electrochemical behavior. The next section summarizes some of the studies performed particularly since the 1970s.

4.2.1
Electrochemical Behavior

The *N*-oxide moiety in heterocyclic amine *N*-oxides, like QDO and PDO, presents a particular electrochemical behavior. On the one hand, the oxygen *N*-oxide atom could suffer loss of one electron, an oxidation process, and on the other hand, the nitrogen *N*-oxide atom, like in nitro compounds or nitrones, could receive one electron in a reduction pathway (Scheme 3). Both processes conduct to different reactive entities that have been studied using different electrochemical and spectroscopic techniques.

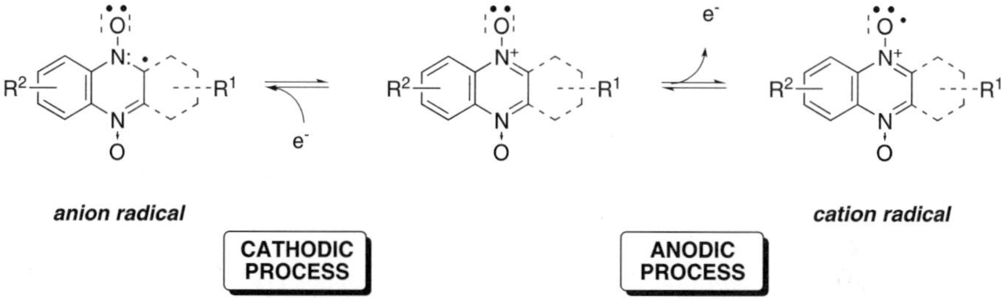

anion radical   cation radical

CATHODIC PROCESS   ANODIC PROCESS

Scheme 3 QDO/PDO electrochemical behavior

The cathodic process (reduction) has been studied in different experimental conditions being the potential of the one-electron reaction correlated with other structural parameters, i.e., with the reduced-product lowest vacant orbital energy [86], with the substituent Hammett σ parameter [87], or with the biological activity (see below). Table 4 shows reduction potentials for some QDO and PDO derivatives in different experimental conditions [88, 89].

The anodic process (oxidation) has been studied by electrochemical experiments combining with certain spectroscopies, i.e., electron spin resonance (ESR) [87, 90], ultraviolet [91], and visible [92] spectroscopies. Table 4 shows oxidation potentials for some QDO and PDO derivatives in different experimental conditions. From the ESR studies, the free radicals were characterized in terms of hydrogen and nitrogen hyperfine coupling constants and *g*-values [93, 94]. The *g*-values of the QDO cation radicals are higher than the values for PDO cation radicals and they are considerably larger than those of the corresponding anion radicals. These differences were explained by the spin density on the oxygen *N*-oxide and the contribution from lone pair electrons. The cation radicals from PDO have been studied for its reactivity both in dimerization process, generating 1,1'-dimers [95], and as mediators in the hydrocarbons and alcohols oxidation [92, 94, 96, 97].

Table 4 Electrochemical data of QDO, 2-methylquinoxaline 1,4-dioxide, 2,3-dimethyl-quinoxaline 1,4-dioxide, 1,2,3,4-tetrahydrophenazine 5,10-dioxide, and PDO

Derivative	Electrochemical findings	
	First reduction potential (V) vs sce	First oxidation potential (V) vs. indicated electrode
Quinoxaline 1,4-dioxide	$E_{1/2}$ – 1.34[a] – 1.22[b]	+ 1.72[c]
2-methylquinoxaline 1,4-dioxide	$E_{1/2}$ – 1.45[a] – 1.32[b]	–
2,3-dimethylquinoxaline 1,4-dioxide	$E_{1/2}$ – 1.55[a] – 1.39[b]	–
1,2,3,4-tetrahydrophenazine 5,10-dioxide	$E_{1/2}$ – 1.55[a] – 1.36[b]	–
Phenazine 5,10-dioxide	– 1.35[d,e]	+ 1.53[c,f] + 1.3[g]

[a] In acetonitrile, working electrode: platinum spiral, sweep rate: 0.1 V s^{-1}, supporting electrolyte: tetrabutylammonium hexafluorophosphate (0.1 M) [88]
[b] In DMF, working electrode: silver wire, sweep rate: 0.1 V s^{-1}, supporting electrolyte: tetraethylammonium perchlorate (0.1 M) [89]
[c] In benzonitrile, working electrode: platinum, supporting electrolyte: tetrabutylammonium perchlorate, reference electrode: Ag/AgCl [90]
[d] From ESR experiments
[e] In acetonitrile, working electrode: platinum wire, supporting electrolyte: tetrabutylammonium perchlorate (0.1 M) [94]
[f] Sweep rate: 0.6 V s^{-1}
[g] In acetonitrile, working electrode: platinum wire, supporting electrolyte: LiClO$_4$ (0.1 M), reference electrode: sce [94]

4.3
Other Studies

A great number of studies related to thermochemical properties of QDO and PDO derivatives have been recently described by Ribeiro da Silva et al. [98–103]. These studies, which have involved experimental and theoretical determinations, have reported standard molar enthalpies of formation in the gaseous state, enthalpies of combustion of the crystalline solids, enthalpies of sublimation, and molar (N – O) bond dissociation enthalpies. Table 5 shows the most relevant determined parameters. These researchers have employed, with excellent results, calculations based in density functional theory in order to estimate gas-phase enthalpies of formation and first and second N – O dissociation enthalpies [103].

Table 5 Characteristic thermochemical properties of QDO, 2-methylquinoxaline 1,4-dioxide, 2,3-dimethylquinoxaline 1,4-dioxide, and 2-hydroxyphenazine 5,10-dioxide, determined by Ribeiro da Silva et al.

| Derivative | Termochemical findings [a] | | |
	$\Delta_f H_m^o$ (g) [b]	$\langle DH_m^o(N-O)\rangle$ [c]	$DH_{(N-O),DFT}^d$
Quinoxaline 1,4-dioxide	227.1 ± 2.4	255.8 ± 2.0	261.1
2-methylquinoxaline 1,4-dioxide	169.9 ± 7.2	268.3 ± 4.9	263.4
2,3-dimethylquinoxaline 1,4-dioxide	149.4 ± 4.5	260.9 ± 2.7	265
2-hydroxyphenazine 5,10-dioxide	120 ± 6	263 ± 4	265

[a] $kJ \cdot mol^{-1}$
[b] $\Delta_f H_m^o$(g): standard molar enthalpy of formation in the gaseous state
[c] $\langle DH_m^o(N-O)\rangle$: molar (N−O) bond dissociation enthalpy
[d] $DH_{(N-O),DFT}$: molar (N−O) bond dissociation enthalpy theoretically calculated using density functional theory (B3LYP/6-311+G(2d,2p)//B3LYP/6-31G(d))

5
Bioactivity

The QDO and PDO derivatives have shown very interesting biological properties and its interest in medicinal chemistry has grown in the last three decades. Maybe the first reported bioactivities have been described in the mid-1940s involving the QDO veterinary feed growth promotion capacity [104], i.e., compounds **4–7** (Fig. 9), and the animal and human antibacterial action of certain synthetic QDO and natural PDO derivatives [105, 106], i.e., compounds **1–3** (Fig. 1). Currently, the interest in both kinds of bioactivities is still in progress [107] and there are a great number of patent registrations and commercial formulations concerned with the QDO or PDO use as antibacterial agents and animal feed additives [108–112]. In this sense in the final of 1980s Olaquindox (**7**, Fig. 9), which has been marketed as BayoNox® by Bayer AG (Germany), was authorized throughout the European Community as a veterinary growth promoter [113].

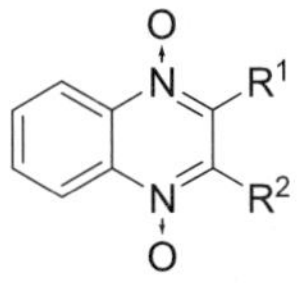

4, Quindoxin, Grofas, Celbar, $R^1 = R^2 = -H$
5, Carbadox, Mecadox, $R^1 = -CH=NNHCO_2CH_3$, $R^2 = -H$
6, Dioxidine, $R^1 = R^2 = -CH_2OH$
7, Olaquindox, BayoNox, $R^1 = -CONHCH_2CH_2OH$, $R^2 = -CH_3$

Fig. 9 First QDO patented as antibacterial agent and animal-feed additive

The next sections outline other bioactivities described in the most recent years.

5.1
Antibacterial and Antifungal Activity

The findings that demonstrated the activity of compound **4** (Fig. 9) under anaerobic conditions opened the studies of QDO as specific anti-micro-organism agents. Anaerobic antibacterial mechanism of compound **4** involves DNA synthesis inhibition without effects on the RNA and proteins synthe-ses [114]. It has been demonstrated that compound **6** inhibits *Staphylococcus aureus* DNase biosynthesis and plasma coagulase [115].

One of the most recent biological results in the field of QDO as an an-tibacterial and antifungal agents come from Carta et al. These researchers found 2-phenylsulfonyl-substituted QDO to be an excellent scaffold against the nosocomial *Enterococcus faecalis* and *Enterococcus faecium*, being com-pounds **8** and **9** the most representative examples (Fig. 10) [116]. On the other hand, they have investigated the QDO action against fungus; specifically they have employed clinically isolated *Candida albicans*, *Candida glabrata*, *Candida krusei* and *Candida parapsilosis* as models for their biological evalu-ations. Compounds **10** and **11** (Fig. 10) were the most actives against *C. krusei* fungus [117, 118].

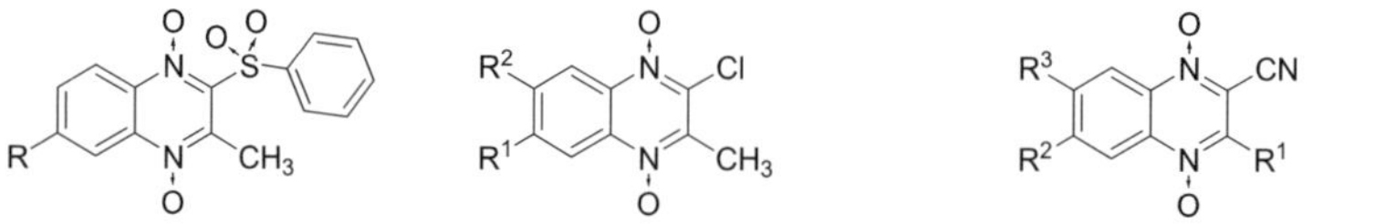

8, R= -F
9, R= -Cl

Actives against
E. faecalis and E. faecium

10, R^1= -F, R^2= -OEt
11, R^1= -H, R^2= -Cl

Actives against
C. krusei

12, R^1= -NHCOCH$_3$, R^2= R^3= -Cl
13, R^1= -NHPh, R^2= -H, R^3= -Cl
14, R^1= 4-(*p*-NO$_2$Ph)piperazin-1-yl, R^2= R^3= -CH$_3$

Actives against
M. tuberculosis

Fig. 10 Noteworthy QDO with antibacterial and antifungal activities

Recently, Takabatake et al. described the anti-*Bacteroides fragilis* activity of a series of 2-hydroxy-1/3-substituted PDO [33].

As a result of the urgency to find new agents for the anaerobic microorgan-ism *Mycobacterium tuberculosis* (*M. tuberculosis*), considerable research with QDO has been amassed in this field. The next section summarizes the most significant results in these approaches.

5.1.1
QDO as Antimycobacteria Agents

Tuberculosis (TB), an infection of *M. tuberculosis*, still remains the leading cause of death worldwide among infectious diseases. The statistics indicate that 3 million people throughout the world die annually from TB and there are an estimated 8 million new cases each year, of which 95% occur in developing countries. One-third of the population is infected with *M. tuberculosis* and the World Health Organization (WHO) estimates that within the next 20 years, about 30 million people will be infected with the bacillus [119]. The current frontline therapy for TB consists on administering three different drugs: isoniazid, rifampin, and pyrazinamide, over an extended period of time. On the other hand, important therapeutic problems have been recognized with the presence of multidrug-resistant TB (MDR-TB).

At the end of the 1990s, Monge et al. described QDO derivatives series with extensively substitution patterns possessing excellent anti-*M. tuberculosis* $H_{37}Rv$ activity. In a first approach 3-amino-2-cyano-substituted QDO (i.e. **12–14**, Fig. 10) was identified as antimycobateria lead system [120–122]. Not only this structural architecture resulted significant for the antibacterial activity but also both *N*-oxide moieties play a role in the bioactivity finding that the Qx analogues were inactives [122–124]. Furthermore, toxicity against mammal-VERO cells was assessed possessing compound **14** one of the best selectivity index (SI) (SI = $IC_{50,VERO}/MIC$) (MIC: minimum inhibitory concentration). Further QDO-antimycobacteria-actives new generations were obtained by changing 3-amino-2-carbonitrile motives by 2-carboxamide, 2-acetyl/benzoyl or 2-carboxylate ones [125–127]. From these studies, compounds **15–17** (Fig. 11) have been outstanding, between the 2-carboxamide derivatives, for its good EC_{90}/MIC ratio (0.89, 0.42, and 0.14, respectively), being EC_{90} the lowest concentration effecting a 90% reduction at 7th treatment day of TB-infected macrophages. In the 2-acetyl/benzoyl-substituted series, the best results were obtained with 2-acetyl derivatives highlighting compounds **18–21** activity (Fig. 11) but without improvement in the EC_{90}/MIC ratios (0.87, 3.13, 0.80, and 4.29, respectively). Finally, at the moment, the best results are obtained with the 2-carboxylate-substituted derivatives. Compounds **22–25** (Fig. 11) exhibited excellent bioactivity profiles with moderate to excellent EC_{90}/MIC ratios (0.56, 2.30, 1.50, and 0.01, respectively). In addition, compounds **26** and **27** (Fig. 11) were active against

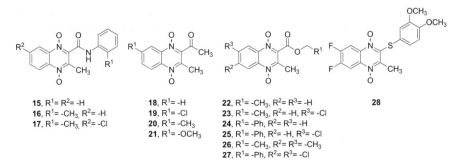

15, $R^1= R^2= $ -H
16, $R^1= $ -CH$_3$, $R^2= $ -H
17, $R^1= $ -CH$_3$, $R^2= $ -Cl

18, $R^1= $ -H
19, $R^1= $ -Cl
20, $R^1= $ -CH$_3$
21, $R^1= $ -OCH$_3$

22, $R^1= $ -CH$_3$, $R^2= R^3= $ -H
23, $R^1= $ -CH$_3$, $R^2= $ -H, $R^3= $ -Cl
24, $R^1= $ -Ph, $R^2= R^3= $ -H
25, $R^1= $ -Ph, $R^2= $ -H, $R^3= $ -Cl
26, $R^1= $ -CH$_3$, $R^2= R^3= $ -CH$_3$
27, $R^1= $ -Ph, $R^2= R^3= $ -Cl

28

Fig. 11 Antimycobacterial QDO

all the seven of the studied resistant strains, being an excellent chemical candidate to develop new agents for MDR-TB.

Carta et al., working with 2-phenylthio/2-phenylsulfonyl-substituted QDO [117], have evidenced the 3-methyl-2-phenylthio derivatives as the best antimycobacteria agents [118]. Recently, the authors emphasized the activity of compound 28 (Fig. 11), which was evaluated against clinical isolated MDR-TB and non-tubercular MDR-TB [128].

5.2
Antiparasite Activity

Parasitic diseases, such as malaria, leishmaniasis, and trypanosomiasis, affect hundreds of millions people around the world, mainly in underdeveloped countries. They are also the most common opportunistic infections affecting patients with acquired immunodeficiency syndrome (AIDS). Malaria occupies the first place but in Latin America, Chagas' disease (American Trypanosomiasis) is the most relevant parasitic disease that produces morbidity and mortality in low-income individuals. Significant advances for the cure of bacterial, fungal, and viral diseases have been made in the last century. However, compared to bacterial diseases, development of chemotherapy for the treatment of parasitic diseases has been hindered since parasitic protozoa are eukaryotic, so they share many common features with their mammalian host making the development of effective and selective drugs a hard task. What is more, the spread of drug-resistant strains and the differences in drug susceptibility among different parasite strains lead to varied parasitological cure rates according to the geographical area [129].

QDO derivatives have been evaluated against the parasite *Plasmodium falciparum* (*P. falciparum*) responsible of Malaria and *Trypanosoma cruzi* (*T. cruzi*) responsible of American Trypanosomiasis.

In a first approach, Monge et al. have synthesized a series of QDO and its Qx-reduced analogues, using traditional methodologies (Fig. 12), and evaluate them in vitro and in vivo as anti-*P. falciparum* agents. For these evaluations different strains including drug-resistant ones were employed [53]. The best biological results were observed in 1,4-dioxide derivatives 29 and 30 (Fig. 12). From these results, a second [130] and a third generation [131] of QDO derivatives with potential antimalarial activity were developed. From these studies, compounds 31 and 32 (Fig. 12) were highlighted for their activity on cultures of FcB1 (chloroquine-resistant strain) of *P. falciparum*.

In other approaches, close to 30 QDO derivatives were carefully selected from Monge's more than 200-quinoxaline library in order to evaluate its anti-*T. cruzi* activity. The study, performed in two different *T. cruzi* strains, showed that compounds 32–35 (Fig. 12) are an excellent starting point for further structural modifications [132, 133]. Some reduced derivatives, Qx, displayed relevant activities but these compounds could be also toxic in mammal

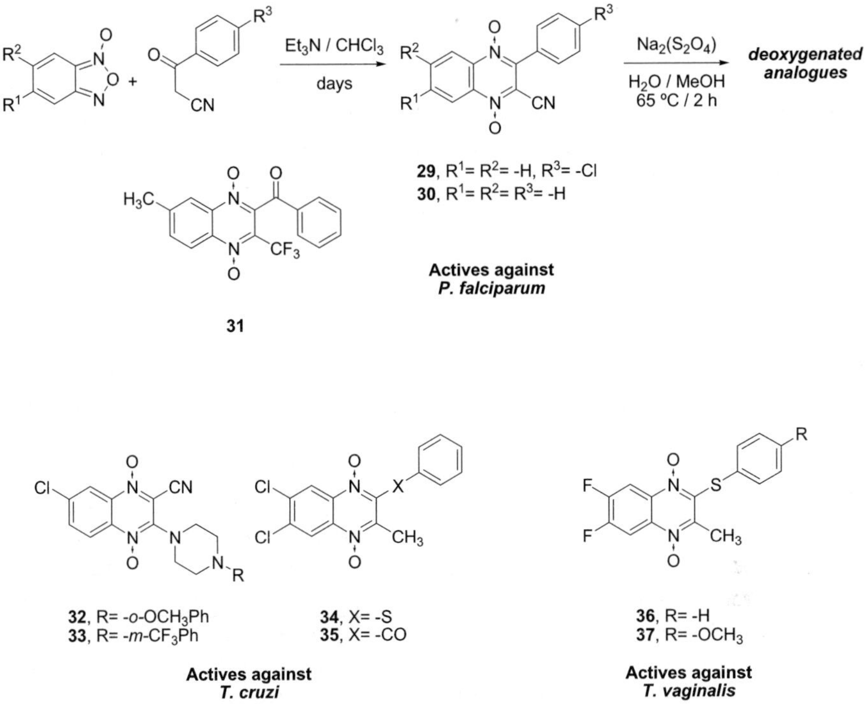

$Et_3N / CHCl_3$
days

$Na_2(S_2O_4)$
$H_2O / MeOH$
65 °C / 2 h

deoxygenated analogues

29, $R^1 = R^2 = $ -H, $R^3 = $ -Cl
30, $R^1 = R^2 = R^3 = $ -H

Actives against
P. falciparum

31

32, R= -o-OCH$_3$Ph
33, R= -m-CF$_3$Ph

Actives against
T. cruzi

34, X= -S
35, X= -CO

36, R= -H
37, R= -OCH$_3$

Actives against
T. vaginalis

Fig. 12 *Top*: Preparation of 2-quinoxalinecarbonitrile 1,4-dioxide and its deoxygenated analogues actives against *P. falciparum*. *Bottom*: Remarkable QDO with anti-*T. cruzi* and anti-*T. vaginalis* activities

cells due to their structural features. In these experiments, the 3-amino-2-cyano QDO derivatives were very insoluble in water and therefore poorly active against the parasite in these conditions. Consequently, in order to improve water solubility, the authors prepared vanadyl complexes of the type $[V^{IV}O(L)_2]$, where L are 3-amino-2-cyano QDO (Fig. 7) [134]. Complexation to vanadium leads to excellent antiparasite activity higher than that of the corresponding free ligands. Structural requirements for anti-*T. cruzi* activity were studied in terms of QSAR analysis (see below).

Trichomoniasis is a protozoan infection of the human and bovine genitourinary tracts being the responsible anaerobic protozoan *Trichomonas vaginalis*. Recently, Carta et al. reported the synthesis and anti-*T. vaginalis* (SS22) activity of a series of 6,7-difluoro-3-methyl QDO derivatives [118]. Several 2-phenylthio derivatives were 20 to 30-fold more potent than the reference drug used in the assay, i.e., compounds **36** and **37** (Fig. 12). 2-Phenylsulfinyl analogues maintain the biological activity whereas 2-phenylsulfonyl-substituted QDOs were fewer actives.

5.3
Selective Hypoxia-Cytotoxic Activity

Because of their rapid growth, cancerous cells can become relatively isolated from the blood supply, and it becomes increasingly difficult for nu-
trients, especially oxygen, to diffuse to them, resulting in hypoxia. There is
evidence that viable hypoxic cells may contribute up to 20% of the tumor
mass [135]. Hypoxic cells are very resistant to radiation damage, and the
same diffusional limitations for oxygen can apply to the cytotoxic drugs used
in chemotherapy, furthermore they appear to accelerate malignant progres-
sion and increase metastasis. However, hypoxic cellular population has an
advantage because a type of compounds known as bioreductive drugs [136].
These drugs are able to undergo metabolism to cytotoxic species exclusively
in the hypoxic conditions. Consequently, this refractory tumoral population
selectively died. Bioreductive drugs belong to different structural entities,
one of them being the *N*-oxide derivatives [137]. The prototype drug of this
family of compounds is Tirapazamine (Scheme 4), currently in Phase III clinical trials. Tirapazamine produces differential hypoxic cytotoxicity due to
a first step of bioreduction, reversible in oxic conditions, where it is converted into an intermediate that releases a hydroxyl radical, OH·, damaging
DNA [138].

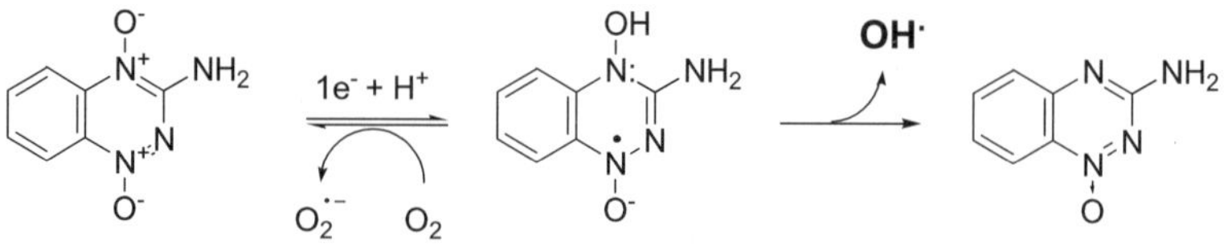

Tirapazamine

Scheme 4 Bioreductive cytotoxic mechanism for Tirapazamine

On the basis of both structural correlation between benzotriazine and
Qx nucleus and mode of action reported for QDO, Monge et al. described
at a first time 3-amino-2-cyano-substituted QDO as selective hypoxic cytotoxins, bioreductive compounds, i.e., compounds **38–41** (Table 6) [27]. In
this first approach, the best compounds were the 7-electron-withdrawing
substituted derivatives. Electrochemical properties, assessed via voltammetric studies, showed that as the electron-withdrawing nature of the 6-(7)-
substituent increases, the reduction potential becomes more positive. Compounds with reduction potential more positive are more hypoxia-cytotoxic.
However, due to the poor solubility in water of these derivatives, a new
generation of compounds was designed and synthesized, i.e., compounds
42–45 (Table 6) [28]. In this second generation of compounds it is possible
to highlight derivatives **42** and **45** (Table 6) with excellent hypoxic potency

Table 6 3-Amino-2-quinoxalinecarbonitrile 1,4-dioxide derivatives described as hypoxic selective cytotoxins

Derivative	R^1	R^2	NR^3R^4	HCR[a,b], P[a,c]	$E_{pc}^{d,e}$ (V) vs sce
38	H	CF_3	–	75, 7.0	– 0.65
39	Cl	Cl	–	80, 1.0	– 0.62
40	H	H	–	80, 30.0	– 0.88
41	H	Cl	–	150, 9.0	– 0.74
42	H	Cl	$NH(CH_2)_2NMe_2 \cdot HCl \cdot 0.25\,H_2O$	100, 0.3	nr[f]
43	H	Cl	$NH(CH_2)_3NMe_2 \cdot HCl$	250, 0.4	nr
44	H	H	$NH(CH_2)_3NMe_2 \cdot HCl$	300, 1.0	nr
45	H	CF_3	$NH(CH_2)_3NMe_2 \cdot HCl \cdot H_2O$	340, 0.3	nr
46	OCH_3	Cl	4-(p-NO_2Ph)piperazin-1-yl	> 50, 2.0	nr
47	–	–	–	nr, 35.1	nr
48	–	–	–	> 15, 3.0	– 0.78
Tirapazamine	–	–	–	75, 30.0	– 0.90

[a] Using V79 cells
[b] HCR: Hypoxia cytotoxicity selectivity relationship, determined as the ratio between the concentration of drug in air and the concentration of drug in hypoxia that produce 1% of cell killing
[c] P: potency under hypoxia, defined as the dose in μM which gives 1% of cell survival with respect to the control
[d] E_{pc}: Reduction potential
[e] Using cyclic voltammetry
[f] nr: not reported

(P, see definition in Table 6) and hypoxia cytotoxicity selectivity relationship (HCR, see definition in Table 6), where the basic side chain in the 2-position improves the water solubility of these compounds. Compound **43**, one of the most active in vitro derivative, was submitted to in vivo pharmacokinetic studies. In these experiments it was found that it has too short half-life to be used as an anticancer drug. Further structural modifications were studied by these authors that modified the basic chain in the 3-position of QDO, i.e., compound 46 (Table 6) [30]. Some of these showed better P than that of the parent compounds, however, the HCRs were not improved. Other approaches with the parent 3-amino-2-cyano QDO compounds as bioreductive agents have involved metal complexation (Fig. 7) [69, 141]. For example, copper (II) or vanadyl complexes, i.e., compounds 47 and 48 (Table 6), were developed in order to improve the bioavailability and the pharmacological and toxicological properties of the QDO derivatives. Particularly, the biological interest of Cu(II) as a coordinating metal relays on its capacity to be bioreduced into the solid tumor releasing the ligand, in this case QDO, and consequently producing the killing of tumor cells. On the other hand, if radioactive Cu isotopes were used in the coordination process, the generated radiopharmaceutical could act via a dual mechanism of action into the tumoral tissue, namely via radiation damage and the via-bioreduction process [142]. These complexes were shown to be as potent as the corresponding ligands and selective for the hypoxic conditions (see examples in Table 6).

In 1995, a series of QDO were prepared from the corresponding Bfx and evaluated as hypoxic selective cytotoxins and radiosensitizer [143]. For example, hydroxamic acid 49 (Fig. 13) is an excellent hypoxia cytotoxic agent and radiosensitizer in vitro. Recently Gali-Muhtasib et al. investigated the in vitro ability of three QDO, compounds 18 (Fig. 11), 50 and 51 (Fig. 13) and one PDO, tetrahydro derivative 52 (Fig. 13), to inhibit cell growth and induce cell-cycle changes in human colon cancer cell line T-84 under normoxic and hypoxic conditions [144]. Under normoxia, the compounds displayed mild cytotoxic effects whereas under hypoxia compound 51 was the most potent cytotoxin and hypoxia-selective agent ($P_{T-84} = 1.0\,\mu M$, $HCR_{T-84} = 100.0$). Compounds 18 and 50 were less cytotoxic but arrested almost 50% of the cells in the G2M phase of the cell cycle. The PDO derivative, 52, was unable to affect the growth and cycling of cells cultured in normoxia and was the least potent cytotoxin under hypoxic conditions.

In 2006, another approach in QDO as bioreductive agents was described [66]. Figure 13 shows some relevant derivatives of this work (53–55) being compound 53 the most potent and selective cytotoxin with a HCR higher than that of compound 40 (Table 6) used for the authors as the standard. Compounds 54 and 55 were also more selective than the standard. On the other hand, some derivatives were evaluated as in vitro antitumoral agents in normoxia against liver carcinoma and brain tumor cells finding significant activities that justify further in vivo studies.

Recently, González et al. described the synthesis and evaluation as hypoxic selective cytotoxins of a series of PDO and its reduced analogues, Pz, as the generated metabolic products into the hypoxic tissues [54]. Com-

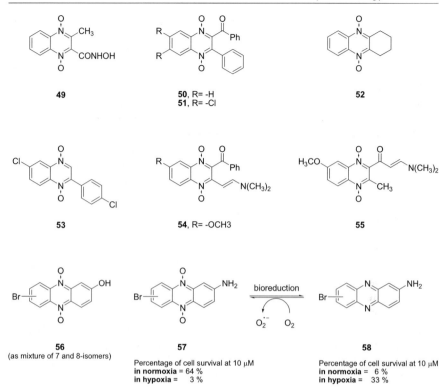

Fig. 13 *Top*: QDO with selective hypoxia cytotoxic activity. *Bottom*: PDO, and Pz related derivative, described as selective hypoxia cytotoxins

pound **56** (Fig. 13) was the most active developed derivative ($P_{V79} = 10\,\mu$M, $HCR_{V79} > 10.0$). Furthermore, the reduced analogues were, in general, non-cytotoxic in hypoxia. Consequently, this biological behaviour indicates that the corresponding parent PDO could be acting as hypoxic selective biore-ductive compounds, such as Tirapazamine. The unique reduced derivative that showed cytotoxic effects on hypoxia and on oxia was compound **58** (Fig. 13), and observing its oxic cytotoxicity some interesting feature for par-ent compound **57** was outlined by the authors. The authors thought that compound **57** is irreversibly reduced in hypoxic conditions to **58**, generates toxic damage in the hypoxic cells. Whereas its metabolite **58** can migrate to the surrounding oxic cells and produce damage in this tissue. These re-sults indicate that the parent PDO **57** could act as selective hypoxic cytotoxic agent and after the bioreductive trigger a normoxia-antitumoral agent, com-pound **58**. In a second approach, structural modifications were performed in the 2-hydroxy/amino moiety generating compounds without improved ac-tivities [145]. Also, DNA-PDO-interaction studies were performed observing no PDO DNA-interacting capability. Electrochemical and QSAR studies were performed in order to explain the substituent-depending differential activities (see below).

Gates et al. investigated if QDO, non-substituted and 3-amino-2-cyano-substituted, biological properties were similar to those of TPZ [146, 147]. The results showed that QDOs act in vitro as redox-activated DNA-cleaving agent under hypoxic conditions. Compound **4** (Fig. 9) was the bioreductive substrate of xanthine/xanthine oxidase reducing system. For compound **40** (Table 6), the findings indicate that one-electron reduction promoted by NADPH:cytochrome P450 reductase produces, like Tirapazamine, DNA damage under low-oxygen levels.

On the basis that numerous genes are activated under hypoxia, the most important of which is hypoxia-inducible factor 1 α (HIF-1 α), Gali-Muhtasib et al. studied the QDO capacity to reduce the expression of HIF-1 α in T-84 cells [148]. Compound **51** (Fig. 13) reduced the levels of HIF-1 α transcript and protein expressing compounds **18** (Fig. 11) and **52** (Fig. 13) in a much lesser proportion.

Furthermore, these researchers have found that compound **51** induces growth inhibition in T-84 cells due to its ability to induce cell-cycle arrest and/or apoptosis, while compound **18** blocked more than 60% of cells at the G2/M phase without inducing apoptosis [149]. Compound **50** (Fig. 13) also produced G2/M cell-cycle arrest and induction of apoptosis. Studying the QDO effects on molecules that regulate apoptosis and the G2 to M transition, **18** and **50** inhibited the expression of cyclin B while **51** decreased the levels of Bcl-2, gene that codifies for the antiapoptotic protein bcl-2, and increased Bax expression, gene with proapoptotic functions.

In 2005, López de Ceráin et al. found that by using comet assays and flow cytometry analysis, compound **43** (Table 6) damages DNA under hypoxia and normoxia conditions [150]. In another study, it has been demonstrated that compound **43** produces reactive oxygen species (ROS) and oxidative DNA damage both in normoxic and hypoxic conditions into Caco-2 cells [151]. In normoxia, a significant increase of ROS was evidenced in a manner dose-dependent, while in hypoxia high ROS levels were observed at all the studied concentrations. Furthermore, modified comet assay demonstrated that oxidative DNA damage in normoxia and hypoxia was promoted by compound **43**.

5.4
Others Relevant Bioactivities

In addition to the biological properties described in the previous section, other significant bioactivities have been reported for QDO and PDO derivatives. In Table 7 and Fig. 14, listed examples, bioactivities, comments, and references for the most recent works are shown.

Table 7 QDO and PDO derivatives miscellaneous bioactivities

Family	Examples[a]	Bioactivity	Comments	Refs.
QDO	59	antitumoral	in vitro assays against MCF7 (breast), NCI-H460 (lung), SF-268 (CNS) and leukemia cells	[152]
QDO	51[b]	angiogenesis inhibition	in vivo studies	[153]
PDO	60	site-specific DNA cleavage	in the first approach DNA-oligonucleotides covalently bonded to PDO were used	[154, 155]
QDO	61, 62	herbicidal	–	[65, 156, 157]
PDO	63	xanthine oxidase inhibition	–	[158]
QDO	48[c]	insulin-mimetic activity	–	[70]

[a] See structure in Fig. 14
[b] See structure in Fig. 13
[c] See structure in Table 6

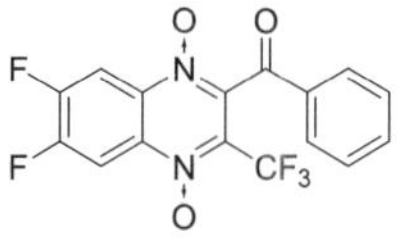

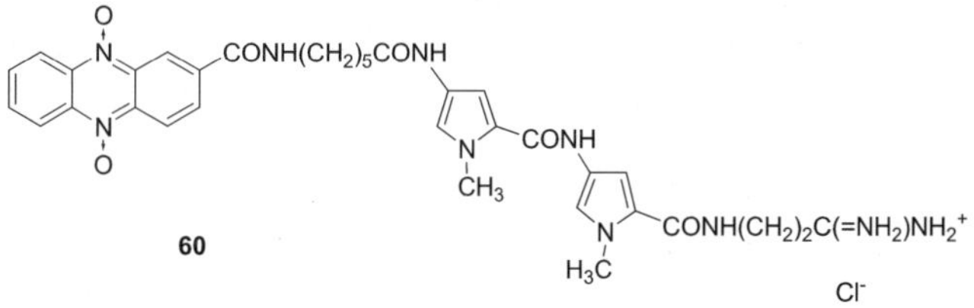

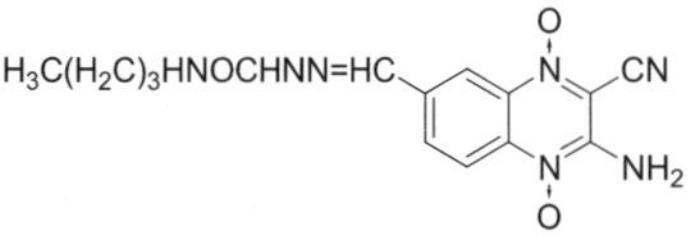

Fig. 14 QDO and PDO derivatives miscellaneous bioactivities

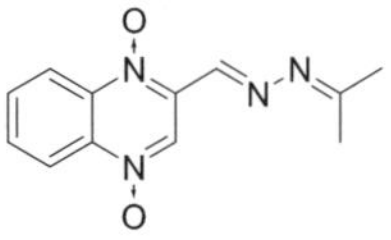

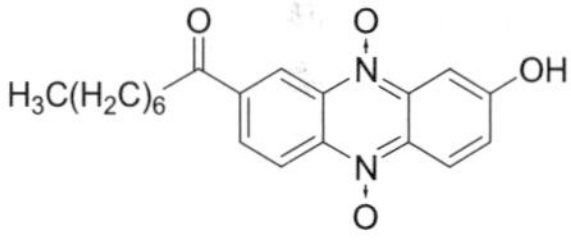

6
Structure-Activity Relationships

In the 1980s, the first works studying QDO and PDO structure-activity relationship were published describing the relations between antibacterial or radiosensitizer bioactivity and experimental and theoretical physicochemical descriptors [159–162]. Afterwards, very few jobs relating to QDO and PDO structure-activity relationships were described.

It can be mentioned the work of Cerecetto et al. [133, 134] where the anti-*T. cruzi* activities of a series of QDO and its vanadyl complexes were correlated with theoretical physicochemical descriptors. On the one hand, a clear two-variable correlation was found between both QDO-antiparasite activity and theoretical lipophilicity and LUMO energy [133]. The equation confirms some previously observed aspects, i.e., that the presence of an electron-withdrawing substituent, like halogen, in 5-8-positions produces more active compounds. On the other hand, it was found that the anti-*T. cruzi* response of a family of QDO-vanadyl complexes depends on the lipophilic properties and the electronic properties, σ_m, of 6-7-substituents [134]. Another QSAR study involving PDO derivatives with selective hypoxic cytotoxicity was performed by these researchers [145]. In this case, electrochemical properties, expressed as E_{pc}, together with lipophilicity correlated to the fraction of cell survival in hypoxia finding that when the redox potential of the *N*-oxide moiety decreases the hypoxic cytotoxicity increase.

Recently, Ribeiro da Silva et al. have described another QSAR study. The authors studied the substitution influence on the N – O bond dissociation enthalpies of a series of QDO and its relationship to antitumor activity [163].

7
Pharmacological Behaviors

In this section two aspects related to QDO and PDO future use as drugs will be discussed: in vivo metabolic behavior and toxic effects.

7.1
In Vivo Metabolic Studies

Recently, only the in vivo pharmacokinetic studies performed on QDO **43** (Table 6) have been described [140]. Pharmacokinetic parameters without metabolic data were reported.

7.2
Toxicity Studies

A great number of toxicity studies have been performed that analyze QDO mutagenic effects [164–167]. For example, compounds **4–7** (Fig. 9) and 2,3-dimethyl QDO were mutagenic to *Salmonella typhimurium* (*S. typhimurium*) TA 98 and TA 100 in the presence and absence of the rat liver S-9 fraction (Ames test). Mutagenicity is dependent on the presence of *N*-oxide moiety since Qx is not mutagenic whereas QMO exhibits lower mutagenic activity than compound **4**. Furthermore, the mutagenicity of QDO is enhanced under anaerobic conditions. On the other hand, the clastogenicity of QDO was also examined analyzing the metaphase of chromosome aberrations in bone marrow of mice. Compound 5 (Fig. 9) induces chromosome aberrations at a dose of $100\ mg\ kg^{-1}$, compound 7 (Fig. 9) has the same toxic effects at a dose of $800\ mg\ kg^{-1}$ whereas compound 64 (Fig. 15), structurally closer to compound 5, does not produce chromosome-damaging effects, even at a dose of $1200\ mg\ kg^{-1}$ [168, 169]. Furthermore, compound 5 has a teratogenic effect and should contribute to fetal malformations and embryo-fetal deaths in rats [170]. Some deep studies on the mechanism of compound 7 mutagenicity were performed [171], which determined that 7-induced mutations show sequence-specificity in which most point mutations occurred at site N in a 5'-NNTTNN-3' sequence.

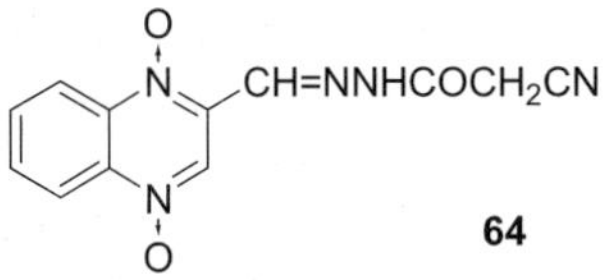

Fig. 15 QDO structurally related to carbadox, 5, without clastogenicity

Several authors have reported about the photoallergy of some QDO derivatives. The oxaziridines generated upon UV irradiation (see Fig. 8) are very reactive with proteins [73]. Compound 40 (Table 6) produces DNA alkalilabile lesions selectively at 2'-deoxyguanosine positions upon irradiation in the UV-A region suggesting that this molecule is potentially phototoxic [172]. In another study, DNA photodamage by Tirapazamine, 4 (Fig. 9) and 40 were compared [173]. Under anaerobic conditions, NADH increased photoinduced strand breaks in pBR322 plasmid DNA caused by Tirapazamine or 40. For Tirapazamine, the reactive species is probably nitroxide radical or the hydroxyl radical generated from its decomposition. In contrast, DNA damage by 4 was not affected by NADH, suggesting a different mechanism, possibly involving a photogenerated oxaziridine intermediate.

8
Others Applications

QDO, PDO, and related compounds are the subject of a great number of invention patents, particularly for its uses in material sciences. For example, QDOs were included in the formulation of modified unsaturated polymers and rubbers [174, 175]. Polymers with a QDO substructure as monomeric unit were used to produce fibers, films, electrochromic elements, electrodes, semiconductors, and electrolyte solutions for secondary batteries [176–179].

References

1. Katritzky AR, Lagowski JN (1971) Chemistry of heterocyclic N-oxides. Academic Press, New York
2. Albini A, Pietra A (1991) Heterocyclic N-oxides. CRC Press, Boston
3. Dvoryantseva GG, Kaganskii MM, Musatova IS, Elina AS (1974) Khim Geterotsikl Soedin 1554
4. Pushkareva ZV, Varyukhina LV, Kokoshko Z (1956) Dokl Akad Nauk SSSR 108:1098
5. Aaron JJ, Maafi M, Párkányi C, Boniface C (1995) Spectrochim Acta 51A:603
6. Laursen JB, Nielsen J (2004) Chem Rev 104:1663
7. McIlwain H (1943) J Chem Soc 322
8. Haddadin MJ, Issidorides CH (1965) Tetrahedron Lett 36:3253
9. Douglass ML (1973) US Patent 3 733 323
10. Mahajanshetti CS, Balse MN (1978) Indian J Chem 16B:830
11. Bowie RA (1971) DE Patent 2 127 902
12. Kobayashi Y, Kumadaki I, Sato H, Sekine Y, Hara T (1974) Chem Pharm Bull 22:2097
13. Crommelynck F, Basso MT (1975) FR Patent 2 265 740
14. Sandtnerova R, Repas M, Bilik V (1987) CZ Patent 240 379
15. Carmeli M, Rozen S (2006) J Org Chem 71:5761
16. Cooper MS, Heaney H, Newbold AJ, Sanderson WR (1990) Synlett 9:533
17. Haddadin MJ, Issidorides CH (1968) US Patent 3 398 141
18. Edwards ML, Bambury RE (1975) J Heterocycl Chem 12:835
19. Kluge AF, Maddox ML, Lewis GS (1980) J Org Chem 45:1909
20. Issidorides CH, Haddadin MJ (1982) US Patent 4 343 942
21. Monge A, Gil MJ, Pascual M (1983) Anal Real Acad Farm 49:199
22. Bowie RA, Jones G (1972) DE Patent 2 136 962
23. Usta JA, Haddadin MJ, Issidorides CH, Jarrar AA (1981) J Heterocycl Chem 18:655
24. Ludwig GW, Baumgartel H (1982) Chem Ber 115:2380
25. Monge A, Palop JA, Del Castillo JC, Calderó JM, Roca J, Romero G, Del Rio J, Lasheras B (1993) J Med Chem 36:2745
26. Monge A, Palop JA, Piñol A, Martínez-Crespo FJ, Narro S, González M, Sáinz Y, López de Ceráin A, Hamilton E, Barker AJ (1994) J Heterocycl Chem 31:1135
27. Monge A, Palop JA, López de Ceráin A, Senador V, Martínez-Crespo FJ, Sáinz Y, Narro S, García E, de Miguel C, González M, Hamilton E, Barker AJ, Clarke ED, Greenhow DT (1995) J Med Chem 38:1786
28. Monge A, Martínez-Crespo FJ, López de Ceráin A, Palop JA, Narro S, Senador V, Martín A, Sáinz Y, González M, Hamilton E, Barker AJ (1995) J Med Chem 38:4488
29. Barker AJ, Monge A, Hamilton E (1996) UK Patent 2 297 089
30. Ortega MA, Morancho MJ, Martínez-Crespo FJ, Sainz Y, Montoya ME, López de Ceráin A, Monge A (2000) Eur J Med Chem 35:21
31. Haroun M, Helissey P, Giorgi-Renault S (2001) Synth Commun 31:2329
32. Takabatake T, Miyazawa T, Kojo M, Hasegawa M (2000) Heterocycles 53:2151
33. Takabatake T, Miyazawa T, Takei A, Hasegawa M (2001) Igaku to Seibutsugaku 142:5
34. Takabatake T, Ito H, Takei A, Miyazawa T, Hasegawa M, Miyairi S (2003) Heterocycles 60:537
35. Sekimura N, Saito H, Miyairi S, Takabatake T (2005) Heterocycles 65:1589
36. Lima LM, Zarranz B, Marin A, Solano B, Vicente E, Pérez S, Aldana I, Monge A (2005) J Heterocycl Chem 42:1381
37. Lukasiewicz Z, Wrotek J (1976) PL Patent 86 391
38. Monge A, Llamas A, Pascual MA (1977) An Quim 73:912

39. Monge A, Llamas A, Pascual MA (1977) An Quim 73:1208
40. Li J, Ji M, Hua W, Hu H (2001) Indian J Chem 40B:1230
41. Boulton AJ, Katritzky AR, Sewell MJ, Wallis B (1967) J Chem Soc B 914
42. Boulton AJ, Halls PJ, Katritzky AR (1970) J Chem Soc B 636
43. Cerecetto H, González M, Lavaggi ML, Porcal W (2005) J Braz Chem Soc 16:1290
44. Haddadin MJ, Issodorides CH (1976) Heterocycles 4:767
45. Edwards ML, Bambury RE, Kim HK (1976) J Heterocycl Chem 13:653
46. Issidorides CH, Haddadin MJ (1970) UK Patent 1215815
47. Boiani M, Boiani L, Denicola A, Torres de Ortiz S, Serna E, Vera de Bilbao N, Sanabria L, Yaluff G, Nakayama H, Rojas de Arias A, Vega C, Rolán M, Gómez-Barrio A, Cerecetto H, González M (2006) J Med Chem 49:3215
48. Mason JC, Tennant G (1970) J Chem Soc B 911
49. Seng F, Ley K (1972) Angew Chem Int Ed 11:1009
50. Suzuki H, Kawakami T (1997) Synthesis 8:855
51. Abushanab E (1970) DE Patent 1929541
52. Abushanab E (1970) J Org Chem 35:4279
53. Zarranz B, Jaso A, Aldana I, Monge A, Maurel S, Deharo E, Jullian V, Sauvain M (2005) Arzneim-Forsch 55:754
54. Cerecetto H, González M, Lavaggi ML, Azqueta A, López de Ceráin A, Monge A (2005) J Med Chem 48:21
55. Wang S, Miller W, Milton J, Vicker N, Stewart A, Charlton P, Mistry P, Hardick D, Denny WA (2002) Bioorg Med Chem Lett 12:415
56. Balicki R, Chmielowiec U (2000) Monats Chem 131:1105
57. Hahn WE, Lesiak J (1985) Pol J Chem 59:627
58. Balicki R, Cybulski M, Maciejewski G (2003) Synth Commun 33:4137
59. Balicki R (1990) Chem Ber 123:647
60. Kaczmarek L, Malinowski M, Balicki R (1988) Bull Soc Chim Belg 97:787
61. Malinoswaki M (1987) Synthesis 732
62. Novácek L, Nechvátal M (1988) Collect Czech Chem Commun 53:1302
63. Ahmed Y, Qureshi MI, Habib MS, Farooqi MA (1987) Bull Chem Soc Jpn 60:1145
64. Kotovskaya SK, Perova NM, Charushin VN, Chupakhin ON (1999) Mendeleev Commun 9:76
65. Kim HS, Hur JH (2004) J Korean Chem Soc 48:385
66. Amin KM, Ismail MMF, Noaman N, Soliman DH, Ammar YA (2006) Bioorg Med Chem 14:6917
67. Karayannis NM, Speca AX, Chasan DE, Pytlewski LL (1976) Coord Chem Rev 20:37
68. Matrosov EI, Starikova ZA, Khodak AA, Nifant'ev EE (2003) Zh Neorg Khim 48:2008
69. Torre MH, Gambino D, Araujo J, Cerecetto H, González M, Lavaggi ML, Azqueta A, López de Ceráin A, Monge A, Abram U, Costa-Filho AJ (2005) Eur J Med Chem 40:473
70. Noblía P, Vieites M, Torre MH, Costa-Filho AJ, Cerecetto H, González M, Lavaggi ML, Adachi Y, Sakurai H, Gambino D (2006) J Inorg Biochem 100:281
71. Martínez-Crespo FJ, Palop JA, Sáinz Y, Narro S, Senador V, González M, López de Ceráin A, Monge A, Hamilton E, Barker AJ (1996) J Heterocycl Chem 33:1671
72. Haddadin MJ, Issidorides CH (1967) Tetrahedron Lett 8:753
73. de Vries H, Bojarski J, Donker AA, Bakri A, Beyersbergen van Henegouwen GM (1990) Toxicol 63:85
74. Jarrar AA (1978) J Heterocycl Chem 15:177
75. Kawata H, Kikuchi K, Kokubun H (1983) J Photochem 21:343
76. Namba Y, Oda T, Watanabe T (1963) Bull Chem Soc Jpn 36:79

77. Cong Q, Lin S, Wang H (1989) Jiegou Huaxue 8:31
78. Jiang FQ, Hu YZ (2006) Acta Crystallogr E62:o649
79. Greer ML, Duncan JR, Duff JL, Blackstock SC (1997) Tetrahedron Lett 38:7665
80. Kraessig R, Bergmann D, Fliegen N, Kummer F, Seiffert W, Zimmermann H (1970) Berichte Bunsen-Ges 74:617
81. Andreev VP, Ryzhakov AV (1993) Khim Geterotsikl Soedin 1662
82. Limbach HH, Seiffert W, Ohmes E, Zimmermann H (1970) Berichte Bunsen-Ges 74:966
83. Stefaniak L (1976) Spectrochim Acta 32A:345
84. Guenther H, Gronenborn A (1978) Heterocycles 11:337
85. Panasyuk PM, Melnikova SF, Tselinskii IV (2005) Chem Heterocycl Comp 41:909
86. Brodskii AE, Gordienko LL, Degtyarev LS (1968) Electrochim Acta 13:1095
87. Kubota T, Nishikida K, Miyazaki H, Iwatani K, Oishi Y (1968) J Am Chem Soc 90:5080
88. Barqawi KR, Atfah MA (1987) Electrochim Acta 32:597
89. Ames JR, Houghtaling MA, Terrian DL (1992) Electrochim Acta 37:1433
90. Stüewe A, Weber-Schäefer M, Baumgäertel H (1972) Tetrahedron Lett 13:427
91. Stüewe A, Weber-Schäefer M, Baumgäertel H (1974) Berichte Bunsen-Ges 78:309
92. Koldasheva EM, Strelets VV, Tse YK, Geletii YV, Shestakov AF (1996) Izvestiya Akademii Nauk, Seriya Khimicheskaya 1992
93. Nishikida K, Kubota T, Miyazaki H, Sakata S (1972) J Mag Reson 7:260
94. Kulakovskaya SI, Kulikov AV, Berdnikov VM, Ioffe NT, Shestakov AF (2002) Electrochim Acta 47:4245
95. Stuewe A, Baumgaertel H (1974) Berichte Bunsen-Ges 78:320
96. Koldasheva EM, Shestakov AF, Geletii YV, Shilov AE (1992) Izvestiya Akademi Nauk, Seriya Khimicheskaya 845
97. Razskazovskiy Y, Close DM (2006) Res Chem Intermed 32:625
98. Acree WE, Powell JR, Tucker SA, Ribeiro da Silva MDMC, Matos MAR, Goncalves JM, Santos LMNBF, Morais VMF, Pilcher G (1997) J Org Chem 62:3722
99. Ribeiro da Silva MDMC, Gomes JRB, Gonçalves JM, Sousa EA, Pandey S, Acree WE (2004) Org Biomol Chem 2:2507
100. Ribeiro da Silva MDMC, Gomes JRB, Goncalves JM, Sousa EA, Pandey S, Acree WE (2004) J Org Chem 69:2785
101. Gomes JRB, Sousa EA, Gonçalves JM, Monte MJS, Gomes P, Pandey S, Acree WE, Ribeiro da Silva MDMC (2005) J Phys Chem B 109:16188
102. Gomes JRB, Sousa EA, Gomes P, Vale N, Gonçalves JM, Pandey S, Acree WE, Ribeiro da Silva MDMC (2007) J Phys Chem B 111:2075
103. Gomes JRB, Ribeiro da Silva MDMC, Ribeiro da Silva MAV (2006) Chem Phys Lett 429:18
104. Novacek L, Polasek L (1982) Biol Chem Zivocisne Vyroby – Vet 18:17
105. Oda M, Sekizawa Y, Watanabe T (1966) Appl Environ Microbiol 14:365
106. McIlwain H (1943) Biochem J 37:265
107. Carta A, Corona P, Loriga M (2005) Curr Med Chem 12:2259
108. Elina AS, Musatova IS, Padejskaya EN, Pershin GN (1993) RU Patent 677 320
109. Ono M, Chen YSE, Marks DF (1997) JP Patent 09 100 231
110. Wu H (2005) CN Patent 1 687 038
111. Zhao R, Wang Y, Xue F, Xu Z, Li J, Li J, Yan X, Du X, Miao X (2006) CN Patent 1 785 979
112. Takahata T, Sekimura N, Sumiyoshi Y (2006) JP Patent 2 006 241 111
113. Council Directive. OJ No L160/34, 20.6.1987:1–2

114. Suter W, Rosselet A, Knuesel F (1978) Antimicrob Agents Chemother 13:770
115. Fadeeva NI, Padeiskaya EN, Degtyareva IN, Pershin GN (1978) Farmakologiya i Toksikologiya 41:613
116. Usai D, Sanna P, Sechi LA, Carta A, Paglietti G, Zanetti S (2004) L'Igiene Moderna 121:289
117. Carta A, Paglietti G, Rahabar Nikookar ME, Sanna P, Sechi L, Zanetti S (2002) Eur J Med Chem 37:355
118. Carta A, Loriga M, Paglietti G, Mattana A, Fiori PL, Mollicotti P, Sechi L, Zanetti S (2004) Eur J Med Chem 39:195
119. OMS. WHO annual report on global TB control summary (2003) Weekly Epidemiol Rec 78:121
120. Sainz Y, Montoya ME, Martínez-Crespo FJ, Ortega MA, López de Ceráin A, Monge A (1999) Arzneim-Forsch 49:55
121. Ortega MA, Sáinz Y, Montoya ME, López de Ceráin A, Monge A (1999) Pharmazie 54:24
122. Ortega MA, Sáinz Y, Montoya ME, Jaso A, Zarranz B, Aldana I, Monge A (2002) Arzneim-Forsch 52:113
123. Montoya ME, Sáinz Y, Ortega MA, López de Ceráin A, Monge A (1998) Farmaco 53:570
124. Ortega MA, Montoya ME, Jaso A, Zarranz B, Tirapu I, Aldana I, Monge A (2001) Pharmazie 56:205
125. Zarranz B, Jaso A, Aldana I, Monge A (2003) Bioorg Med Chem 11:2149
126. Jaso A, Zarranz B, Aldana I, Monge A (2003) Eur J Med Chem 38:791
127. Jaso A, Zarranz B, Aldana I, Monge A (2005) J Med Chem 48:2019
128. Zanetti S, Sechi L, Molicotti P, Cannas S, Bua A, Deriu A, Carta A, Paglietti G (2005) Int J Antimicrob Agents 25:179
129. WHO Thirteenth Program Report: UNDP/World Bank/World Health Organization program for research and training in tropical diseases (1997) Geneva
130. Aldana I, Ortega MA, Jaso A, Zarranz B, Oporto P, Giménez A, Monge A (2003) Pharmazie 58:68
131. Zarranz B, Jaso A, Lima LM, Aldana I, Monge A, Maurel S, Sauvain M (2006) Braz J Pharm Sci 42:357
132. Cerecetto H, Di Maio R, González M, Risso M, Saenz P, Seoane G, Denicola A, Peluffo G, Quijano C, Olea-Azar C (1999) J Med Chem 42:1941
133. Aguirre G, Cerecetto H, Di Maio R, González M, Montoya ME, Jaso A, Zarranz B, Ortega MA, Aldana I, Monge A (2004) Bioorg Med Chem Lett 14:3835
134. Urquiola C, Vieites M, Aguirre G, Marín A, Solano B, Arrambide G, Noblía P, Lavaggi ML, Torre MH, González M, Monge A, Gambino D, Cerecetto H (2006) Bioorg Med Chem 14:5503
135. Overgaard J (1989) Int J Radiat Biol 56:801
136. Lin AJ, Cosby LA, Shansky CW, Sartorelli AC (1972) J Med Chem 15:1247
137. Cerecetto H, González M (2001) Mini-Rev Med Chem 1:219
138. Daniels JS, Gates KS (1996) J Am Chem Soc 118:3380
139. Zamalloa E, Aldana I, Bachiller CM, Monge A (1997) Arzneim-Forsch 47:873
140. Zamalloa E, Dios-Viéitez C, González-Peñas E, Monge A, Aldana I (1997) Arzneim-Forsch 47:1044
141. Vieites M, Noblía P, Torre MH, Cerecetto H, González M, Lavaggi ML, Da Costa Filho AJ, López de Ceráin A, Monge A, Parajón-Costa B, Gambino D (2006) J Inorg Biochem 100:1358
142. Lin PS, Ho KC (1998) J Nucl Med 39:677

143. Zhang W, Wu Y, Lu W, Li W, Sun X, Shen X, Li X, Tang L (1995) Zhongguo Yaowu Huaxue Zazhi 5:242
144. Gali-Muhtasib HU, Haddadin MJ, Rahal DN, Younes IH (2001) Oncol Rep 8:679
145. Cerecetto H, González M, Lavaggi ML, Aravena MA, Rigol C, Olea-Azar C, Azqueta A, López de Ceráin A, Monge A, Bruno AM (2006) Med Chem 2:511
146. Ganley B, Chowdhury G, Bhansali J, Daniels JS, Gates KS (2001) Bioorg Med Chem 9:2395
147. Chowdhury G, Kotandeniya D, Daniels JS, Barnes CL, Gates KS (2004) Chem Res Toxicol 17:1399
148. Diab-Assef M, Haddadin MJ, Yared P, Assaad C, Gali-Muhtasib HU (2002) Mol Carcinog 33:198
149. Gali-Muhtasib HU, Diab-Assaf M, Haddadin MJ (2005) Cancer Chem Pharmacol 55:369
150. Azqueta A, Pachón G, Cascante M, Creppy EE, López de Ceráin A (2005) Mutagenesis 20:165
151. Azqueta A, Pachón G, Cascante M, Creppy EE, López de Ceráin A (2007) Chem-Biol Interact 168:95
152. Zarranz B, Jaso A, Aldana I, Monge A (2004) Bioorg Med Chem 12:3711
153. Gali-Muhtasib H, Sidani M, Geara F, Mona AD, Al-Hmaira J, Haddadin MJ, Zaatari G (2004) Int J Oncol 24:1121
154. Nagai K, Hecht SM (1991) J Biol Chem 266:23994
155. Helissey P, Giorgi-Renault S, Colson P, Houssier C, Bailly C (2000) Bioconjugate Chem 11:219
156. Cerecetto H, Dias E, Di Maio R, González M, Pacce S, Saenz P, Seoane G, Suescun L, Mombrú A, Fernández G, Lema M, Villalba J (2000) J Agric Food Chem 48:2995
157. Jingzhong M, Chaonan H, Shengwei Z, Jianhong L, Hong J (2005) Xiandai Huagong 25:34
158. Kang IY, Kim SY, Kim HS, Huh K (1990) Yakhak Hoechi 34:112
159. Leach SC, Weaver RD, Kinoshita KL, William W (1981) J Electroanal Chem Interf Electrochem 129:213
160. Ryan MD, Scamehorn RG, Kovacic P (1985) J Pharm Sci 74:492
161. Crawford PW, Scamehorn RG, Hollstein U, Ryan MD, Kovacic P (1986) Chem-Biol Interact 60:67
162. Schoenfelder D, Stumm G, Bohle M, Niclas J (1988) Pharmazie 43:837
163. Gomes JRB, Ribeiro da Silva MDMC, Ribeiro da Silva MAV (2006) Chem Phys Lett 429:18
164. Hashimoto T, Negishi T, Namba T, Hayakawa S, Hayatsu H (1979) Chem Pharm Bull 27:1954
165. Voogd CE, Van der Stel JJ, Jacobs JJJAA (1980) Mut Res 78:233
166. Negishi T, Tanaka K, Hayatsu H (1980) Chem Pharm Bull 28:1347
167. Beutin L, Preller E, Kowalski B (1981) Antimicrob Agents Chemother 20:336
168. Cihak R, Srb V (1983) Mut Res 116:129
169. Cihak R, Vontorkova M (1983) Mut Res 117:311
170. Yoshimura H (2002) Toxicol Lett 129:115
171. Hao L, Chen Q, Xiao X (2006) Mut Res Fund Mol Mech Mut 599:21
172. Fuchs T, Gates KS, Hwang JT, Greenberg MM (1999) Chem Res Toxicol 12:1190
173. Inbaraj JJ, Motten AG, Chignell CF (2003) Chem Res Toxicol 16:164
174. Crosby J, Milner JA (1979) UK Patent 2 010 850
175. Graves DF, Engels HW (1988) EU Patent 289 752
176. Yamamoto T (1997) Int Appl Patent 9 715 607

177. Morioka Y, Sato M, Iwasa S, Iriyama J, Nakahara K, Suguro M (2003) JP Patent 2 003 115 297
178. Iriyama J, Morioka Y, Iwasa S, Nakahara K, Sato M, Suguro M (2003) JP Patent 2 003 178 801
179. Morioka Y, Iwasa S, Iriyama J, Yamamoto H, Nakahara K, Suguro M, Sato M (2004) JP Patent 2 004 192 829

Top Heterocycl Chem (2007) 11: 213–229
DOI 10.1007/7081_2007_087
© Springer-Verlag Berlin Heidelberg
Published online: 6 September 2007

Quinoline Analogs as Antiangiogenic Agents and Telomerase Inhibitors

Mahmud Tareq Hassan Khan[1,2]

[1]School of Molecular and Structural Biology, and Department of Pharmacology,
Institute of Medical Biology, University of Tromsø, 9037 Tromsø, Norway
mahmud.khan@fagmed.uit.no

[2]*Present address:*
Pharmacology Research Lab., Faculty of Pharmaceutical Sciences,
University of Science and Technology, Chittagong, Bangladesh

Abstract Quinoline and quindoline analogs have been shown to exhibit a wide variety of pharmacological activities, including on different types of cancers. This chapter begins by reviewing some general aspects of quindoline and its analogs, including their potential pharmacological and other general uses. The chapter then details several of the latest findings on the activities of quindoline analogs against a number of specific cancers and tumor subtypes, as well as angiogenesis. Methods of synthesizing several of these analogs, which were proposed by subsequent authors, are also discussed.

Keywords Angiogenesis · Cancer · Isoquinoline · Quindoline · Quinoline · Skraup synthesis · Telomerase

Abbreviations
ALT   alternative lengthening of telomere
CAM   chicken chorioallantoic membrane
CD   circular dichroism
EMSA   electrophoretic mobility shift assay

HUVEC human umbilical vein endothelial cell
MS   multiple sclerosis
TRAP   telomerase repeat amplification protocol

1
Introduction

Quinoline, 1-azanaphthalene and isoquinoline are aromatic nitrogen compounds characterized by a double-ring structure (as shown in Fig. 1) that
contains a benzene ring fused to pyridine at two adjacent carbon atoms [1].
These compounds are stable. Quinoline is a high-boiling liquid and isoquinoline is a low-melting solid, each with sweetish odor [2].

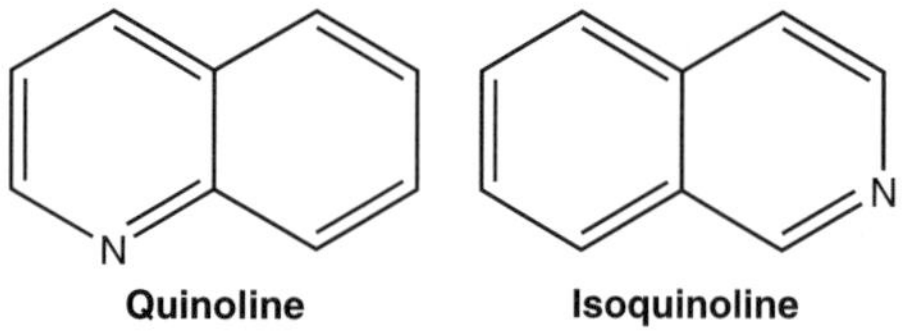

Quinoline   **Isoquinoline**

Fig. 1 Basic structures of quinoline and isoquinoline

The simplest member of the quinoline family is quinoline itself. It is a hygro-
scopic, yellowish oily liquid that is slightly soluble in water, soluble in alcohol,
ether, carbon disulfide, and readily soluble in many other organic solvents.
It can be obtained through the distillation of coal tar [2]. Quinoline can be
prepared from aniline with acrolein under heated sulfuric acid (the Skraup synthesis [3–6], shown in Scheme 1) [1]. Isoquinoline was also isolated from coal
tar in 1885 [2]. Both of these bases have been known about for a long time.

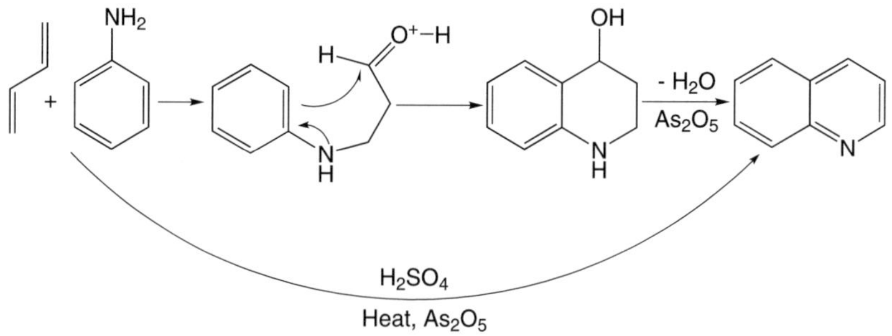

Scheme 1 Synthesis of quinoline, starting from acrolein and aniline (Skraup synthesis) [7]

Various quinoline compounds can be prepared by a series of Skraup syntheses utilizing different oxidizing agents. Isoquinoline differs from quinoline
at the N position (at 2) [1].

Quinoline derivatives can also be synthesized by [4 + 2] cycloaddition reactions (shown in Scheme 2) [8].

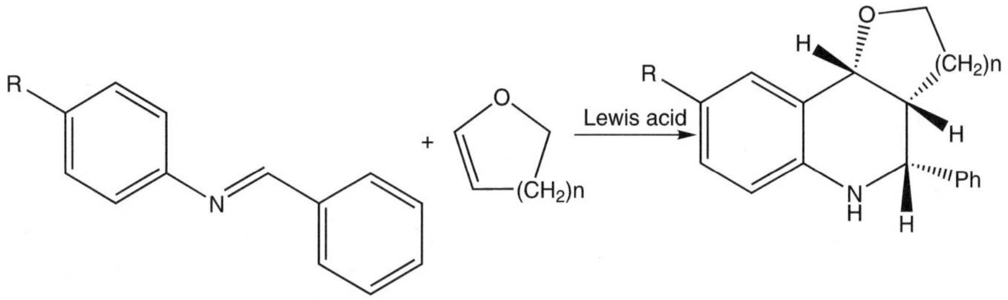

Scheme 2 Synthesis of quinoline derivatives by [4 + 2] cycloaddition reaction [8]

The synthesis of the quinoline ring system by an intermolecular [4 + 2] cycloaddition involves the Schiff base acting as a diene, with the styrene derivative acting as a dienophile. The above cycloaddition can also be performed using a cyclic enol ether as a dienophile in order to construct a fused quinoline ring system; this process is interesting not only from the point of view of extending the ring, but also because of the occurrence of a number of physiologically interesting quinoline alkaloids as natural products [8].

A new synthetic route to quinoline N-oxides was proposed by Wrobel et al. in 1993 (shown in Scheme 3), via cyclization of alkylidene O-nitroarylacetonitriles [9].

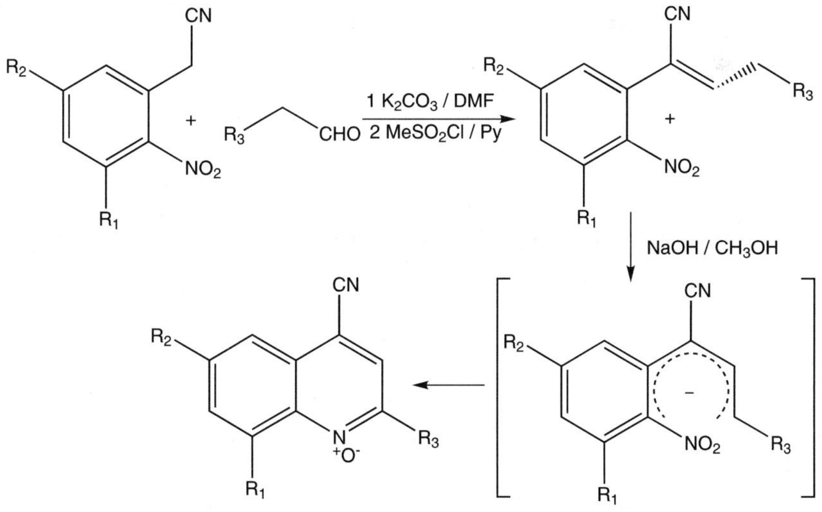

Scheme 3 Synthetic route to quinolines N-oxides via cyclization of alkylidene O-nitroarylacetonitriles [9]

Substituted quinoline N-oxides have been prepared via the base-induced cyclization of alkylidene derivatives of O-nitroarylacetonitriles, which are readily available via the vicarious nucleophilic substitution cyanomethylation of nitroarenes followed by Knoevenagel condensation [9].

2
Applications of Some Quinolines

Quinoline compounds are widely used as parent compounds to synthesize other molecules of pharmaceutical interest, especially antimalarial drugs, fungicides, biocides, alkaloids, dyes, rubber chemicals and flavoring agents. They have antiseptic, antipyretic, and antiperiodic properties. They are also used as catalysts, corrosion inhibitors, preservatives, and as solvents for resins and terpenes. They are used in transition metal complex catalysis chemistry for uniform polymerization, in luminescence chemistry, and as antifoaming agents in refinery field. Quinaldine, 2-methylquinoline, is used as an antimalarial agent and to prepare other antimalarial drugs. It is used in the manufacturing of oil-soluble dyes, food colorants, pharmaceuticals, pH indicators and other organic compounds. Quinaldic acid, a quinoline with a carboxylic acid at position 2, is a catabolite of tryptophan (an aromatic side chain amino acid). Quinazoline, 1,3-diazanaphthalene, is used as a chemical intermediate when synthesizing medicines and other organic compounds. The quinazoline structure is present in some antihypertensive agents, such as prazosin and doxazosin, which are peripheral vasodilators. Quinoxaline, 1,4-diazanaphthalene, is used as a chemical intermediate to synthesize fungicides and other organic compounds [1].

The 3-(2-furoyl)quinoline-2-carbaldehyde has been used as a fluorogenic reagent for the analysis of primary amines by liquid chromatography with laser-induced fluorescence detection [10, 11].

3
Pharmacological Activities of Some Quinoline Analogs

Tens of thousands (around 71 720) of biological uses of quinoline and its derivatives have been described in research articles in PubMed (www.pub med.gov) from 1893 to today.

Imiquimod (Aldara™, a heterocyclic imidazoquinoline amide (the structure of which is shown in Fig. 2), is an immune response modifier that has potent antiviral and antitumor activities [12–16]. Imiquimod and resiquimod (the structure of which is also shown in Fig. 2) have been shown to be topical immune response modifiers [17]. Imiquimod is approved for the treatment of actinic keratoses, superficial basal cell carcinomas and warts; it has also

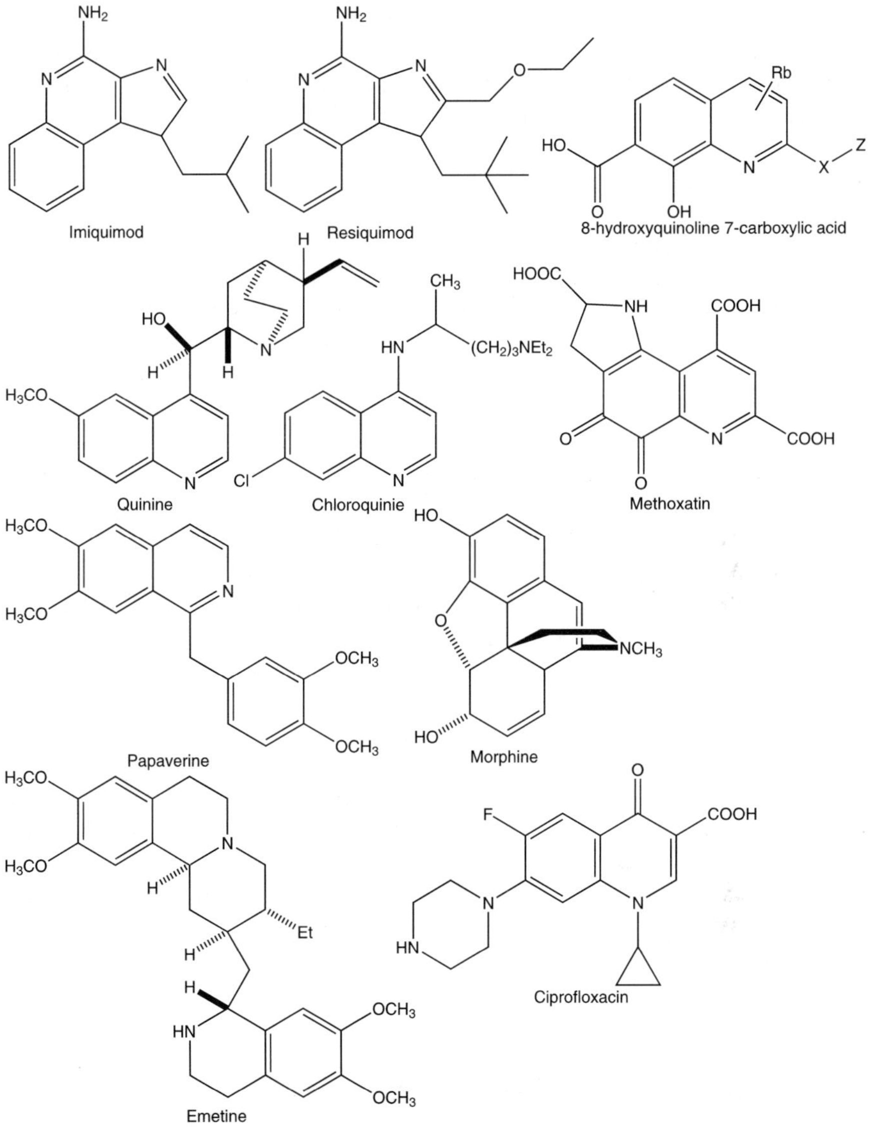

Fig. 2 Molecular structures of some commonly found quinoline analogs with diverse pharmacological activities

been documented as being used to successfully treat other forms of skin cancer, such as squamous cell carcinoma in situ, melanoma, and extramammary Paget's disease [17]. Both also proved to be novel immunomodulators [18].

8-Hydroxyquinoline 7-carboxylic acid (structure shown in Fig. 2) derivatives have been shown to be useful for producing integrase-inhibiting medicaments, to be capable of blocking viral replication in the stages preceding integration, and if appropriate at this integration stage, these medicaments can be used to treat retroviral pathologies, in AIDS [19]. Figure 2 shows the structural features of some commonly found quinoline analogs that are pharmacologically active and find clinical use.

4
General Methods of Synthesizing Some Quinolines Used Pharmacologically

Chloroquine (structure shown in Fig. 2) is a commonly used synthetic antimalarial drug. In 1946, Surrey and Hammer [20] proposed a method of synthesizing it, which is shown in Scheme 4 [2].

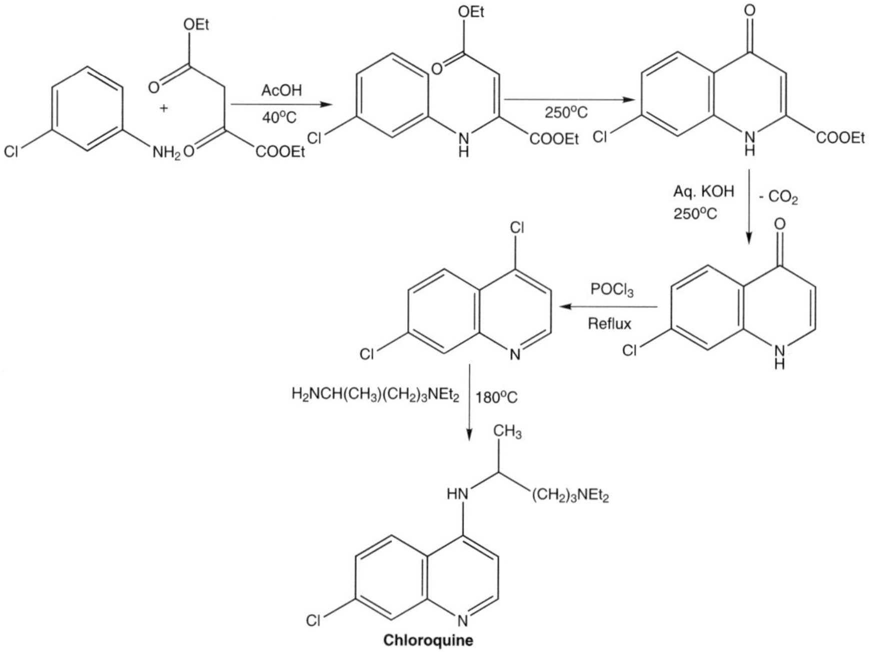

Scheme 4 Steps involved in the synthesis of chloroquine proposed by Surrey and Hammer [2, 20]

Papaverine is a commonly known opium alkaloid that is a smooth muscle relaxant and is used as a coronary vasodilator [2]. Scheme 5 shows the synthetic route to papaverine proposed by Pictet and Gams [21] almost a century ago [2].

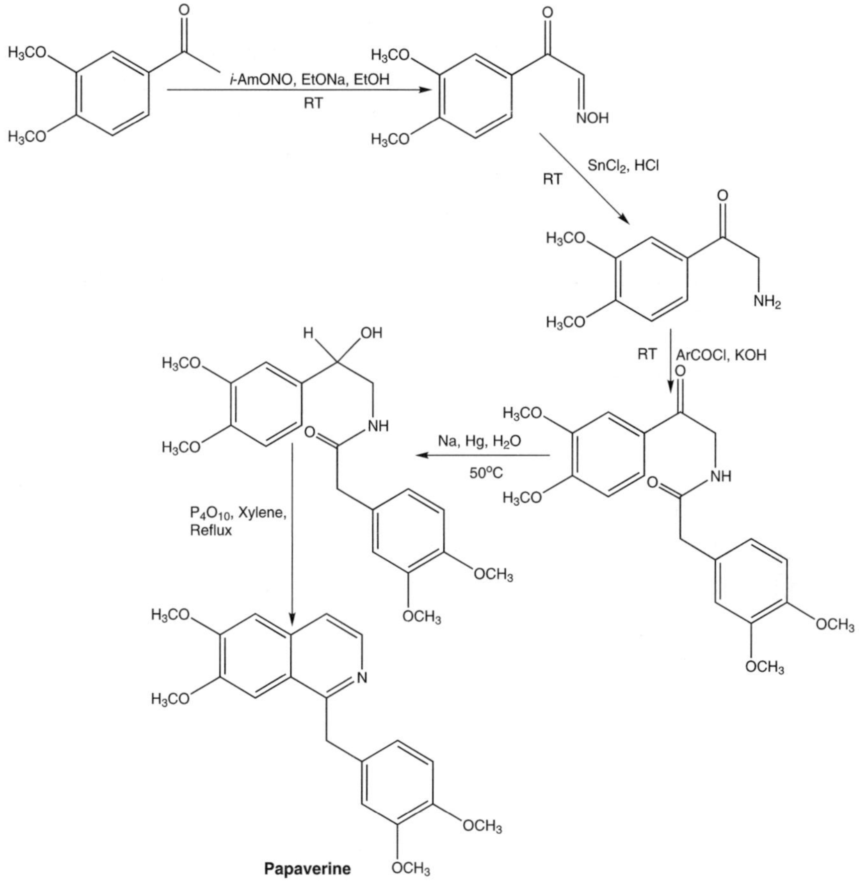

Scheme 5 Steps involved in the synthesis of papaverine proposed by Pictet and Gams [2, 21]

5
Recently Reported Quinolines Derivatives That Show Anticancer Activities

In recent years, large numbers of quinoline derivatives have been synthesized and their various significant biological activities, including against different types of cancers, have been reported. The following section provides some examples of these very important novel quinoline derivatives and their anticancer properties, as well as some discussion about synthetic pathways to them.

5.1
Human Telomerase Inhibitors

Telomeres are the specialized structures at the ends of eukaryotic chromosomes. Their role is to prevent chromosomal degradation, end to-end fusion and rearrangement. However, they shorten after each cellular division because of incomplete DNA replication, and so they act as a mitotic clock for permanent proliferation arrest or entry into senescence in normal somatic cells. The short telomeres are perceived as being damaged DNA that lead to p53/ATM pathway activation. In tumor cells, a ribonucleoprotein complex termed telomerase enables telomere elongation. This complex—which is composed of two main components, the telomerase RNA component or hTR, the RNA template for telomere synthesis, and telomerase reverse transcriptase, the catalytic subunit—is reactivated in the majority of cancers, including those of the lung [22]. Telomerase stabilisation may be required for cells to escape replicative senescence and to enable them to proliferate indefinitely. Because of a very strong association between telomerase and malignancy, both clinicians and pathologists expect this molecule to be a useful diagnostic and prognostic marker and a new therapeutic target [23].

Current standard cancer therapies (chemotherapy and radiation) often cause serious adverse off-target effects. Drug design strategies are therefore being developed that will target cancer cells more precisely for destruction while leaving surrounding normal cells relatively unaffected. Telomerase, which is widely expressed in most human cancers but almost undetectable in normal somatic cells, provides an exciting and broad-spectrum therapeutic target for cancer treatment [24, 25]. Therapeutic strategies for inhibiting telomerase activity have involved both targeting components of telomerase (the protein component, TERT, or the RNA component, TERC) or directly targeting telomere DNA structures. A combination telomerase inhibition therapy has also been studied recently. The TERT promoter has been used to selectively express cytotoxic gene(s) in cancer cells, and a TERT vaccine for immunization against telomerase has been tested. The 10% to 15% of immortalized cancer cells that do not express telomerase use a recombination-based mechanism to maintain telomere structure, which has been called the "alternative lengthening of telomeres" (ALT). In view of the increasing study of telomerase inhibitors as anticancer treatments, it is crucially important to determine whether inhibition of telomerase will select for cancer cells that activate ALT mechanisms of telomere maintenance [25].

Agents that stabilize G-quadruplexes have the potential to interfere with telomere replication by blocking the elongation step catalyzed by telomerase or the telomerase-independent mechanism, and could therefore act as antitumor agents [26–30].

Caprio et al. (in 2000) reported the synthesis and inhibitory profile of a novel quinoline derivative against human telomerase [27]. The molecule bis-dimethylaminoethyl is a derivative of quindoline (10H-indolo[3,2-b]quinoline) (for structure see Fig. 3), an alkaloid from the West African shrub *Cryptolepis sanguinolenta*, as reported by the authors [27]. The compound exhibits moderate cytotoxicity and inhibitory activity against telomerase. Utilizing molecular modelling technology, the authors hypothesized that the inhibitory

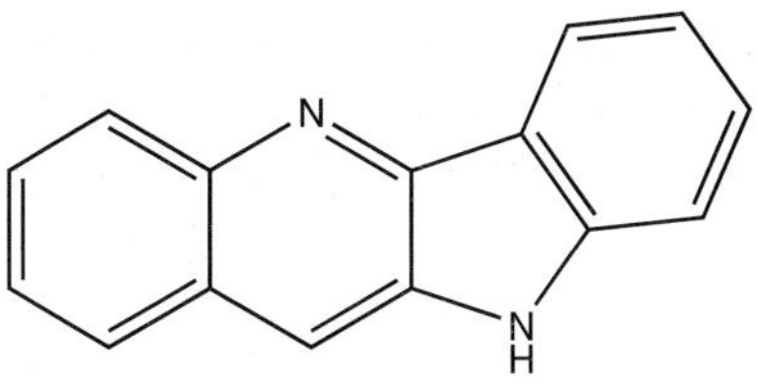

Quindoline

Fig. 3 Molecular structure of quindoline

activity against telomerase was due to the stabilization of an intermediate guanine–quadruplex complex [27].

The synthetic method reported by Caprio et al. is shown in Scheme 6.

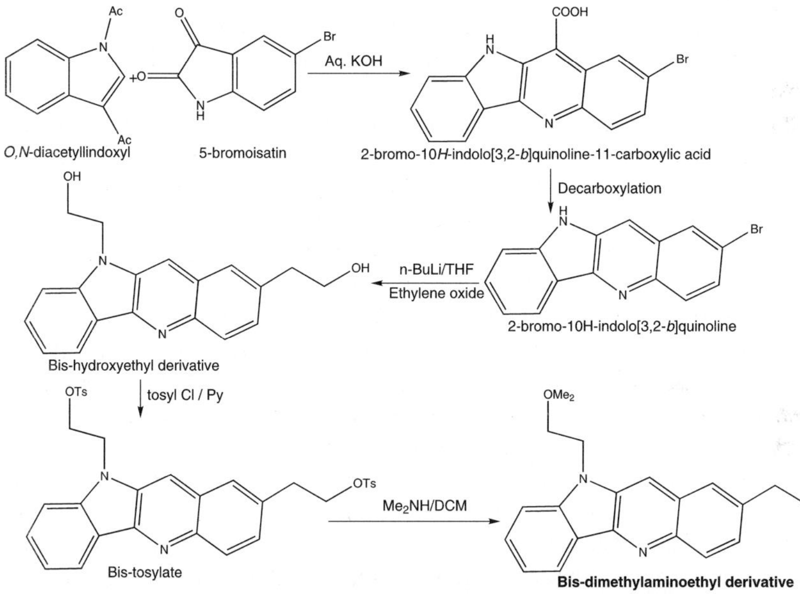

Scheme 6 Steps involved in the synthesis of bis-dimethylaminoethyl derivative, a novel human telomerase inhibitor, as reported by Caprio et al. (in 2000) [27]

Zhou and coworkers reported a new series of quindoline derivatives (**1–10**) that could be used to develop novel and potent telomerase inhibitors [31]. The structure of quindoline is shown in Fig. 3.

The interactions of the G-quadruplex of human telomere DNA with these newly designed molecules have been examined via CD spectroscopy and electrophoretic mobility shift assay (EMSA). The selectivity between the quindoline derivative and G-quadruplex or duplex DNA has been investigated by competition dialysis. These new compounds have also been investigated as inhibitors of telomerase through the utilization of the modified telomerase repeat amplification protocol (TRAP) assay [31]. The synthetic steps involved are shown in Scheme 7.

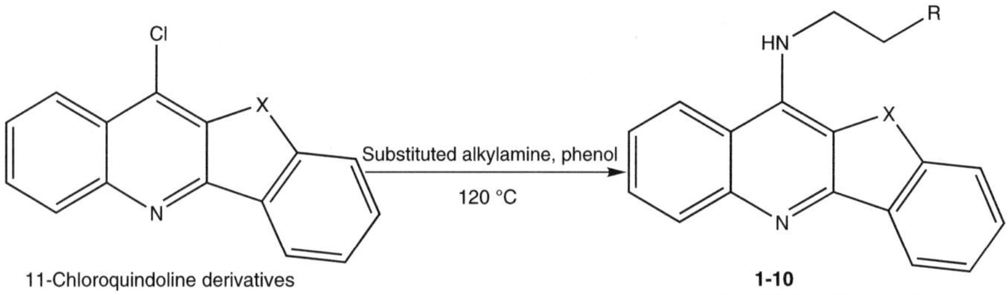

11-Chloroquindoline derivatives

1-10
1 X = NH, R = Morphoryl
2 X = NH, R = $N(CH_3)_2$
3 X = NH, R = $CH_2N(CH_3)_2$
4 X = NH, R = OH
5 X = NH, R = CH_2OH
6 X = O, R = Morphoryl
7 X = O, R = $N(CH_3)_2$
8 X = O, R = $CH_2N(CH_3)_2$
9 X = O, R = OH
10 X = O, R = CH_2OH

Scheme 7 Steps involved in the synthesis of a new series of quindoline derivatives (**1–10**) that yield potent telomerase inhibition [31]

The results from Zhou et al. revealed that the introduction of electron-donating groups such as substituted amino groups at the C-11 position of the quindoline significantly enhanced the ability of the molecule to inhibit telomerase activity ($IC_{50} > 138\ \mu M$ for quindoline, 0.44–12.3 μM for quindoline derivatives **1–10**). The quindoline derivatives not only stabilized the G-quadruplex structure but also induced the G-rich telomeric repeated DNA sequence to fold into a quadruplex [31].

The authors suggested that the quindoline derivatives might be potential lead compounds for the development of new telomerase inhibitors [26, 31].

5.2
Antiangiogenic Agents

Angiogenesis is the formation of new blood vessels from pre-existing ones. It has been shown that, without angiogenesis, a tumor can only reach the size of 1–2 mm^3, small enough to be easily resected or treated with conventional cytotoxic chemotherapeutic agents [32]. In vitro, linomide inhibited endothelial cell proliferation and migration [33, 34] as well as invasion through basement membrane at high concentration (> 100 mg/mL) [32, 33].

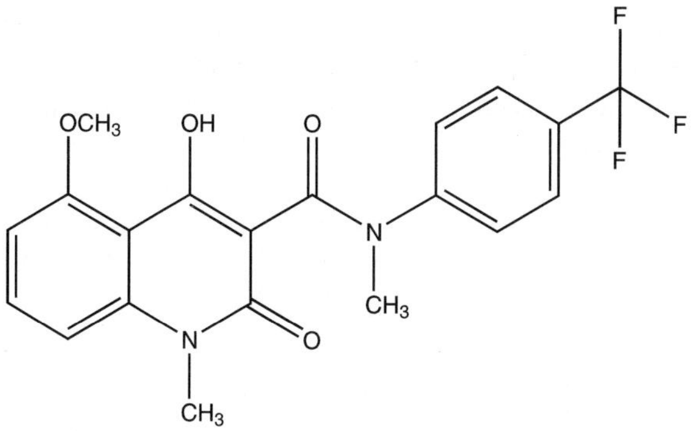

Fig. 4 Molecular structure of tasquinimod, a second-generation orally active quinoline-3-carboxamide analog [36]

Table 1 Molecular features of the linomide analogs studied by Isaac et al. that exhibit activity against prostate cancer [35]

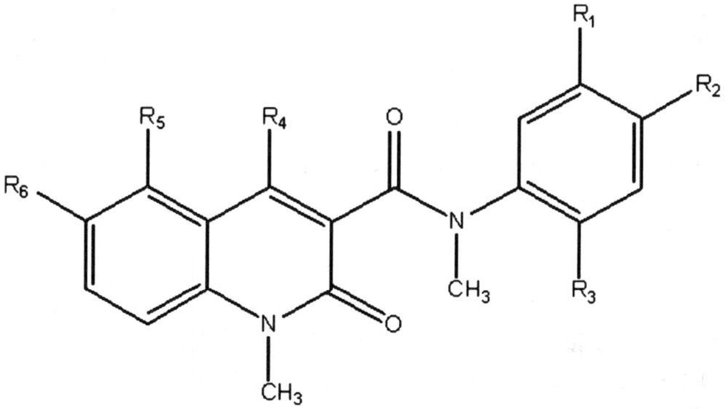

Analogs	R_1	R_2	R_3	R_4	R_5	R_6
Linomide	H	H	H	OH	H	H
ABR-215025	H	H	H	OH	OCH_3	H
ABR-215093	F	H	F	OH	OCH_3	H
ABR-215050	H	CF_3	H	OH	OCH_3	H
ABR-215851	F	F	H	OH	$-OCH_2O$	$-OCH_2O$
ABR-215681	F	F	H	OH	SCH_3	H
ABR-215060	H	Cl	H	OH	Cl	H
ABR-217518	H	CH_3	H	OH	Cl	H
ABR-217365	H	OCH_3	H	OH	$N(CH_3)_2$	H
ABR-216974	H	H	H	NH_2	CH_3	H
ABR-217032	H	H	H	NH_2	OCH_3	H
ABR-217031	H	H	H	$NHCH_3$	OCH_3	H

Tasquinimod (structure shown in Fig. 4) is a second-generation orally active quinoline-3-carboxamide analog with enhanced potency against prostate cancer via its antiangiogenic activity. It is presently undergoing clinical trials [35]. Androgen ablation and taxanes are standard therapies for metastatic prostate cancer [36].

Linomide (N-phenylmethyl-1,2-dihydro-4-hydroxyl-1-methyl-2-oxo-quinoline-3-carboxamide, structure given in Table 1) has been proven to be an immunomodulator [32]. In clinical trials, it has been shown to be a potential treatment for autoimmune diseases such as rheumatoid arthritis, systemic lupus erythematosis and multiple sclerosis [37–39]. It has also been reported to possess antiangiogenic activity [33, 40].

Like tasquinimod, linomide yields robust and consistent in vivo growth inhibition of prostate cancer models due to its antiangiogenic activity, and it inhibits of autoimmune encephalomyelitis models of multiple sclerosis (MS). MS clinical trials were discontinued because of unacceptable toxicity, due to dose-dependent induction of proinflammation [35].

Recently (in 2006) Isaac and coworkers screened linomide analogs to determine their in vivo potency for inhibiting growth of the Dunning R-3327 AT-1 rat prostate cancer model in rats, and their potency for inhibiting angiogenesis in a Matrigel assay in mice [35]. The structures of the studied linomide analogs are shown in Table 1.

Table 2 Structural features and structure–activity relationship of the compounds 11–16 [32]

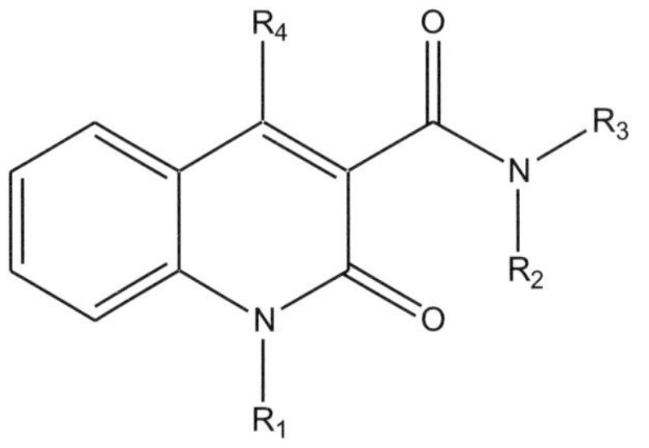

Analogs	R_1	R_2	R_3	R_4	ED$_{50}$ (μg/embryo)	IC$_{50}$ (in HUVEC proliferation, in μM)
Linomide	CH$_3$	CH$_3$	Phenyl	OH	60.1	13.95
11	CH$_3$	CH$_3$	CH$_3$	OH	8.5	16.47
12	CH$_3$	H	Phenyl	OH	18.0	7.2
13	H	CH$_3$	Phenyl	OH	7.5	6.55
14	H	H	Phenyl	OH	0.87	4.0
15	CH$_3$	CH$_3$	Phenyl	H	> 100	20.58
16	H	H	Phenyl	H	> 100	22.26

Based upon its greater potency (i.e., 30- to 60-fold more potent than lino-mide) in assays and the lack of any resulting proinflammation in Beagle dog, ABR-215050 (tasquinimod, structure shown in Fig. 4 as well as in Table 1) was characterized for dose–response ability in the growth inhibition of a series of four additional human and rodent prostate cancer models in mice [35]. Phar-macokinetic analysis following oral dosing indicated that blood and tumor tissue levels of ABR-215050 as low as 0.5–1 µM are therapeutically effect-ive [35].

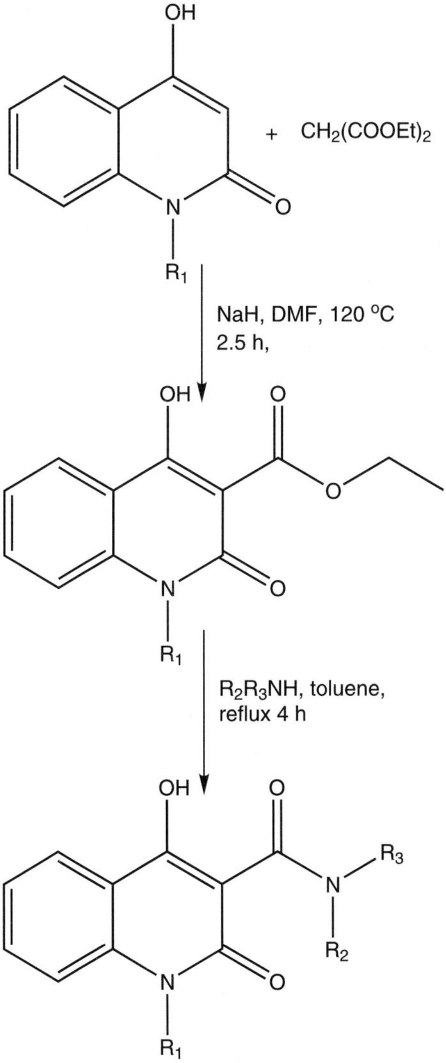

Linomide and compounds 11-14

Scheme 8 Steps involved in the syntheses of linomide and compounds **11–14** [32]

Compound efficacy has been correlated with the inhibition of angiogenesis in a variety of assays (endothelial capillary tube formation, aortic ring assay, chorioallantoic membrane assay, real-time tumor blood flow and PO$_2$ measurements, tumor blood vessel density, and tumor hypoxic and apoptotic fractions) [35].

Shi et al. reported on the synthesis and antiangiogenic activities of linomide and its analogs (compounds 11–16) [32]. Three of the analogs are 3.3–69 times more potent than linomide at inhibiting blood vessel formation in the chicken chorioallantoic membrane (CAM) [41, 42] angiogenesis assay. These compounds possessed considerable antiproliferative activities against isolated human umbilical vein endothelial (HUVEC) cells along with with no

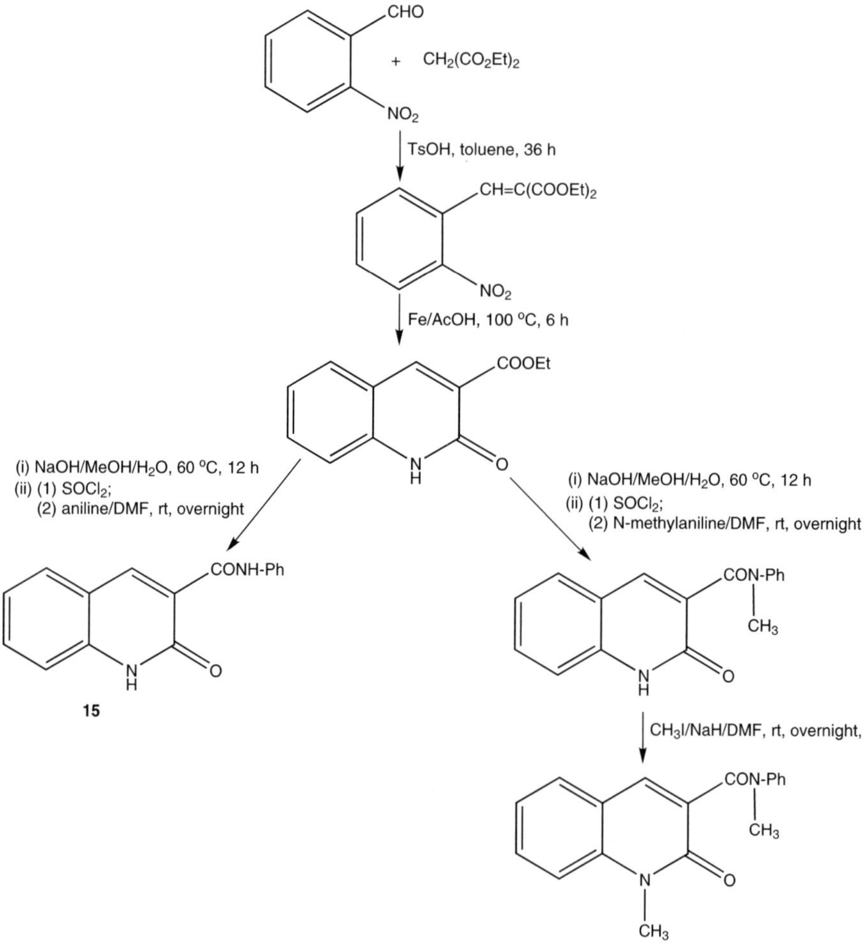

Scheme 9 Steps involved in the syntheses of compounds 15–16 [32]

activity against epithelial-derived prostate tumor cells [32]. The structures of the compounds **11-16**, and their structure–activity relationship, are shown in Table 1.

Syntheses of the compounds **11-14** and **15-16** are described in Schemes 8 and 9, respectively [32].

6
Conclusion

A large number of pharmacologically active molecules that have found clinical use have been synthesized through the derivatization of quinolines, while others are simply quinoline analogs. Some of the most recent studies of quindoline analogs in relation to anticancer activity were discussed above.

References

1. ChemicalLand21.com (2007) Quinoline product identification webpage. http://chemicalland21.com/industrialchem/organic/QUINOLINE.htm. (last visited: 8 July 2007)
2. Joule JA, Mills K (2000) Heterocyclic chemistry, 4th edn. Blackwell, Oxford, pp 121–145
3. Skraup ZH (1880) Eine Synthese des Chinolins. Berichte 13:2086–2087
4. Manske RHF (1942) Chem Rev 30:113
5. Manske RHF, Kulka M (1953) Org React 7:80–99
6. Wahren M (1964) Tetrahedron 20:2773
7. Canov M (2007) http://www.jergym.hiedu.cz/~canovm/mechanic/pravidl2/skr/sk.htm. (last visited: in February 2007)
8. Kametani T, et al. (1985) Synth Commun 15(6):499–505
9. Wrobel Z, Kwast A, Makosza M (1993) New synthesis of substituted quinoline N-oxides via cyclization of alkylidene O-nitroarylacetonitriles. Synthesis 31–32
10. Beale SC, et al. (1990) Application of 3-(2-furoyl)quinoline-2-carbaldehyde as a fluorogenic reagent for the analysis of primary amines by liquid chromatography with laser-induced fluorescence detection. J Chromatogr 499:579–587
11. Liu X, Yang LX, Lu YT (2003) Determination of biogenic amines by 3-(2-furoyl) quinoline-2-carboxaldehyde and capillary electrophoresis with laser-induced fluorescence detection. J Chromatogr A 998(1–2):213–219
12. Jobanputra KS, Rajpal AV, Nagpur NG (2006) Imiquimod. Indian J Dermatol Venereol Leprol 72(6):466–469
13. Eedy DJ (2002) Imiquimod: a potential role in dermatology? Br J Dermatol 147(1):1–6
14. Beutner KR, et al. (1998) Imiquimod, a patient-applied immune-response modifier for treatment of external genital warts. Antimicrob Agents Chemother 42(4):789–794
15. Beutner KR, et al. (1998) Treatment of genital warts with an immune-response modifier (imiquimod). J Am Acad Dermatol 38(2 Pt 1):230–239
16. Ho NT, Lansang P, Pope E (2007) Topical imiquimod in the treatment of infantile hemangiomas: a retrospective study. J Am Acad Dermatol 56(1):63–68

17. Woodmansee C, Pillow J, Skinner RB Jr (2006) The role of topical immune response modifiers in skin cancer. Drugs 66(13):1657–1664
18. Dockrella DH, Kinghorn GR (2001) Imiquimod and resiquimod as novel immunomodulators. J Antimicrob Chemother 48:751–755
19. Mousnier A, et al. (2005) Use of quinoline derivatives with anti-integrase effect and applications thereof (US Patent 20 050 261 336).
    http://www.freepatentsonline.com/20050261336.html. (last visited: 8 July 2007)
20. Surrey AR, Hammer HF (1946) J Am Chem Soc 68:113
21. Pictet A, Gams A (1909) Chem Ber 42:2943
22. Lantuejoul S, et al. (2007) Telomerase expression in lung preneoplasia and neoplasia. Int J Cancer 120(9):1835–1841
23. Sampedro Camarena F, Cano Serral G, Sampedro Santalo F (2007) Telomerase and telomere dynamics in ageing and cancer: current status and future directions. Clin Transl Oncol 9(3):145–154
24. Cunningham AP et al. (2006) Telomerase inhibition in cancer therapeutics: molecular-based approaches. Curr Med Chem 13(24):2875–2888
25. Siddiqa A, Cavazos DA, Marciniak RA (2006) Targeting telomerase. Rejuvenation Res 9(3):378–390
26. Zhou JM, et al. (2006) Senescence and telomere shortening induced by novel potent G-quadruplex interactive agents, quindoline derivatives, in human cancer cell lines. Oncogene 25(4):503–511
27. Caprio V, et al. (2000) A novel inhibitor of human telomerase derived from 10H-indolo[3,2-b]quinoline. Bioorg Med Chem Lett 10(18):2063–2066
28. Guyen B, et al. (2004) Synthesis and evaluation of analogues of 10H-indolo[3,2-b]quinoline as G-quadruplex stabilising ligands and potential inhibitors of the enzyme telomerase. Org Biomol Chem 2(7):981–988
29. Lemarteleur T, et al. (2004) Stabilization of the c-myc gene promoter quadruplex by specific ligands' inhibitors of telomerase. Biochem Biophys Res Commun 323(3):802–808
30. Yamakuchi M, et al. (1997) New quinolones, ofloxacin and levofloxacin, inhibit telomerase activity in transitional cell carcinoma cell lines. Cancer Lett 119(2):213–219
31. Zhou JL, et al. (2005) Synthesis and evaluation of quindoline derivatives as G-quadruplex inducing and stabilizing ligands and potential inhibitors of telomerase. J Med Chem 48(23):7315–7321
32. Shi J, et al. (2003) Structure–activity relationship studies of the anti-angiogenic activities of linomide. Bioorg Med Chem Lett 13(6):1187–1189
33. Vukanovic J, et al. (1993) Antiangiogenic effects of the quinoline-3-carboxamide linomide. Cancer Res 53(8):1833–1837
34. Parenti A, et al. (1996) The effect of linomide on the migration and the proliferation of capillary endothelial cells elicited by vascular endothelial growth factor. Br J Pharmacol 119(4):619–621
35. Isaacs JT, et al. (2006) Identification of ABR-215050 as lead second generation quinoline-3-carboxamide anti-angiogenic agent for the treatment of prostate cancer. Prostate 66(16):1768–1778
36. Dalrymple SL, Becker RE, Isaacs JT (2007) The quinoline-3-carboxamide anti-angiogenic agent, tasquinimod, enhances the anti-prostate cancer efficacy of androgen ablation and taxotere without effecting serum PSA directly in human xenografts. Prostate 67(7):790–797
37. Andersen O, et al. (1996) Linomide reduces the rate of active lesions in relapsing-remitting multiple sclerosis. Neurology 47(4):895–900

38. Andersen O, et al. (1996) Linomide reduces the rate of active lesions in relapsing-remitting multiple sclerosis. Mult Scler 1(6):348

39. Karussis DM, et al. (1996) Treatment of secondary progressive multiple sclerosis with the immunomodulator linomide: a double-blind, placebo-controlled pilot study with monthly magnetic resonance imaging evaluation. Neurology 47(2):341–346

40. Joseph IB, Vukanovic J, Isaacs JT (1996) Antiangiogenic treatment with linomide as chemoprevention for prostate, seminal vesicle, and breast carcinogenesis in rodents. Cancer Res 56(15):3404–3408

41. Sheu JR, et al. (1998) Effect of U-995, a potent shark cartilage-derived angiogenesis inhibitor, on anti-angiogenesis and anti-tumor activities. Anticancer Res 18(6A):4435–4441

42. Brooks PC, Montgomery AM, Cheresh DA (1999) Use of the 10-day-old chick embryo model for studying angiogenesis. Methods Mol Biol 129:257–269

Top Heterocycl Chem (2007) 11: 231–258
DOI 10.1007/7081_2007_070
© Springer-Verlag Berlin Heidelberg
Published online: 9 August 2007

Bioactive Furanosesterterpenoids from Marine Sponges

Yonghong Liu[1] (✉) · Si Zhang[1] · Jee H. Jung[2] · Tunhai Xu[3]

[1]Key Laboratory of Marine Bio-resources Sustainable Utilization,
  South China Sea Institute of Oceanology, Chinese Academy of Sciences,
  510-301 Guangzhou, China
  yonghongliu@scsio.ac.cn

[2]College of Pharmacy, Pusan National University, 609-735 Busan, Korea

[3]School of Chinese Materia Medica, Beijing University of Chinese Medicine,
  100029 Beijing, China

Abstract This review covers the literature published from January 1996 to December 2006 with 133 citations. The emphasis is on furanosesterterpenoids from marine sponges, together with their relevant biological activities, source organisms, and country of origin. The first total syntheses that led to the revision of structures or stereochemistries have been included.

Keywords Bioactive furanosesterterpenoids · Marine sponge

1
Introduction

The sesterterpenoids are a group of pentaprenyl terpenoid substances whose structures are derivable from geranylfarnesyl diphosphate. Although sesterterpenoids are a relatively small group of terpenoids, their sources are widespread, having been isolated from terrestrial fungi, lichens, higher plants, insects, and various marine organisms, especially sponges. Sesterterpenoids exhibit diverse biological properties, such as anti-inflammatory, cytotoxic, antifeedant, platelet aggregation, and antimicrobial effects. The structural conciseness and diverse bioactivity of sesterterpenoids have made them attractive targets for both biomedical and synthetic purposes [1, 2]. This review is of the literature from 1996 to 2006 and describes 245 furanosesterterpenoids from 133 articles. We show structures of furanosesterterpenoids, and previously reported furanosesterterpenoids where there has been a structural revision or a newly established stereochemistry. Previously reported furanosesterterpenoids for which first syntheses or new bioactivities are described are referenced, but separate structures are generally not shown. So far, only a few reviews have dealt with the class of sesterterpenoids: "Sesterterpenoids" [3–5] and "Heterocyclic terpenes: linear furano- and pyrroloterpenoids" [6]; other general reviews include: "Marine natural products" [1, 2]. References to other reviews are more appropriately placed in the following sections.

2
Linear Furanosesterterpenoids

Untenospongin B (1), which was isolated from the marine sponge *Hippospongia communis* collected from the Atlantic Coast of Morocco, possesses a broad and strong antimicrobial activity [7]. A *Hippospongia* sp. from Goa, India, contained *ent*-untenospongin A (2) [8]. A C_{21} difuranoterpene 3 from a Spanish Mediterranean specimen of *Spongia virgultosa* was found to be the enantiomer of (–)-untenospongin B from a Japanese *Hippospongia* sp. [9]. Comparison of spectral data suggested that the structure of tetradehydrofurospongin-1 from an Australian *Spongia* sp. be revised from 4 to 3 [9]. The absolute configuration of the C_{21} difuranoterpene 5, isolated from

Fasciospongia cavernosa from the Arabian Sea, has been determined by synthesis of its enantiomer [10]. A specimen of *Spongia officinalis* from La Caleta, Cádiz, Spain, contained two additional minor C_{21} furanoterpenes, the weakly cytotoxic furospongin-5 (**6**) and cyclofurospongin-2 (**7**) [11]. The C_{21} norsesterterpenoid **8** originally reported with conjugated double bonds has been revised to **9** on the basis of more complete spectroscopic data obtained from a sample isolated from an Australian specimen of *Spirastrella papilosa* [12]. The absolute stereochemistry was determined by degradation and the name (–)-isotetrahydrofurospongin-1 is proposed for this bisfuranoterpene. A C_{21} furanoterpene, nitenin (**10**), has been isolated from the Mediterranean sponge *Spongia virgultosa* [9]. Isonitenin (**11**) is an additional C_{21} difuranoterpene from a Spanish collection of *Spongia officinalis* [13]. *ent*-Kurospongin (**12**) was isolated from a Korean *Sarcotragus* species [14]. A synthesis of ircinin 4 (**13**) employed a palladium-catalyzed reaction as the key step [15].

A furanoterpene designated dehydrofurodendin (**14**) was isolated from two different species of Madagascan sponges of the genus *Lendenfeldia*, both of which seem to be as yet undescribed [16]. Dehydrofurodendin (**14**) has been found to be a potent inhibitor of the HIV-1 RT-associated DNA

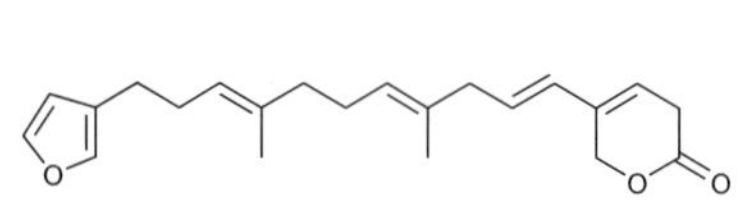

14

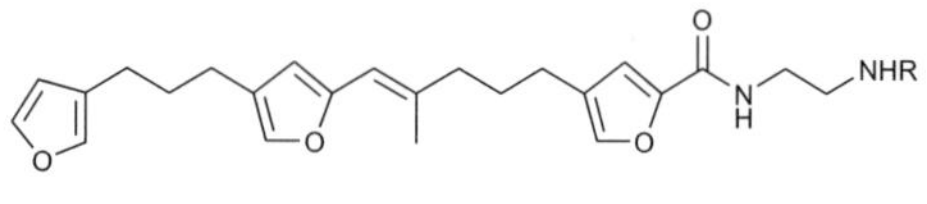

15 R = H
16 R = Ac

polymerase activities. It should be noted that dehydrofurodendin closely resembles furodendin. Whereas in furodendin there are five double bonds, in dehydrofurodendin there is an additional double bond at C-16–C-17, probably resulting from an oxidation occurring at a later stage. Hippospongins E (**15**) and F (**16**) are truncated furanosesterterpenes, which are amides that incorporate diaminoethane, from *Hippospongia* sp. of southern Australian waters [17]. A Korean collection of a sponge of the genus *Sarcotragus* was found to contain sarcotins I (**17**) and J (**18**) with a unique C_{22} trinorsesterterpene, which might be a degradation product of relevant sesterterpenes [18]. A moderately cytotoxic norsesterterpenoid, sarcotin N (**19**), was isolated from a Korean *Sarcotragus* species. The previously reported sarcotin I (**17**) was found to have the 21*R* configuration [14]. Four unprecedented C_{22} sesterterpenes, irciformonins A–D (**20–23**), have been isolated from the marine sponge *Ircinia formosana*, collected off the coast of eastern Taiwan. Irciformonins C and D exhibited significant cytotoxicity against human colon tumor cells [19]. Two trisnorsesterterpenoid lactams, the sarcotragins A (**24**) and B (**25**), were isolated from a *Sarcotragus* species from Jaeju Island in Korean waters [20]. Three norsesterterpenoids **26–28**, isolated from an Okinawa *Ircinia* species, were found to be moderately cytotoxic against KB cells [21]. A new C_{24} furanoterpene, hippospongin D (**29**) was isolated from a southern Australian collection of the marine sponge *Hippospongia* sp. [17]. A moderately cytotoxic norsesterterpenoid, sarcotin O (**30**), was isolated from a Korean *Sarcotragus* sp. [14].

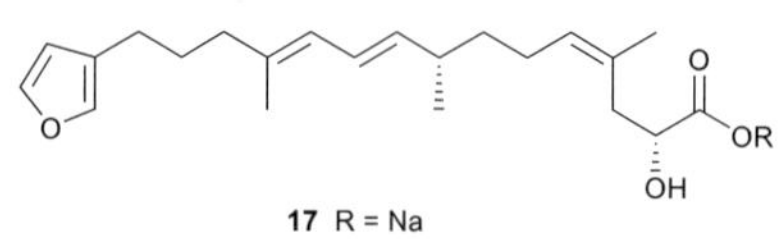

17 R = Na
18 R = CH₃

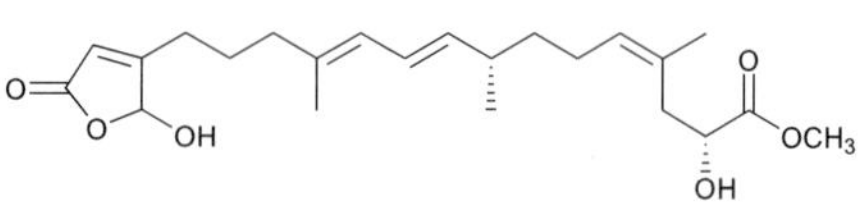

19

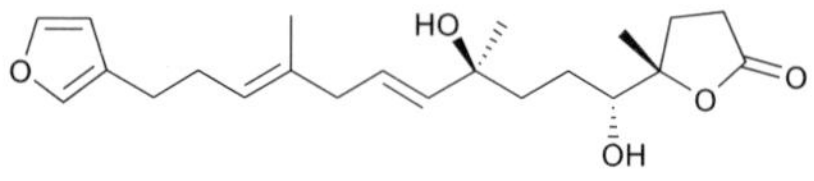

20

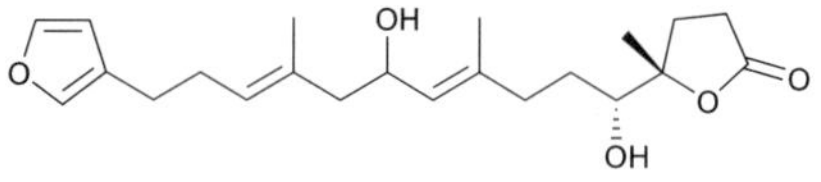

21

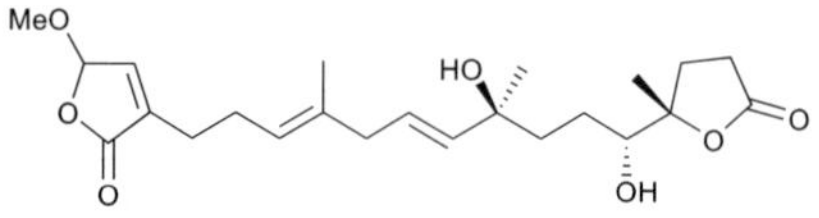

22

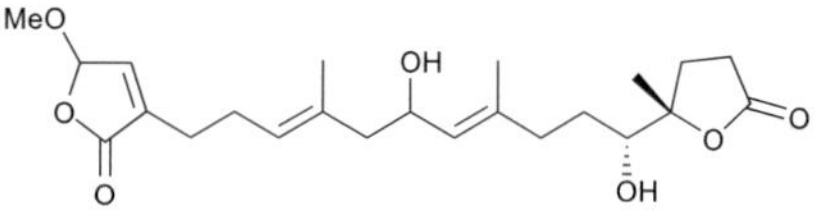

23

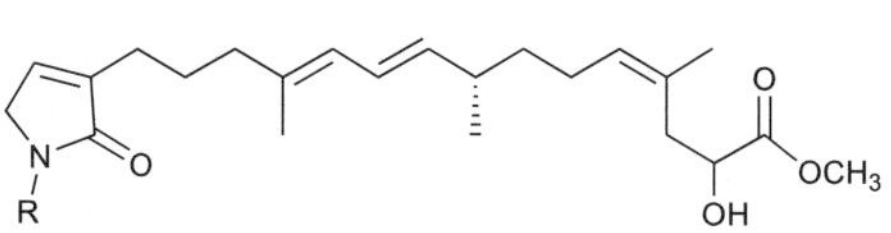

24 R = $\overset{\xi}{\sim}$ COONa

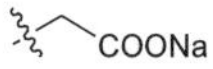

25 R =

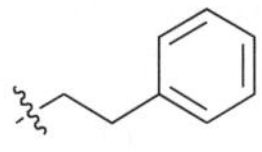

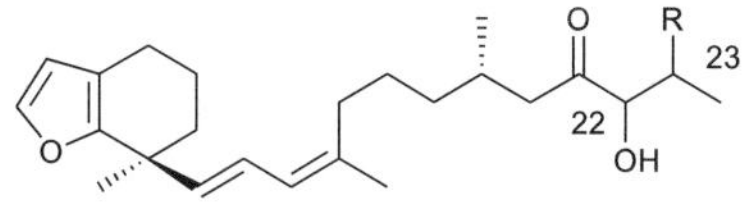

26 R = H

27 R = Cl, 22, 23 anti

28 R = Cl, 22, 23 syn

Both the (8*S*,21*S*,22*S*,23*R*) and (8*R*,21*S*,22*S*,23*R*) isomers of okinonellin B (**31**), which is a cytotoxic and antispasmodic agent from *Spongionella* sp., have been synthesized but neither has the same optical rotation as the natural product [22]. Cytotoxic furanosesterterpene isopalinurin (**32**) was reisolated from a marine sponge *Psammocinia* sp. collected from Ulleung Island, Korea. The configuration at C-21 has not been previously investigated, and it has now been determined to be *R* on the basis of CD analysis [23]. Three cytotoxic furanosesterterpenoids, the sacotins A–C (**33–35**), were reported from a specimen of *Sarcotragus* sp. collected at Cheju Island, Korea [24]. Cytotoxic sesterterpenoids, *epi*-sacotin A (**36**) and sacotins F (**37**) and M (**38**), were isolated from the same Korean *Sarcotragus* sp. [18]. *epi*-Sacotin F (**39**) [14] was found in two Korean *Sarcotragus* sp. Marine organisms, particularly sponges, have continued to provide the source of linear sesterterpenoids. The terminal units often comprise either a furan or

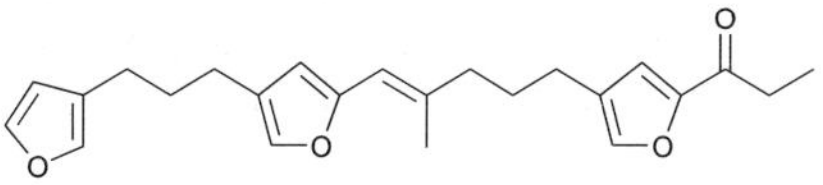

29

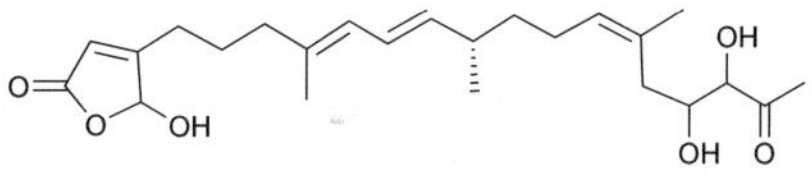

30

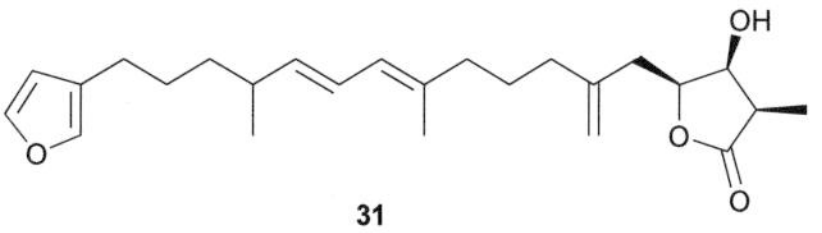

31

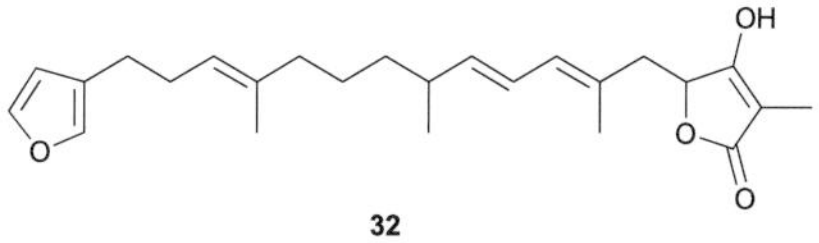

32

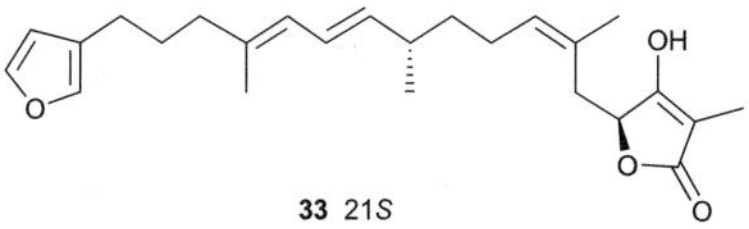

33 21*S*

36 21*R*

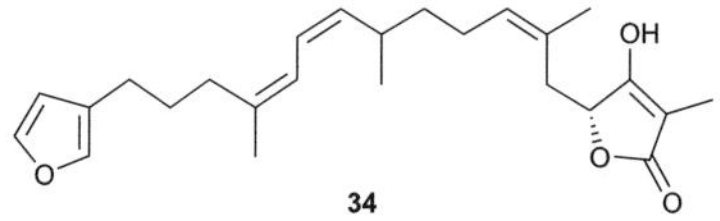

34

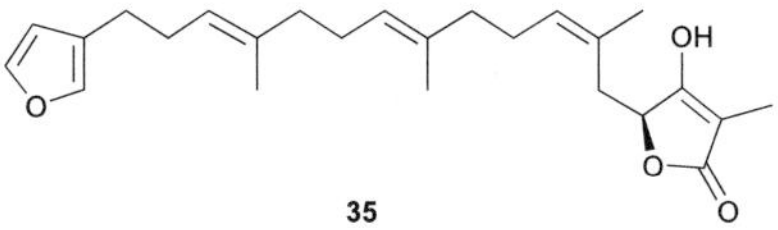

35

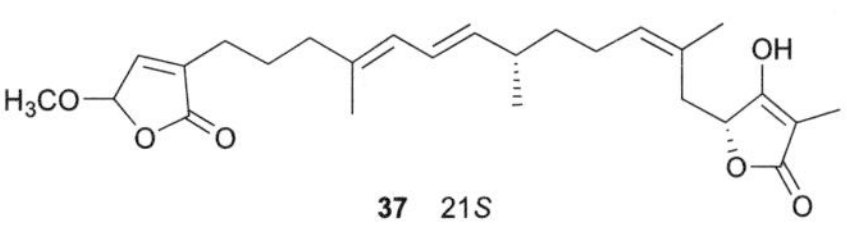

37 21*S*

39 21*R*

a tetronic acid moiety. Ircinin-1 (40) induces cell cycle arrest and apoptosis in SK-MEL-2 human melanoma cells [25]. Two sesterterpenoids 42 and 43, isolated from an Okinawa *Ircinia* species, were found to be moderately cytotoxic [21]. Two additional geometrical isomers, (8Z,13Z,20Z)-strobilinin (44) and (7Z,13Z,20Z)-felixinin (46), were partially characterized after isolation as their 22-O-acetates from Columbian specimens of *I. felix, I. strobilina*, and *I. campana* [26]. (8Z,13Z,18S,20Z)-Strobilinin (45) and (7E,12Z,18S,20Z)-variabilin (47) from a Maltese specimen of *I. oros* were characterized as their corresponding 22-O-methyl derivatives [27]. The first asymmetric synthesis of (18S)-variabilin (41), a furanosesterterpene from the sponge *I. variabilis*, has been reported [28]. An efficient and stereodefined process is described for the first preparation of the marine furanosesterterpene tetronic acid, (18S)-variabilin, featuring lipase-catalyzed asymmetric desymmetrization of two types of propanediol precursors incorporating the terpene skeleton. Three new 48–50 and known cytotoxic furanosesterterpenes, (7E,12E,18R,20Z)-variabilin, (8E,13Z,18R,20Z)-strobilinin, (7E,13Z,18R,20Z)-felixinin (also known as (7E,13Z,18R,20Z)-variabilin), (8Z,13Z,18R,20Z)-strobilinin, and (7Z,13Z,18R,20Z)-felixinin, were isolated from a Korean marine sponge *Psammocinia* sp. These compounds were evaluated for cyto-

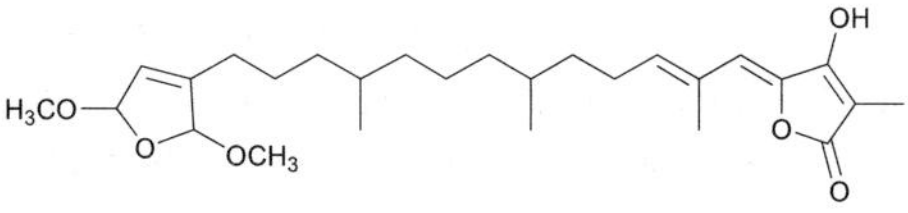

50 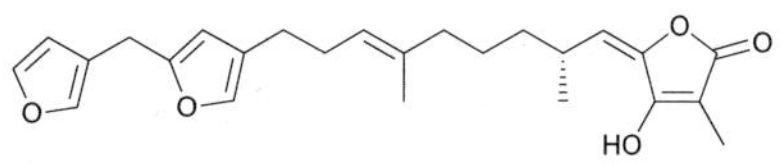   **51**

toxicity against a small panel of five human tumor cell lines, and most of the compounds showed toxicity to SK-MEL-2 [23]. The mixture of (8E,13Z,20Z)-strobilinin/(7E,13Z,20Z)-felixinin displayed significant inhibition of DNA replication and a moderate antioxidant profile [29]. Cytotoxic furanosesterterpenoids, ircinins 1 and 2, sacotins D (**51**) and E (**52**) [24], sacotins G (**53**) and H (**54**) [18], (7E,12E,18R,20Z)-variabilin, and (8E,13Z,18R,20Z)-strobilinin [14], were reported from a specimen of *Sarcotragus* sp. collected off Cheju Island, Korea. 22-Deoxy-23-hydroxymethylvariabilin (**55**) was isolated from the sponge *Fasciospongia* sp. [30].

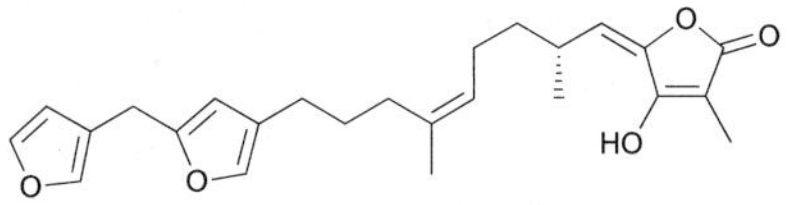

52 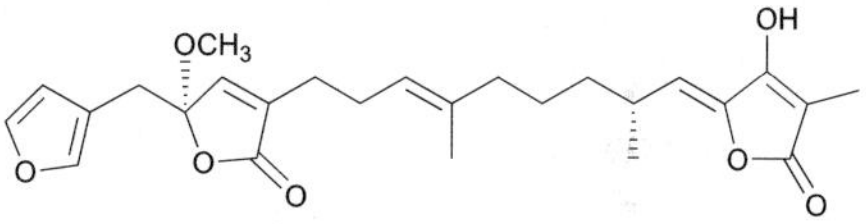   **53**

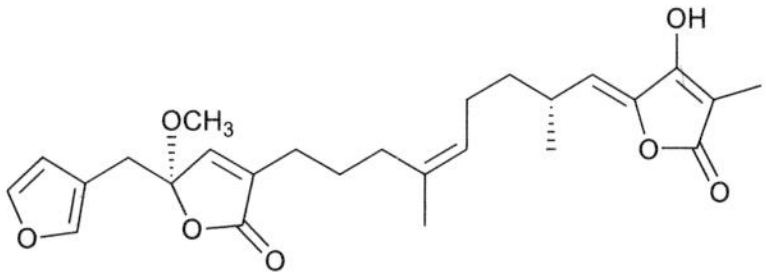

54 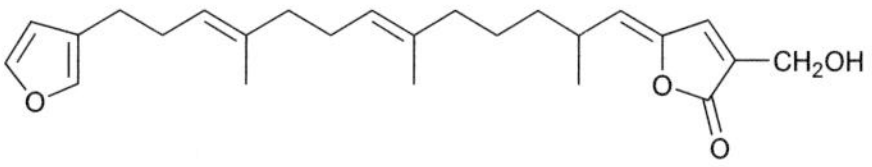   **55**

Idiadione (**56**) was synthesized from citronellal, establishing the stereogenic center as S [31]. Cytotoxic sesterterpenoid cacospongionolide D (**57**) was isolated from *Fasciospongia cavernosa* from the Bay of Naples [32]. Thorectolide monoacetate (**58**), obtained from a New Caledonian marine sponge identified as a *Hyrtios* species [33], in contrast possessed anti-inflammatory properties. The bicyclic lactone astakolactin (**59**) was obtained

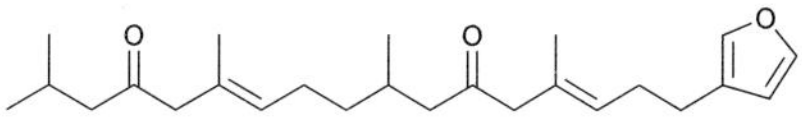

56 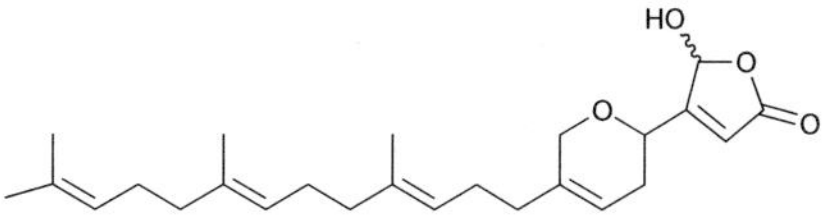   **57**

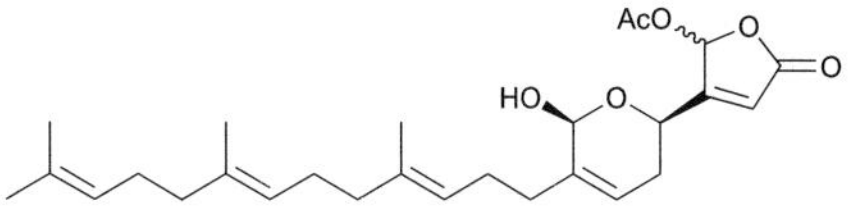

58 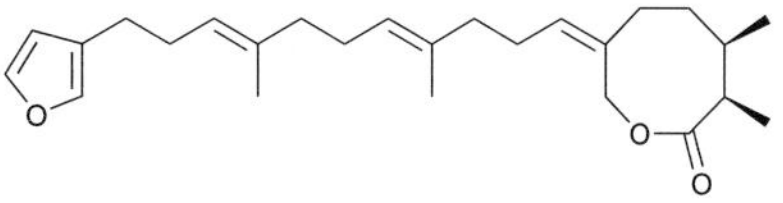   **59**

from specimens of *Cacospongia scalaris* collected from the Gulf of Astakos, Greece [34]. An *Acanthodendrilla* species (Indonesia) provided the cytotoxic acantholide C (**60**) [35]. Cytotoxic sesterterpenoid (6Z)-luffarin-V (**62**) was isolated from *Fasciospongia cavernosa* from the Bay of Naples [32]. Extracts of the marine sponge *Thorectandra* sp. have been found to contain cytotoxic luffarin R (**61**) and luffarin V (**62**) [36]. A new sesterterpene **63**, related to luffarins, has been isolated from the sponge *Fasciospongia cavernosa*, collected in the northern Adriatic Sea. The absolute stereochemistry was determined by application of Mosher's method [37]. Three C_{25} furanoterpenes, hippospongins A–C (**64–66**) were isolated from a southern Australian collection of the marine sponge *Hippospongia* sp. Only hippospongin A (**64**) was found to be a mild antibiotic, inhibiting the growth of *Staphylococcus au-*

reus [17]. Sesterterpenes sarcotrines A (**67**), B (**69**), C (**71**), and D (**73**), and *epi*-sarcotrines A (**68**), B (**70**), and C (**72**) [14, 18] were isolated from the marine sponge *Sarcotragus*. Two microbial metabolites of palinurin (reisolated from a Red Sea sponge *Ircinia echinata*), palinurine A (**74**) and B (**75**), are produced by the fungus *Cunninghamella* sp. NRRL 5695 [38]. The possible transformation of furans to amides (dehydro-3-enepyrrolidin-2- or -5-one) through a biomimetic reaction is certain to have applications in synthetic chemistry. Four unstable sulfate esters **76–79** of furanosesterterpenes were obtained from *I. variabilis* and *I. oros* of the northern Adriatic Sea [39]. The 22-*O*-sulfates **80** and **81** of palinurin and fasciculatin, which were toxic to brine shrimp *Artemia salina*, were isolated from *I. variabilis* and *I. fasciculata*, respectively [40]. A sodium salt of pyrroloterpenoid sarcotragin A (**24**) was isolated from *Sarcotragus* sp. collected from Jaeju Island in Korean waters [20]. A sodium salt of trinorsesterterpene acid sarcotin I (**17**) [18] and two sodium salts of terpenoids sarcotrines E (**82**) and F (**83**) were isolated from Korean sponge *Sarcotragus* sp.

The sponge *Ircinia fefix* from Columbia has yielded ten fatty acid esters: (18*R*)-variabilin-11-methyloctadecanoate, (7*E*,12*E*,18*R*,20*Z*)-variabilin (5*Z*,9*Z*)-22-methyltricosadienoate, (7*E*,12*E*,18*R*,20*Z*)-variabilin (5*Z*,9*Z*)-tetra-

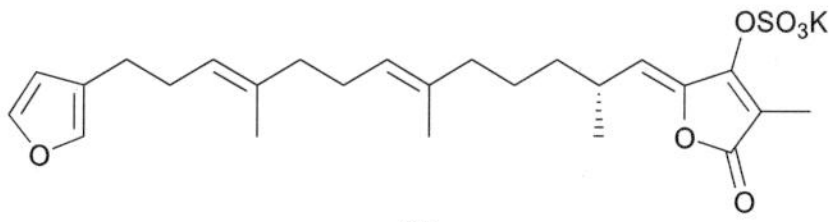

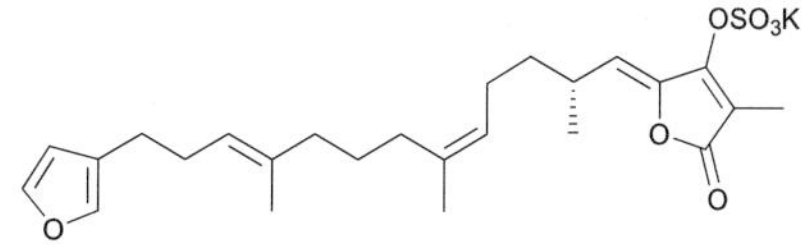

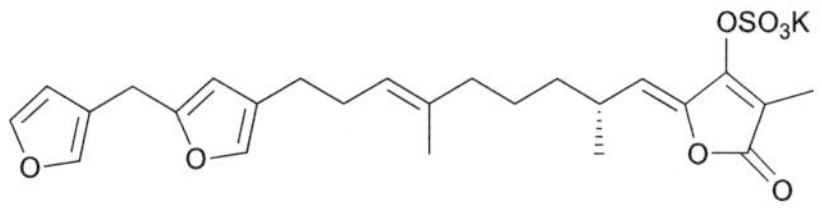

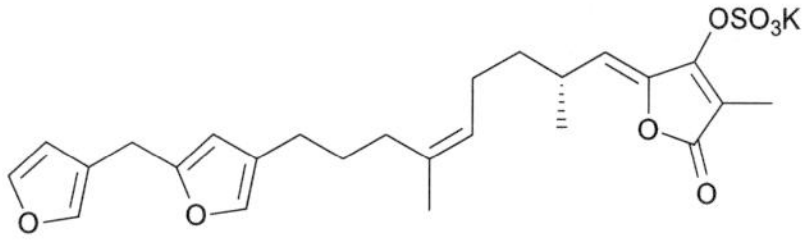

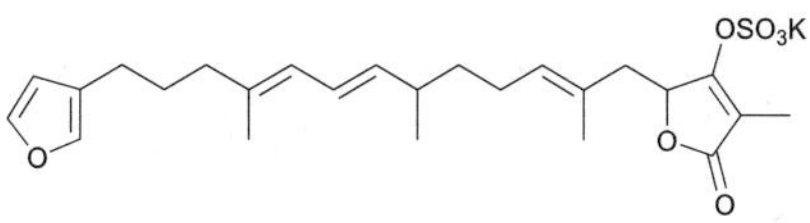

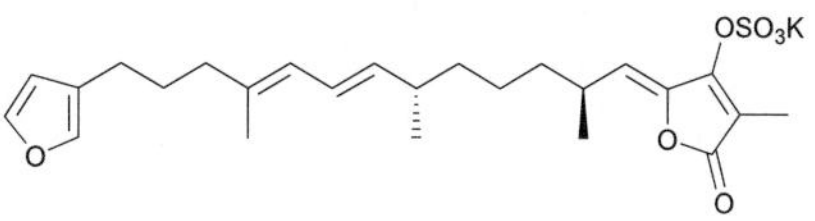

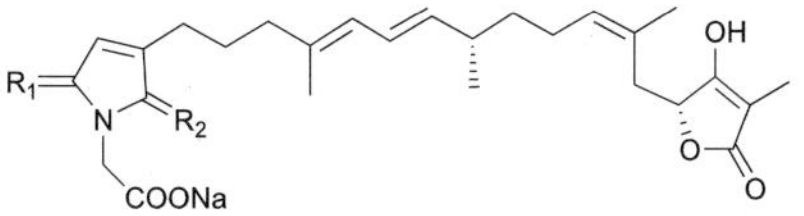

82 R$_1$ = H$_2$, R$_2$ = O

83 R$_1$ = O, R$_2$ = H$_2$

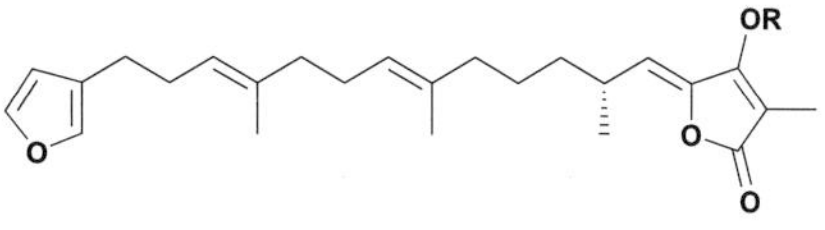

cosadienoate, (7*E*,12*E*,18*R*,20*Z*)-variabilin hexadecanoate, (7*E*,12*E*,18*R*,20*Z*)-variabilin 10-methylhexadecanoate, (7*E*,12*E*,18*R*,20*Z*)-variabilin 15-methyl-hexadecanoate, (7*E*,12*E*,18*R*,20*Z*)-variabilin 14-methylhexadecanoate, (7*E*, 12*E*,18*R*,20*Z*)-variabilin 9-octadecenoate, (7*E*,12*E*,18*R*,20*Z*)-variabilin octa-decanoate, and (7*E*,12*E*,18*R*,20*Z*)-variabilin 2,11-dimethyloctadecanoate (**84–93**) [41].

3
Monocarbocyclic Furanosesterterpenoids

In 1980 de Silva and Scheuer reported the isolation of manoalide (**94**) from the marine sponge *Luffariella rariabilis*, which soon afterwards proved to be a potent analgesic and anti-inflammatory agent. Since then, manoalide has been considered a very useful pharmacological tool, being by far the most well-known phospholipase $A_2(PLA_2)$ inhibitor. Several academic and indus-trial efforts have focused on developing analogues of manoalide to be used in anti-inflammatory therapy, and it has been the target of a number of synthe-ses [42–46]. The absolute stereochemistry of the known anti-inflammatory agent manoalide was determined by comparison of its CD spectrum with those of synthetic model compounds [47]. Cyclolinteinone (**95**), a sesterterpene from sponge *Cacospongia linteiformis,* prevents inducible nitric oxide synthase and inducible cyclo-oxygenase protein expression by blocking nuclear factor kappa B activation in J774 macrophages [48]. Two sesterterpenoids **96** and **97** were isolated from *Hyrtios* cf. *erecta* from Fiji but the major bioactive compounds of this sponge were known [49]. An Okinawan *Luffariella* species yielded two new luffariolides H (**98**) and J (**99**) that were found to be cytotoxic and antimicrobial [50]. An *Acanthodendrilla* species (Indonesia) provided the cytotoxic acantholides A (**100**) and B (**101**). Acantholide B (**101**) was antimicrobial [35]. (6*Z*)-Neomanoalide-24,25-diacetate (**102**) was obtained from a Palauan *Luffariella* species [51]. A *Brachiaster* species (Thailand) yielded the same manoalide congener **102**, the *E* isomer **103** [52]. Cyclolinteinol (**104**) was described as a macrophage activation in-

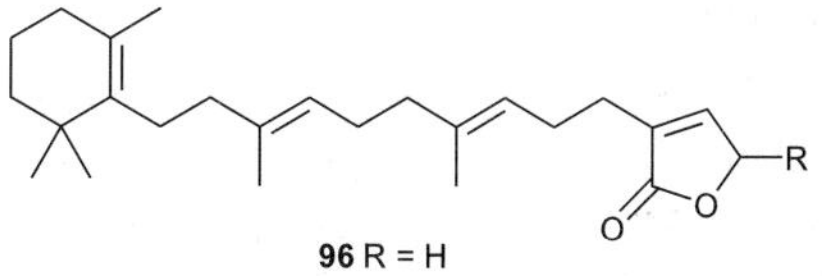

96 R = H

97 R = OEt

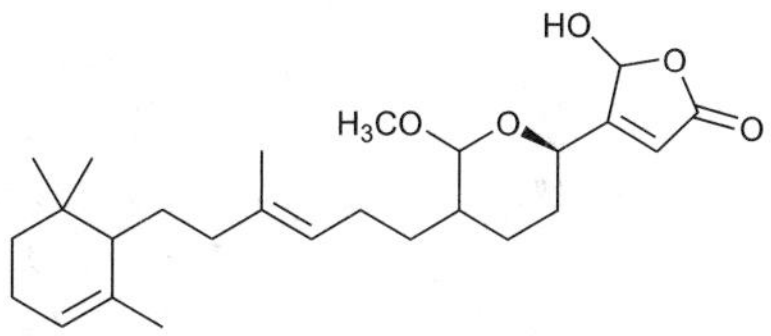

98

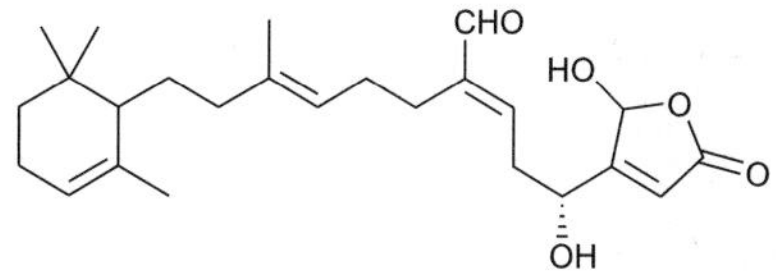

99

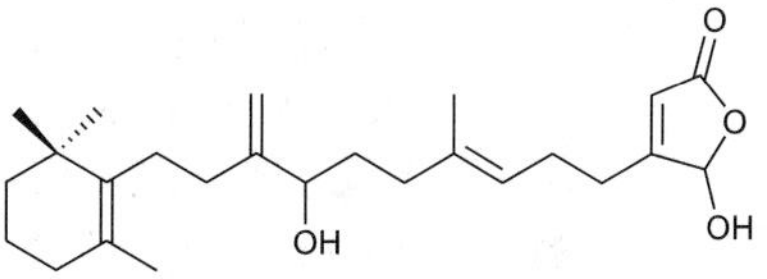

100

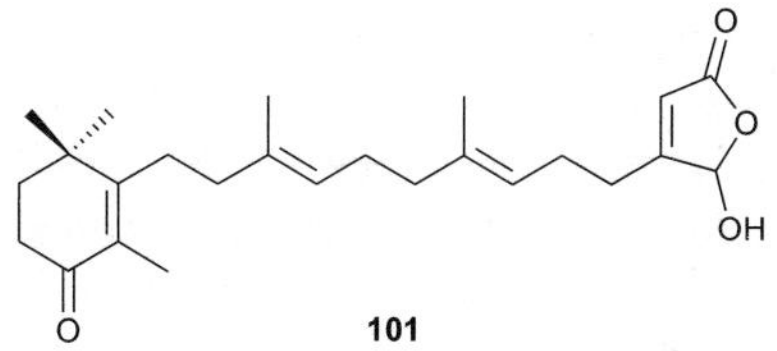

101

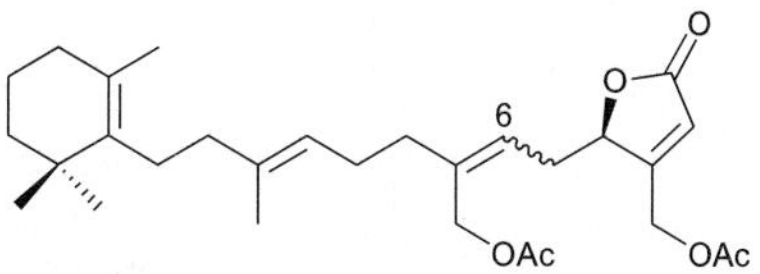

102 6*Z*

103 6*E*

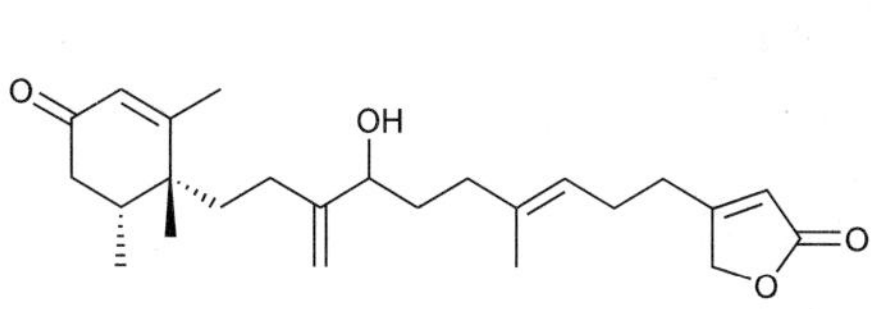

104

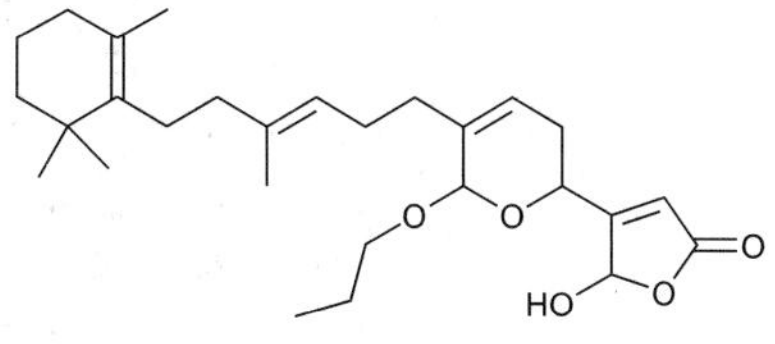

105

106

107 R = β OH
108 R = α OH

hibitor from the Caribbean sponge *Cacospongia* cf. *linteiformis* [53]. A new derivative of manoalide, 24-*n*-propyl-*O*-manoalide (**105**) isolated from *Luffariella* sp., showed significant cytotoxicity against the HCT-116 cell line by MTT assay [54]. A new sesterterpenoid **106**, related to luffarins, has been isolated from the sponge *Fasciospongia cavernosa* collected in the northern Adriatic Sea [37]. An *Acanthodendrilla* species (Indonesia) provided the cytotoxic acantholides D (**107**) and E (**108**) [35].

4
Bicarbocyclic Furanosesterterpenoids

The structure of the former was determined by X-ray crystallographic analysis.

Ircinianin sulfate (**110**) was isolated as a very unstable metabolite of *Ircinia* (*Psammocinia*) *wistarii* from the Great Barrier Reef [55]. The absolute stereochemistry and that of (+)-wistarin **109**, which are metabolites of two *Ircinia* species, was determined by total synthesis from (*S*)-2-methylpropane-1,3-diol mono-THP ether [56]. Wistarin (**109**) has been obtained in 35% yield [57]. (−)-Wistarin (**111**) from a Red Sea *Ircinia* sp. is the enantiomer of a metabolite **109** previously isolated from *I. wisteria* [58]. The absolute configuration of halisulfate 3 (**112**), isolated from a Philippines specimen of *Ircinia* sp., has been determined by application of the chiral amide method coupled with chemical degradation procedures [59]. Halisulfate 7 (**113**) is a sesterterpene sulfate from a species of *Coscinoderma* from Yap, Micronesia [60]. *Darwinella australensis* collected from NW Australia contained sesterterpenoid sulfates **114–116** that inhibited the cell division of sea urchin eggs, but were not cytotoxic to human leukemia cells [61]. An *Ircinia* species collected at − 70 m

109   **110**   **111**

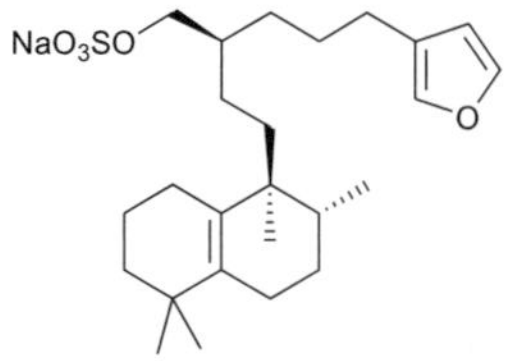

112 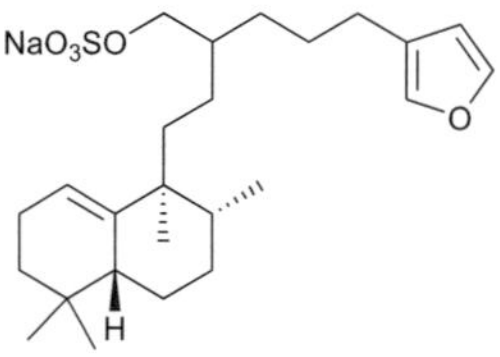   **113** 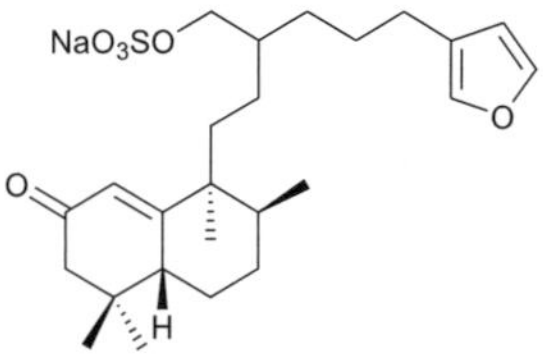   **114**

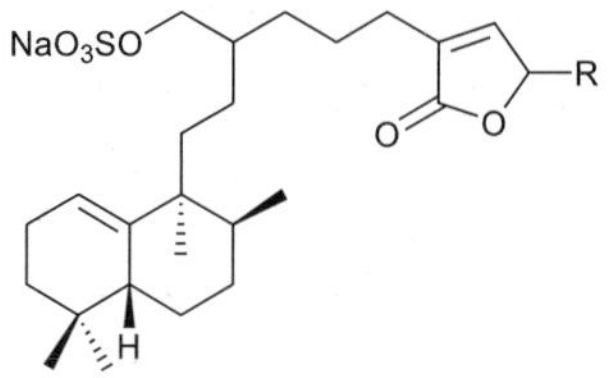

115 R = H
116 R = OMe   **117** 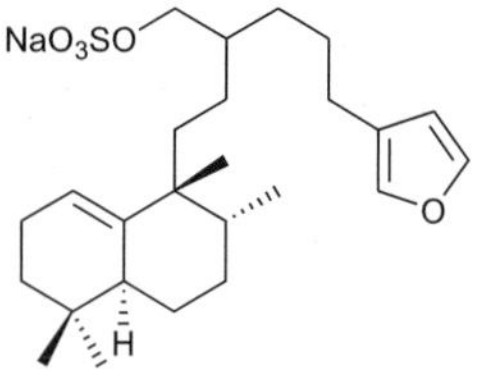   **118**

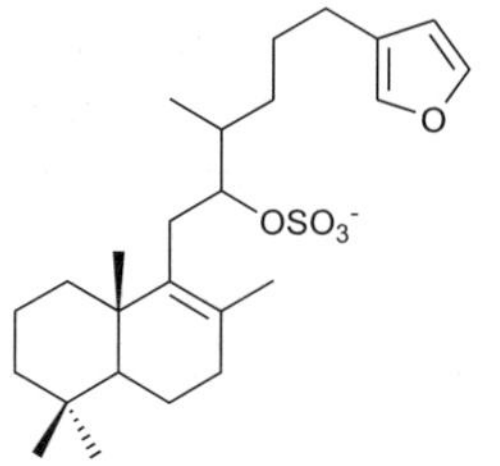

by dredging in the Gulf of Mexico contained a sesterterpenoid **117** [62]. *Stoeba extensa* (Japan) gave the cytotoxic halisulfate 7 (**113**) [63]; the C-8 stereochemistry of halisulfate 7, originally isolated from a *Coscinoderma* species [60], has been revised to that shown (**118**) on the basis of careful NOE analysis [63].

A number of bicarbocyclic sesterterpenoids have a carbon skeleton reminiscent of the clerodane diterpenoids. Cacospongionolide B (**119**) has been isolated from the Adriatic sponge, *Fasciospongia cavernosa*. The absolute stereochemistries of the known anti-inflammatory agent cacospongionolide B (**119**) was determined by comparison of their CD spectra with those of synthetic model compounds [47]. Both the natural (+) and unnatural (–) enantiomers of cacospongionolide B (**119**) have been synthesized; the natural enantiomer is more than twice as active as an inhibitor of sPLA$_2$ [64]. Palaulol (**120**) is an anti-inflammatory sesterterpene which was obtained [65] from a Palauan *Fascaplysinopsis* species of sponge. The number of known norsesterterpene cyclic peroxides has continued to increase. Cacospongionolide E

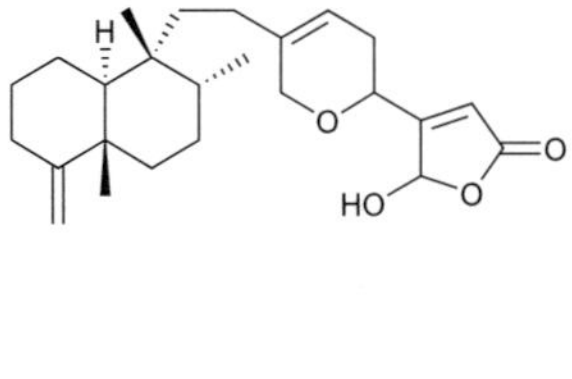

119 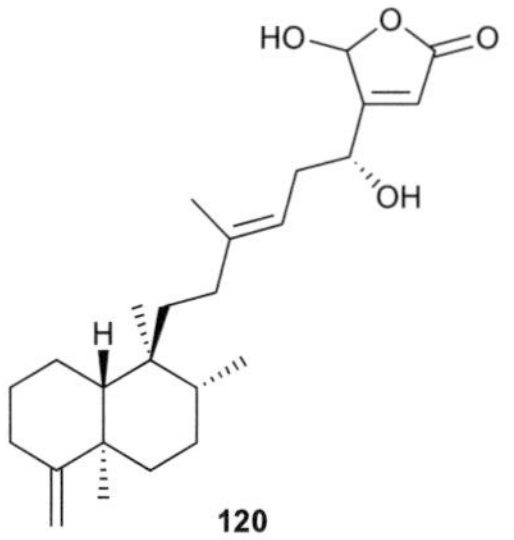   **120** 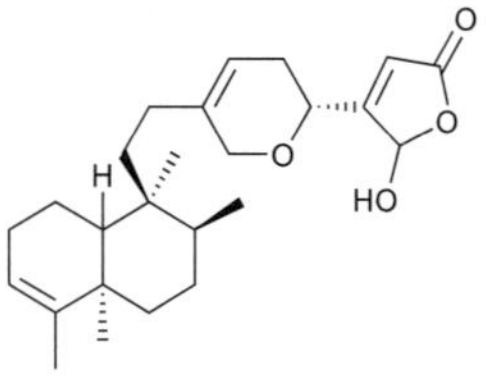   **121**

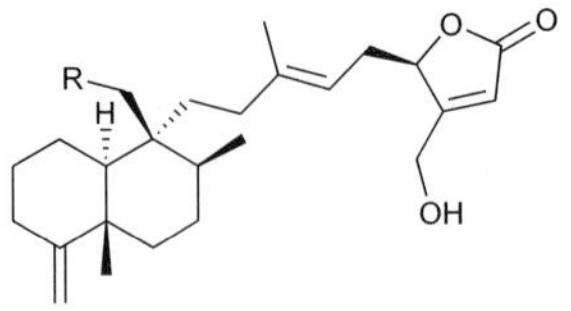

122 R = H
123 R = OAc

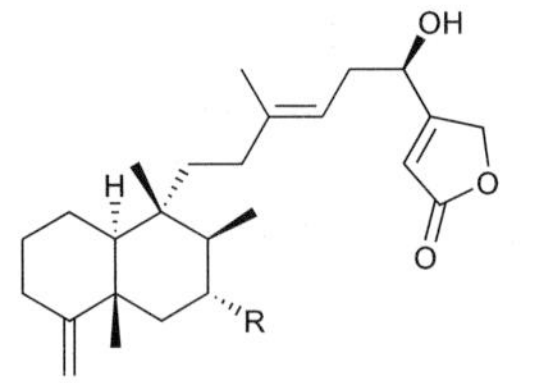

124 R = OAc
125 R = H
126 R = OH

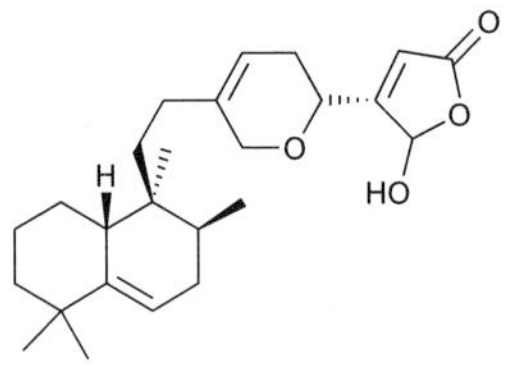

127

(121) is an inhibitor of human secretory phospholipase A_2 that was isolated as a minor constituent of *Fasciospongia cavernosa* from the Adriatic Sea [66]. A Palauan species of *Thorectandra* yielded the cytotoxic thorectandrols A (122) and B (123) [67]. All compounds were found to inhibit the protease activity of human RAS converting enzyme (hRCE1) and are the first natural products reported with this activity. Three further cytotoxic sesterterpenes, thorectandrols C–E (124–126) were isolated from a *Thorectandra* species collected in Palau [36]. An additional sesterterpene, cacospongionolide F (127), was isolated from *Fasciospongia cavernosa* from the northern Adriatic Sea and its absolute stereochemistry was determined using the modified Mosher's method [68]. An asymmetric synthesis of (–)-cacospongionolide F (127), isolated from *Fasciospongia cavernosa* [69], confirmed the original stereochemical assignments [69]. *Dysidea etheria* from the Caribbean contained the sesterterpene dysidiolide (128), which inhibited the cdc25A protein phosphatase. The structure of dysidiolide (128) was determined by single-crystal X-ray diffraction [70]. The absolute configuration of dysidolide (128) has been determined by total synthesis [71]. Syntheses of (+)-dysidiolide (128) have been reported [72–81]. The structures proposed for cladocorans A (129)

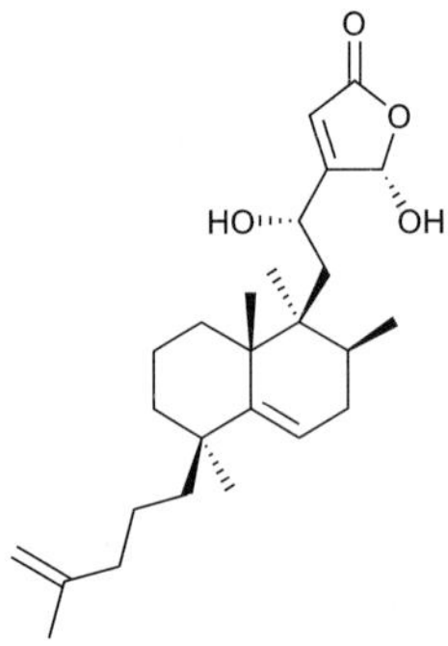

128

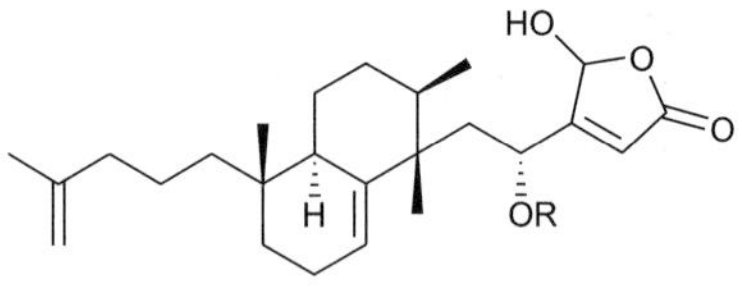

129 R = Ac
130 R = H

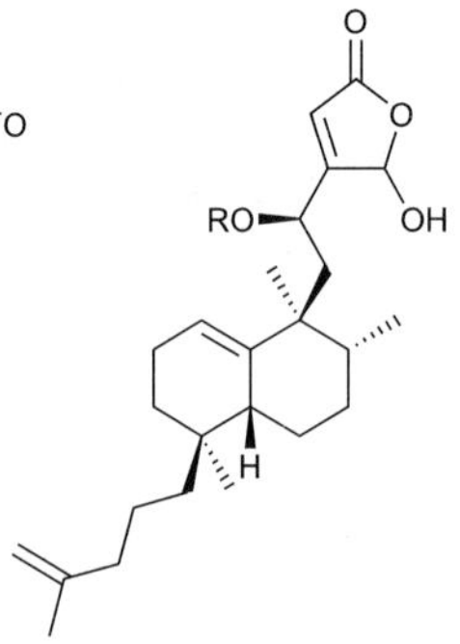

131 R = Ac
132 R = H

and B (**130**), isolated from the Mediterranean anthozoan *Cladocora cespitosa* [82], are in question as spectroscopic data of synthetic materials (**131** and **132**) were not identical to data reported for the natural products [83]. The stereochemistries of sesterterpenes cladocorans A (**131**) and B (**132**) have been revised as **133** and **134** by total synthesis [84], while preparation and testing of related stereoisomers indicated the series exhibits cytotoxicity toward a panel of human tumor cell lines [85]. Inhibition of secretory phospholipase A$_2$ by cladocorans A and B and their diastereomers almost equaled that of manoalide [86]. *Stoeba extensa* (Japan) gave the cytotoxic furanosesterterpenoid shinsonefuran (**135**) [63].

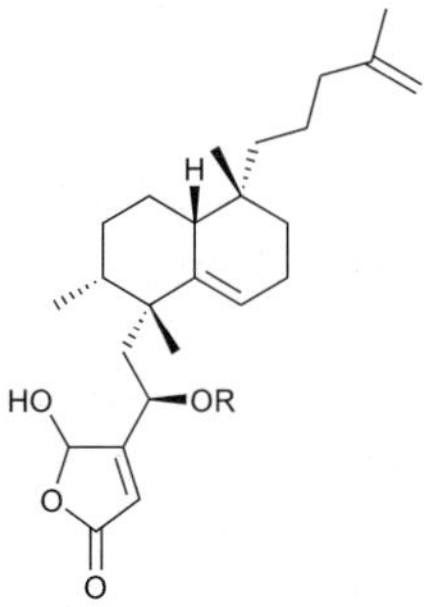

133 R = Ac
134 R = H

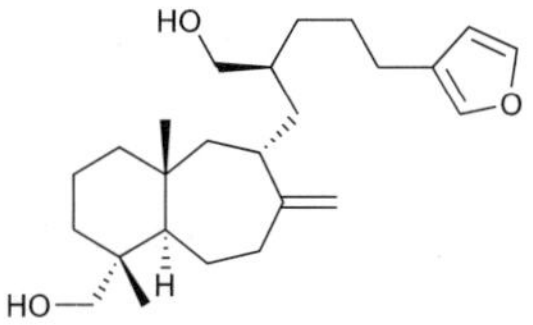

135

5
Tricarbocyclic Furanosesterterpenoids

The synthesis of **136** has been reported [87]. Aplysolide A (**137**) and its isomer aplysolide B, together with aplyolide A (**138**), are hydroxybutenolides obtained from an *Aplysinopsis* species, and their syntheses were reported [88, 89]. Spongianolide A (**139**) was synthesized [90, 91]. Cavernosolide (**140**) has been isolated from the Tyrrhenian sponge *Fasciospongia cavernosa*. Cavernosolide (**140**) showed potent activity in the *Artemia salina* bioassay and a moderate toxicity in a fish (*Gambusia affinis*) lethality assay [92]. The absolute configuration of cavernosolide (**140**) has been determined [47].

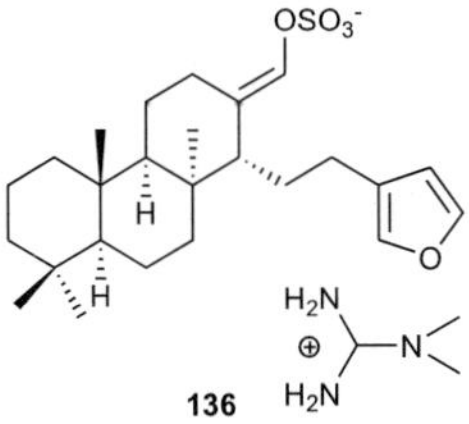

136

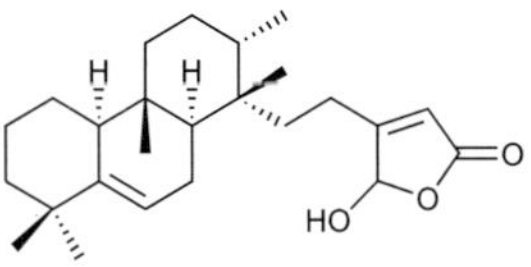

137

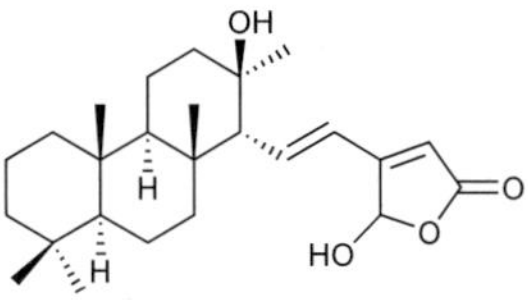

138

139   140   141

Bioassay-guided isolation of serine protease inhibitors from *Coscinoderma mathewsi* yielded the 1-methylherbipoline salt (**141**) [93]. Lintenolides F (**142**) and G (**143**) are two additional antiproliferative sesterterpenes from the Caribbean sponge *Cacospongia* cf. *linteiformis* [94]. Five potent and selective phospholipase A_2 inhibitors, petrosaspongiolides M (**144**), N (**145**), P (**146**), Q (**147**), and R (**148**), were obtained from *Petrosaspongia nigra* from New Aledonia [95]. Petrosaspongiolide M reduces morphine withdrawal in vitro [96] and protects against 2,4,6-trinitrobenzenesulfonic acid-induced colonic inflammation in mice [97]. A structural comparison of the PLA_2 inhibitory activity of the petrosaspongiolides was reported [98]. 21-Hydroxypetrosaspongiolide P (**149**) was isolated as an anti-inflammatory agent from a *Spongia* sp. from Vanuatu [99]. Three tricarbocyclic sesterterpenoids **150–152** of the cheilanthane class isolated from a Queensland *Ircinia* sp. were found to be inhibitors of MSK1 and MAPKAPK-2 protein kinases [100]. Hyr-

142   143   **144** R = H
                                                          145 R = OAc

146 R = H
147 R = OAc   148   149

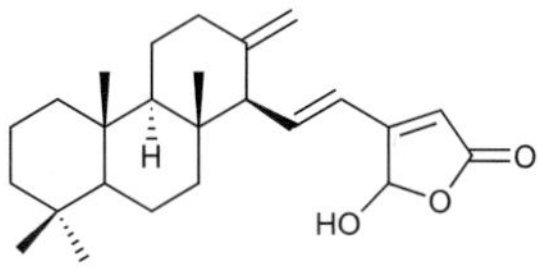

150 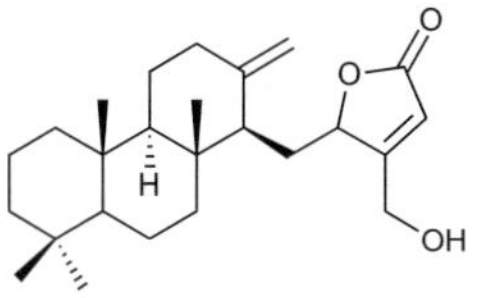   151 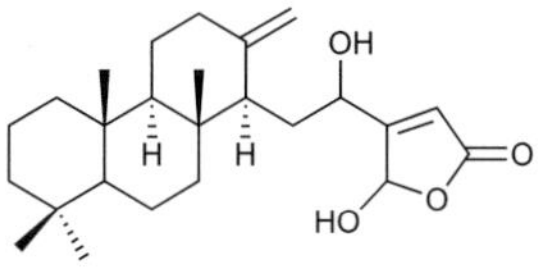   152

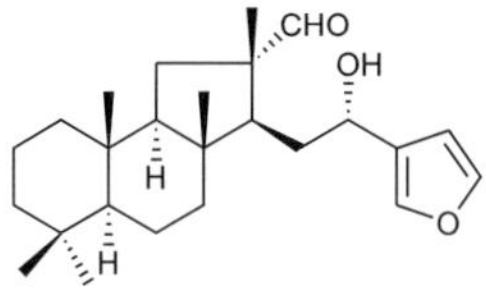

153 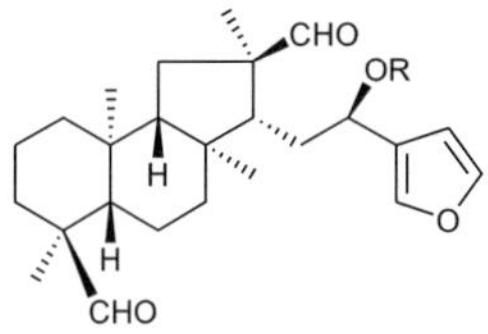   154 R = H   156 R$_1$ = CH$_2$OAc, R$_2$ = OAc
      155 R = Ac   157 R$_1$ = CH$_2$OH, R$_2$ = OAc
             158 R$_1$ = CHO, R$_2$ = OAc
             159 R$_1$ = COOH, R$_2$ = OAc
             160 R$_1$ = CH$_2$OH, R$_2$ = OH
             161 R$_1$ = COOH, R$_2$ = OH

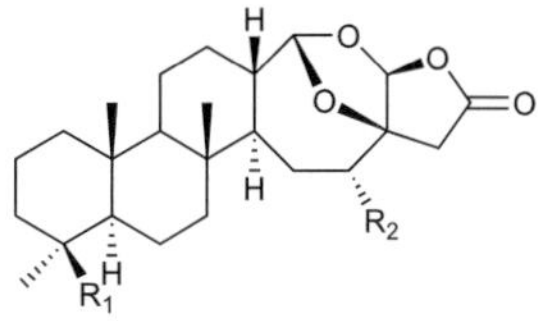

tiosal (**153**) was isolated from the Okinawan marine sponge, *Hyrtios erecta.* Hyrtiosal (**153**) was synthesized [101]. The absolute configuration of (–)-hyrtiosal (**153**) was determined by synthesis from sclareol [102]. *Hyrtios erectus* (Hainan Is., China) contained the two formyl hyrtiosal congeners **154** and **155** [103]. Six sesterterpenes, petrosaspongiolides C–J (**156–161**), and two norsesterterpenes **162**, were isolated as cytotoxic metabolites of *Petrosaspongia nigra* from New Caledonia [104]. Related ketal **163** has been isolated from a sponge originally assigned [105] to the genus *Dactylospongia* but now known [106] to be *Petrosaspongia nigra.* Lintenolides C–E (**164–166**) are additional sesterterpenes from *Cacospongia* cf. *linteiformis* from the Caribbean that inhibit fish feeding [107].

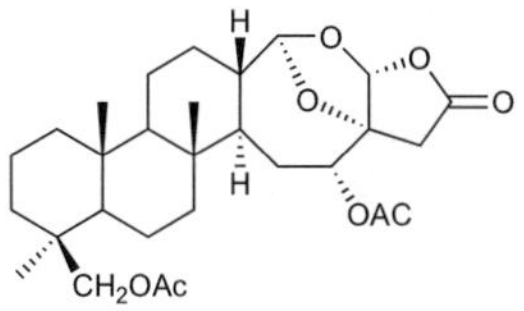

162 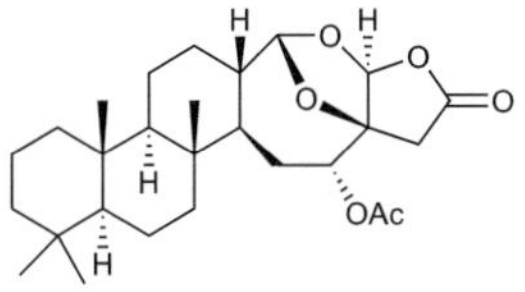   163 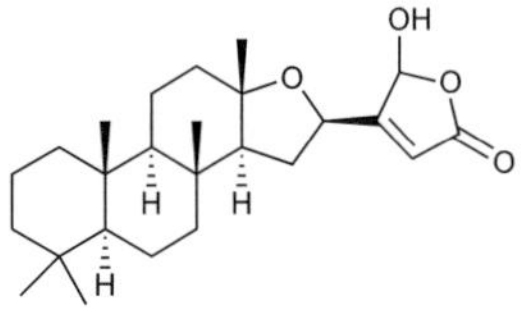   164

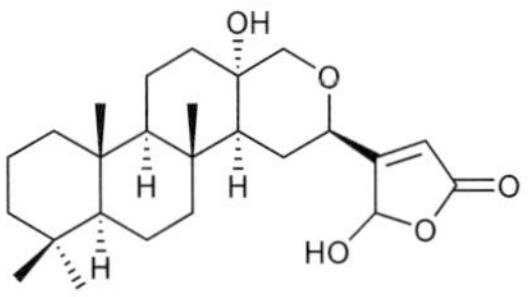

165 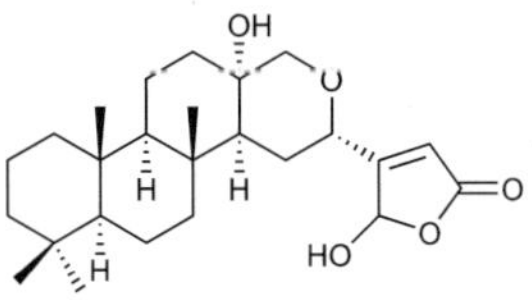   166

6
Tetracarbocyclic Furanosesterterpenoids

The scalaranes are amongst the most common sesterterpenoids and are found in a number of marine sponges, particularly from the family *Dictyoceratida*. They form a closely related series of compounds. In a number of instances these sesterterpenoids have also been isolated from a nudibranch that is associated with a sponge, and hence the sesterterpenoid in the nudibranch may have a dietary origin. Using bioassay-guided fractionation, heteronemin (**167**) was reisolated as a farnesyl transferase inhibitor from *Hyrtios reticulate* [108]. Scalarolide (**168**) was obtained from *Spongia idia*. The X-ray crystallographic data for scalarolides (**168**), which are metabolites of sponges of the family Thorectidae, have been reported [109]. These compounds have an ecological role in preventing predation. *Spongia matamata* from Palau contained 12α-acetoxy-19β-hydroxyscalara-15,17-dien-20,19-olide (**169**) and 12α,16β-diacetoxyscalarolbutenolide (**170**) [110]. A specimen of *Cacospongia scalaris* from Tarifa Island, Spain, contained scalarane sesterterpenes 12-*epi*-acetylscalarolide (**171**) and 16-acetylfuroscalarol (**172**) [111]. A *Cacospongia* sp. from New Zealand contained 12-desacetylfuroscalar-16-one (**173**) [112]. Four cytotoxic scalarane sesterterpenes **174–177** were obtained from a Japanese specimen of *Hyrtios erecta*: the structure of sesterterpene **174** was determined by X-ray crystallography [113]. A specimen of *H. erecta* from the Maldives contained the cytotoxic sesterterpenes sesterstatins 1–5 (**178–182**): the structures of sesterstatins 4 (**181**) and 5 (**182**) were deter-

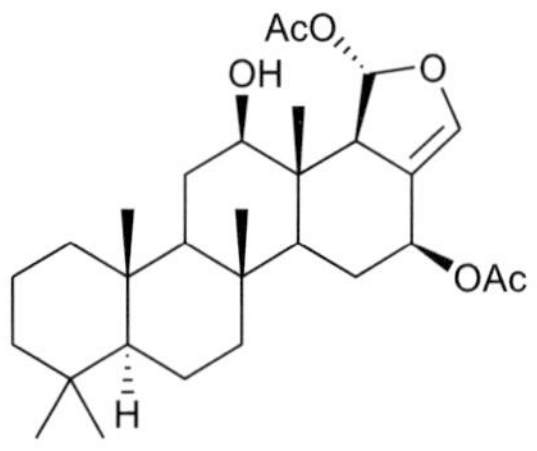

167

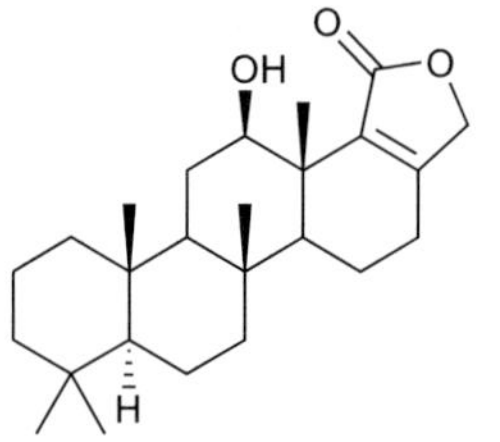

168

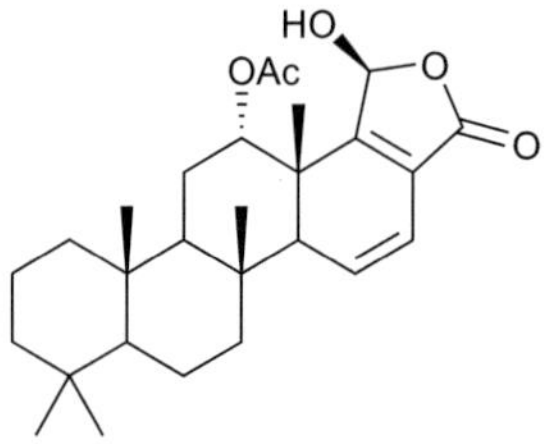

169

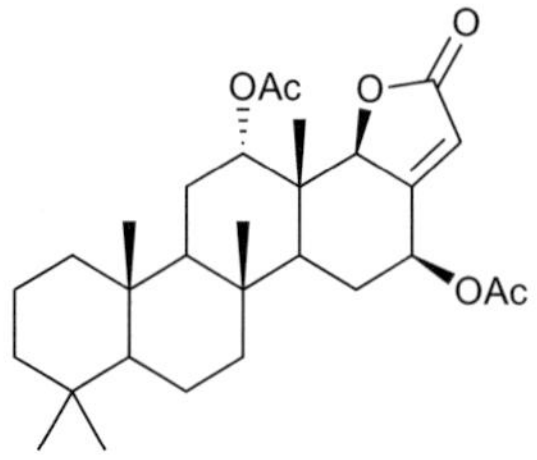

170

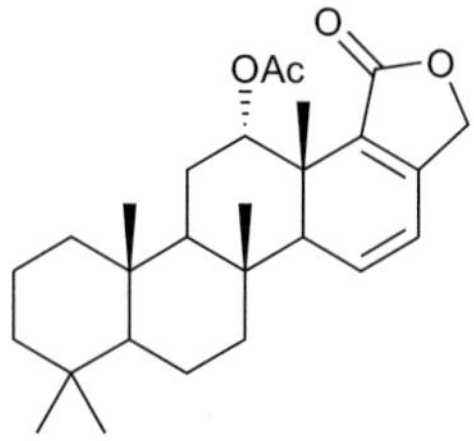

171

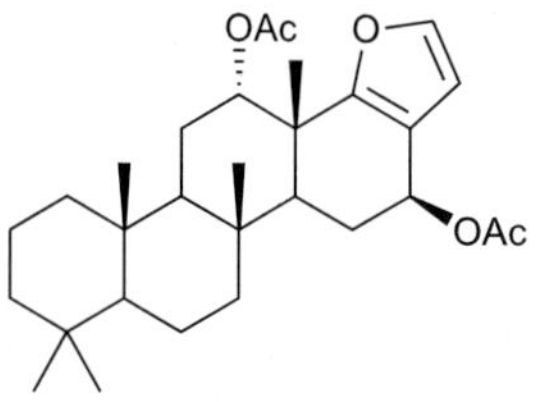

172

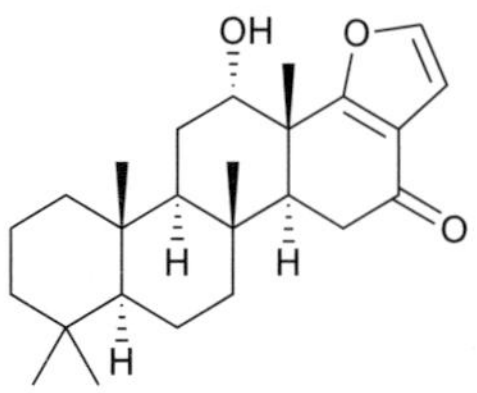

173

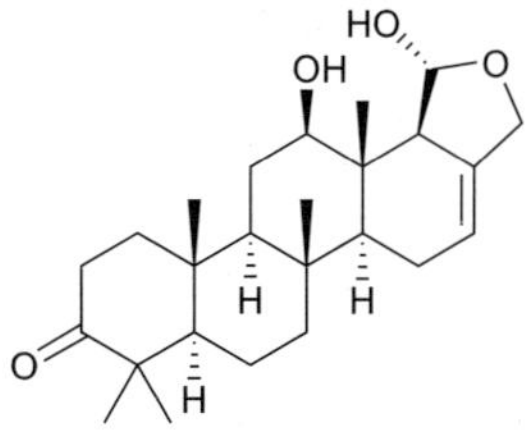

174

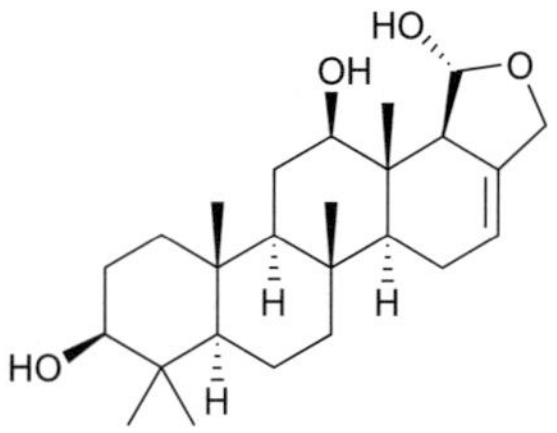

175 R = OH
176 R = H

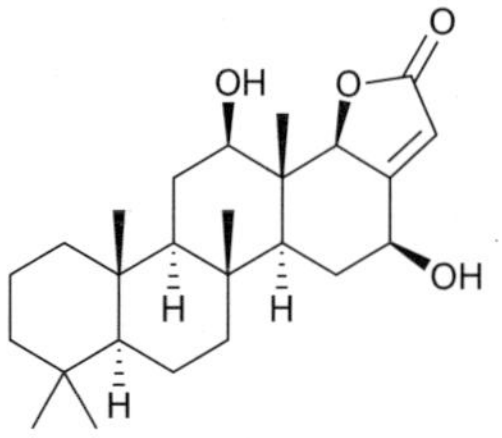

177

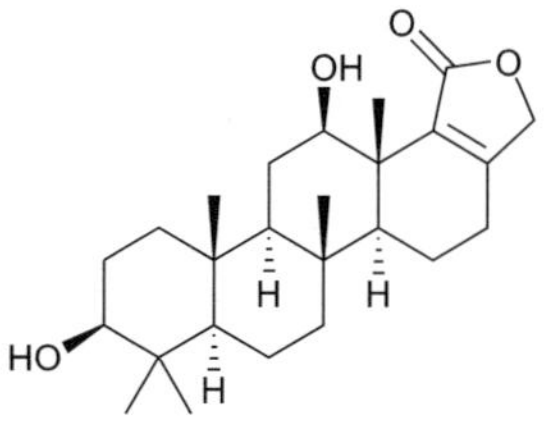

178

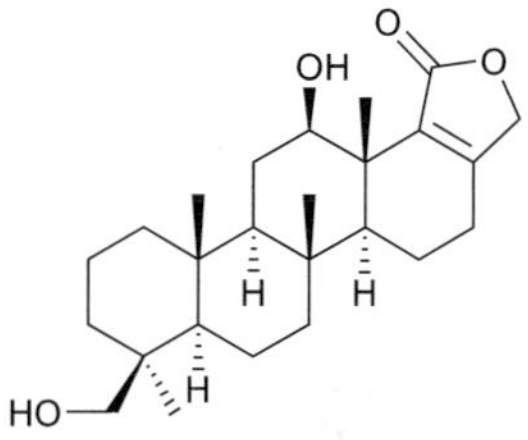

179

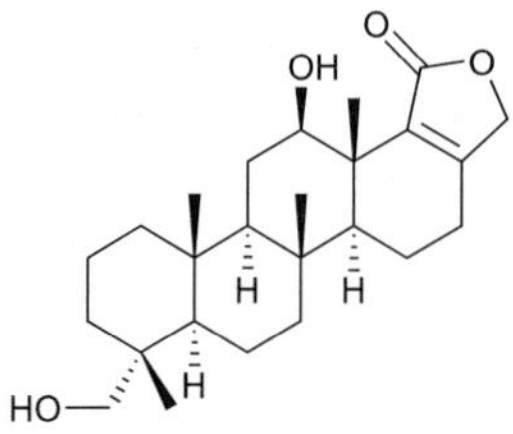

180

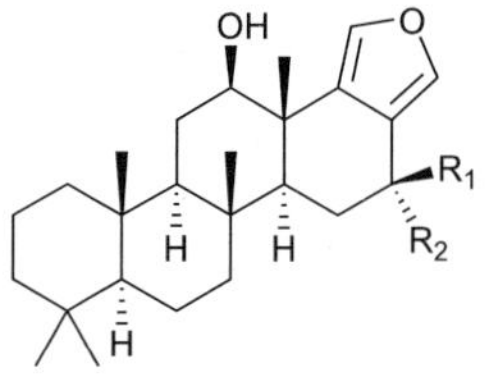

181 R₁ = H, R₂ = OH
182 R₁ = OH, R₂ = H

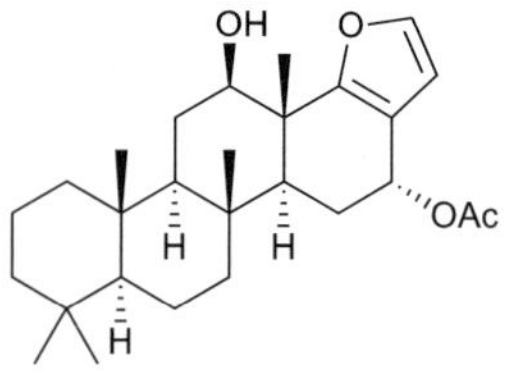

183

mined by X-ray analysis [114, 115]. *Spongia agaricina* from Cádiz, Spain, contained the sesterterpenes 12,16-di-*epi*-12-*O*-deacetyl-16-*O*-acetylfuroscalarol (**183**) and 16-*epi*-scalarolbutenolide (**184**) [116]. The X-ray crystallographic data for 12-*epi*-scalarins (**185**), which are metabolites of sponges of the family Thorectidae, have been reported [109]. A specimen of *Coscinoderma mathewsi* from Pohnpei contained two sesterterpenes **186** and **187** that have an unusual *cis* geometry about the B/C ring junction [117]. Two scalaranes, hyrtiolide (**188**) and 16-hydroxyscalariolide (**189**) were obtained from an Okinawan specimen of *H. erectus* [118]. *Hyrtios erecta* collected from the Egyptian Red Sea was found to contain salmahyrtisol B (**190**) and 3-acetyl- and 19-acetyl-sesterstatin (**191** and **192**), all of which showed significant cytotoxicity in human cancer cell lines [119]. The pentacyclic diacetate 16-acetoxy-dihydrodeoxoscalarin (**193**) was obtained from specimens of *Cacospongia scalaris* collected in Greece [34]. A *Spongia* species collected in Japan yielded three cytotoxic pentacyclic sesterterpenoids **194–196** [120]. *Hyrtios erectus*

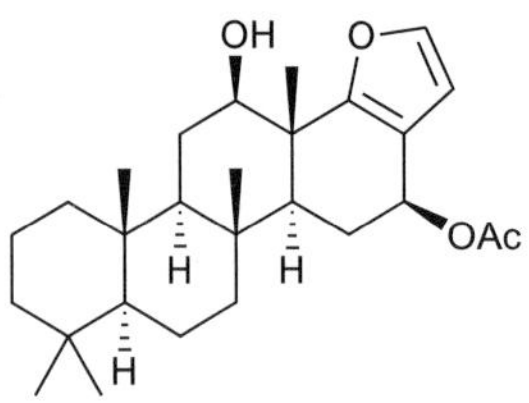

184

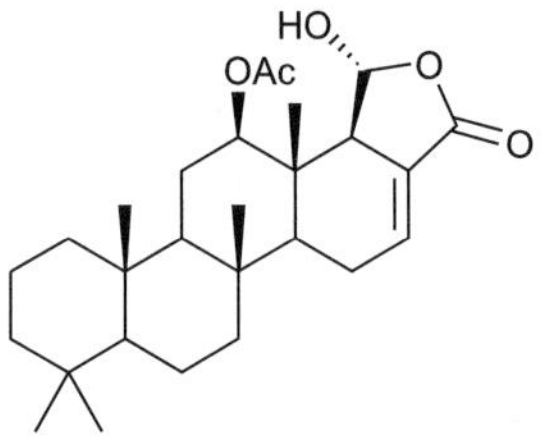

185

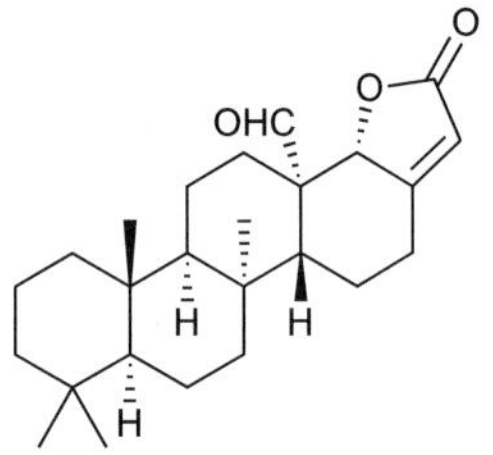

186

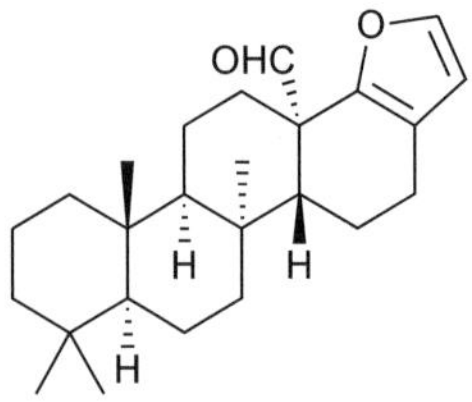

187

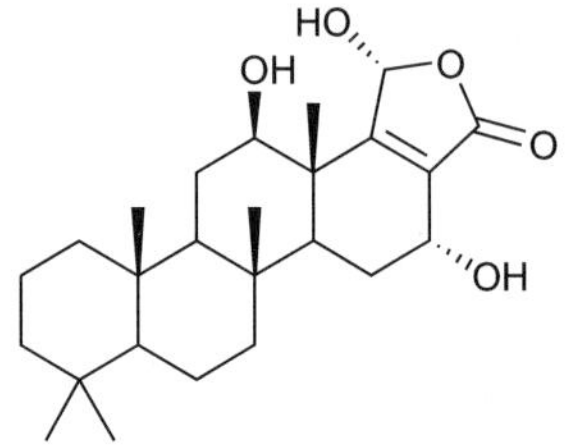

188

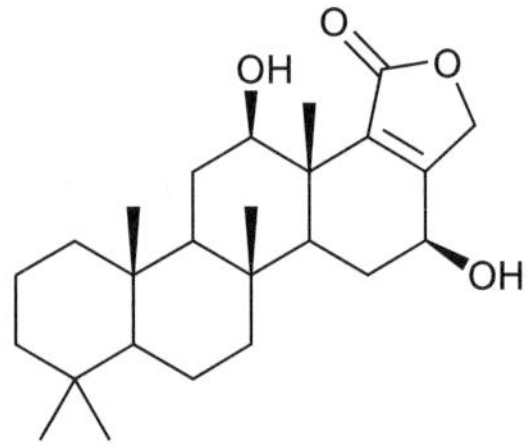

189

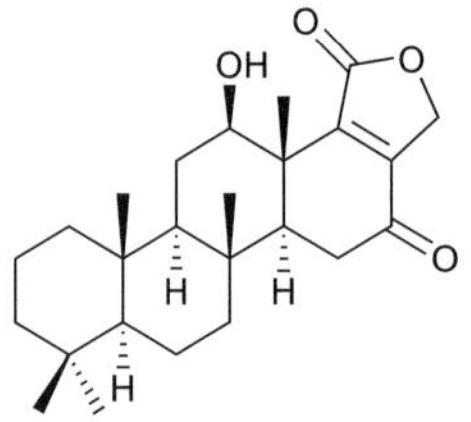

190

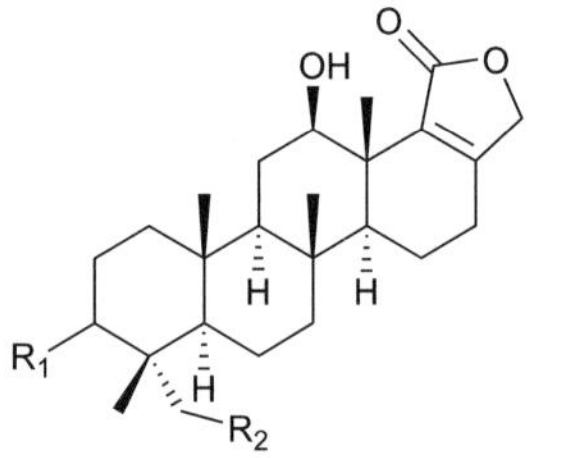

191 R₁ = H, R₂ = OAc

192 R₁ = OAc, R₂ = H

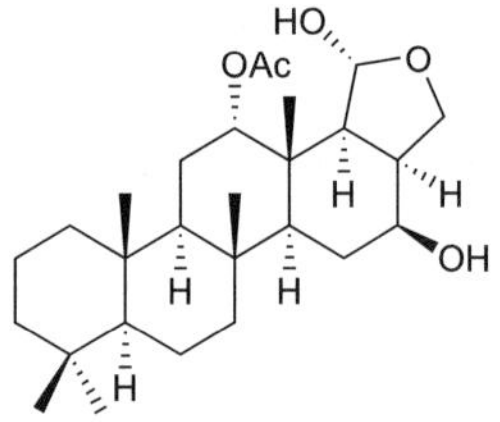

193

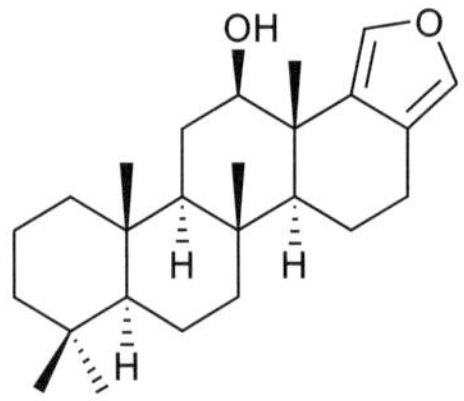

194

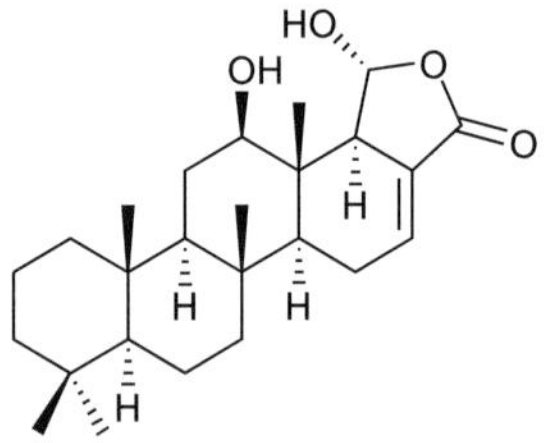

195

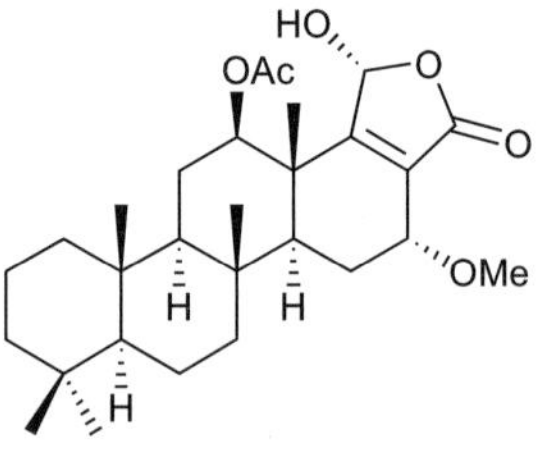

196

(Hainan Is., China) contained 12α-O-acetylhyrtiolide (**197**) [103]. A new scalarane-type pentacyclic sesterterpene, sesterstatin 6 (**198**), was isolated from the Republic of Maldives marine sponge *Hyrtios erecta*. The structure was elucidated by analyses with HRMS and high-field 2-D NMR spec-

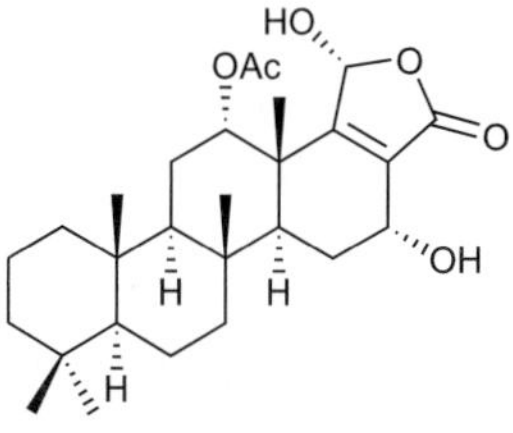

197 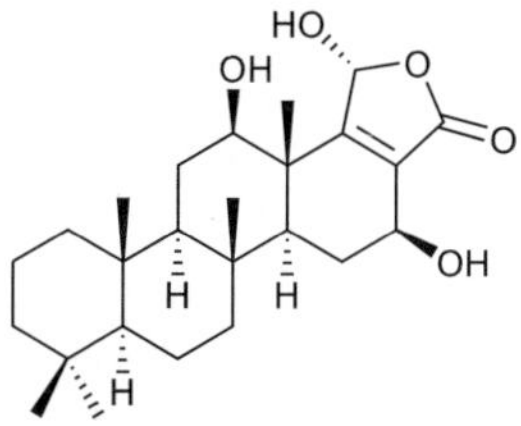 **198** 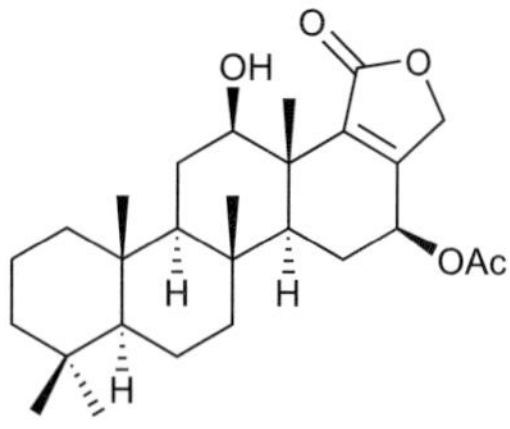 **199**

tra. Sesterstatin 6 showed significant cancer cell growth inhibition against murine P388 lymphocytic leukemia and a series of human tumor cell lines, and proved to be the most inhibitory of the series [121]. A new scalarane-type pentacyclic sesterterpene, sesterstatin 7 (**199**) was isolated from the Red Sea sponge *Hyrtios erecta*. Sesterstatin 7 showed 63% inhibition of *Mycobacterium tuberculosis* (H37Rv) [122]. A new scalarane-type sester-terpene, 12-deacetoxyscalarin 19-acetate (**200**), was isolated from the Thai sponge *Brachiaster* sp., and showed an interesting bioactivity profile in possessing potent antitubercular activity and being practically inactive in the cytotoxicity bioassay [52]. The sponge *Hyatella intestinalis* from the Gulf of California contains the new scalarane-related sesterterpenes hyatelones A–C (**201–203**), together with the new scalaranes hyatolides C–E (**204–206**) and 12-*O*-deacetyl-19-*epi*-scalarin **207**. Hyatelone A has shown activity as

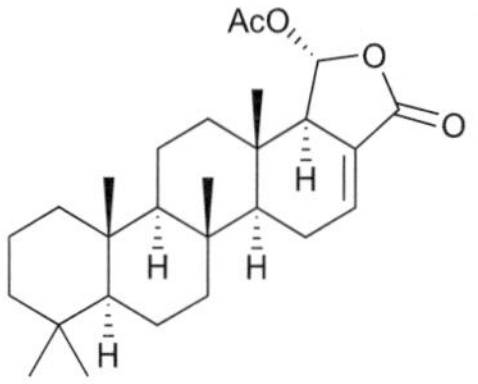

200 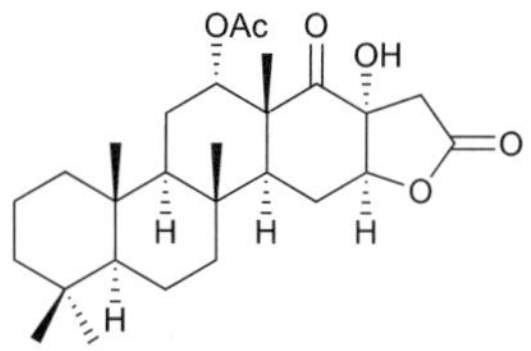 **201**

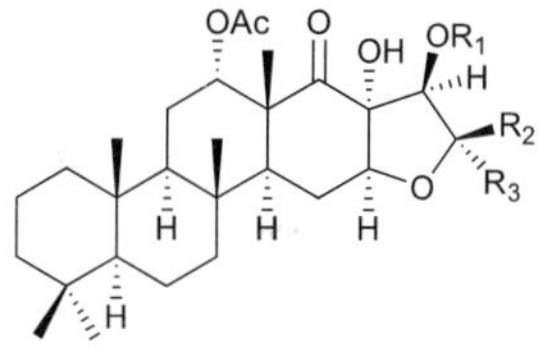

202 R$_1$ = R$_3$ = H, R$_2$ = OH

203 R$_1$= R$_2$ = H, R$_3$ = OH

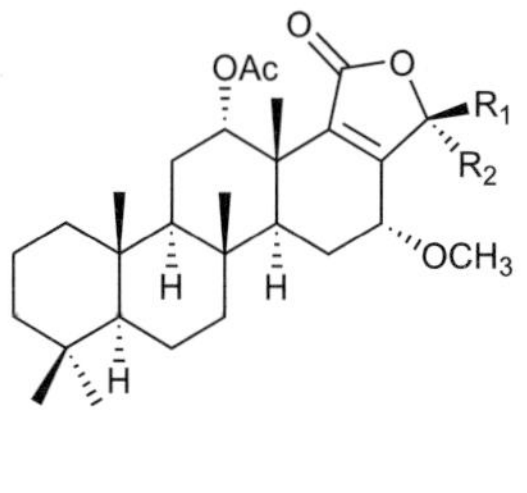

204 R$_1$ = OH, R$_2$ = H

205 R$_1$ = H, R$_2$ = OH

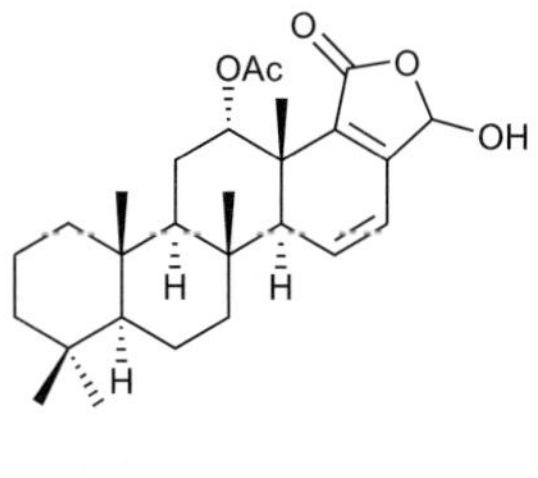

206 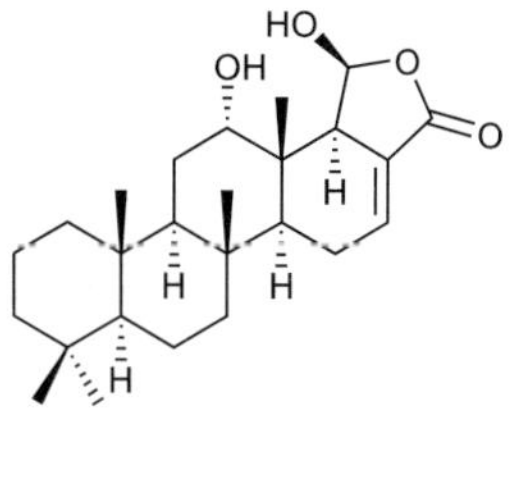 **207**

a growth inhibitor of several tumor cell lines [123]. Hyrtiosins A–E (**208–212**) are five new scalarane sesterterpenes from the South China Sea sponge *Hyrtios erecta* [124]. Three further scalaranetype sesterterpenoids **213–215** were obtained from a Papua New Guinean specimen of *Ledenfeldia frondosa* [125]. Three scalarane sesterterpenes **216–218** were isolated from a marine sponge of the genus *Spongia*. The isolated compounds showed potent inhibition of transactivation for the nuclear hormone receptor, FXR (farnesoid X-activated receptor), which is a promising drug target to treat hypercholesterolemia in humans [126]. Scalarane-type sesterterpene PHC-5 (**219**), which has been isolated from a marine sponge *Phyllospongia chondrodes* collected at Yaeyama Islands, Okinawa, Japan, increased hemoglobin production in human chronic myelogenous leukemia cell line K562. This sesterterpene was found to induce erythroid differentiation in K562 cells [127]. Various bis-homoscalarins that contain extra alkyl groups at C-19 (or C-24) and C-24 have been obtained from sponges. Two homoscalarins, phyllactones H and I, which are C-24 isomers of the bis-homoscalaranes **220** and **221**, have been isolated from a Chinese species of *Phyllospongia* [128]. As part of a chemotaxonomic study of Indo-Pacific foliose sponges, 3-hydroxy-20,22-dimethyl-20-deoxoscalarin (**222**) was identified as a metabolite of a Solomon Islands sponge known either as *Phyllospongia vermicularis* or *Dysidea vermicularis*, while the corresponding ketone, 3-oxo-20,22-dimethyl-20-deoxoscalarin (**223**) was char-

208 R$_1$ = R$_3$ = AcO, R$_2$ = R$_4$ = H

209 R$_1$ = R$_4$ = H, R$_2$ = R$_3$ = AcO

210 R$_1$ = R$_3$ = H, R$_2$ = R$_4$ = AcO

211 R$_1$ = OH, R$_2$ = H

212 R$_1$ = H, R$_2$ = Ac

213

214

215

216 R$_1$ = OAc, R$_2$ = CH$_2$OAc, R$_3$ = OH

217 R$_1$ = OH, R$_2$ = CH$_2$OH, R$_3$ = OH

218 R$_1$ = OH, R$_2$ = Me, R$_3$ = OMe

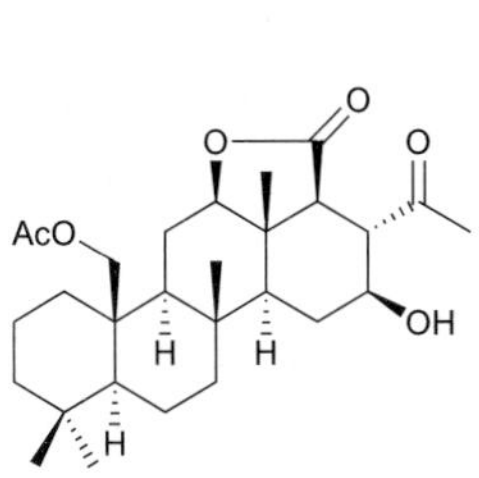

219

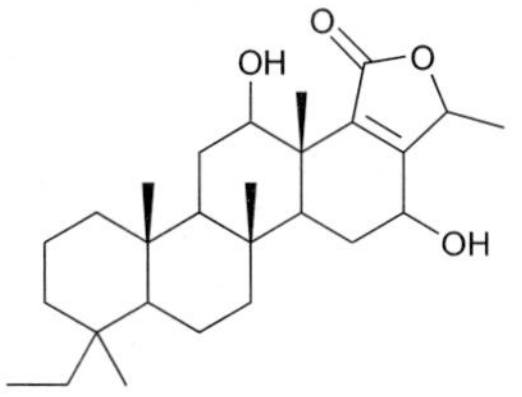

220 Me 26α
221 Me 26β

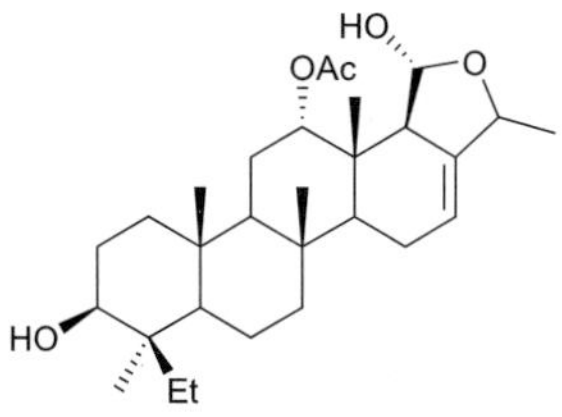

222

acterized as a metabolite of an Indonesian *Carteriospongia* sp., although it had previously been reported without supporting data as a constituent of *Dysidea herbacea* [129]. *Strepsichordaia aliena* from Indonesia contained bis-homoscalaranes honulactones A–L (**224–235**) and phyllofolactones H–K (**236–239**); honulactones A–D (**224–227**), for which the structures **225** and **227** were determined by X-ray analysis, were shown to be cytotoxic [130, 131]. Five inhibitors of in vitro HIV-1 envelope-mediated fusion, the phyllolactones A–E (**240–244**), were obtained from specimens of *Phyllospongia*

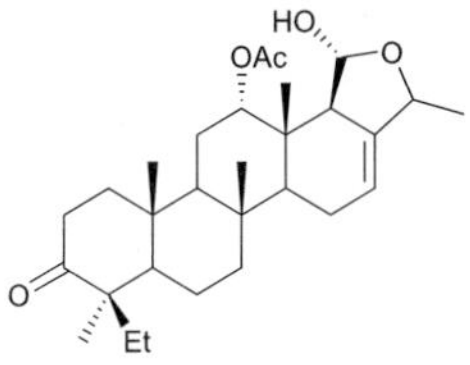

223

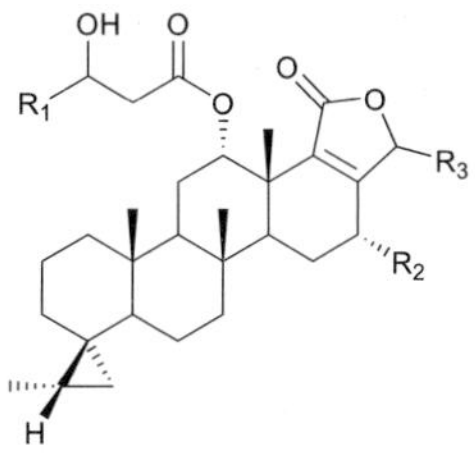

224 R_1 = Me, R_2 = H, R_3 = β-Me
225 R_1 = Me, R_2 = H, R_3 = α-Me
228 R_1 = Et, R_2 = H, R_3 = β-Me
229 R_1 = Et, R_2 = H, R_3 = α-Me
230 R_1 = Me, R_2 = OH, R_3 = β-Me
231 R_1 = Me, R_2 = OH, R_3 = α-Me

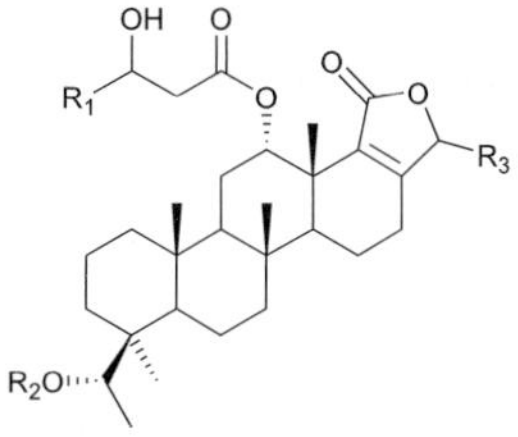

226 R_1 = Me, R_2 = COMe, R_3 = β-Me
227 R_1 = Me, R_2 = COMe, R_3 = α-Me
232 R_1 = Et, R_2 = COMe, R_3 = β-Me
233 R_1 = Et, R_2 = COMe, R_3 = α-Me
234 R_1 = Me, R_2 = COEt, R_3 = β-Me
235 R_1 = Me, R_2 = COEt, R_3 = α-Me

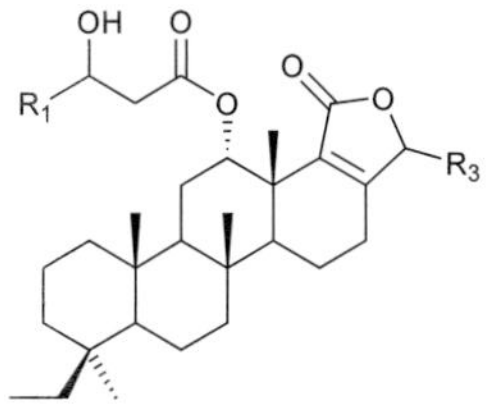

236 R_1 = Me, R_2 = β-Me
237 R_1 = Me, R_2 = α-Me
238 R_1 = Et, R_2 = β-Me
239 R_1 = Et, R_2 = α-Me

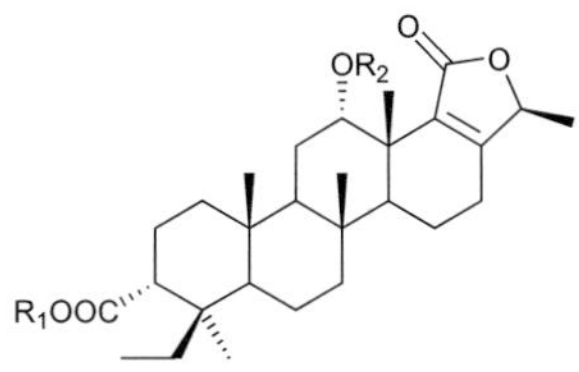

240 R_1 = CH$_2$CH$_2$CH$_3$, R_2 = H
241 R_1 = CH$_2$CH$_3$, R_2 = H
242 R_1 = Me, R_2 = H
243 R_1 = CH$_2$CH$_3$, R_2 = Ac

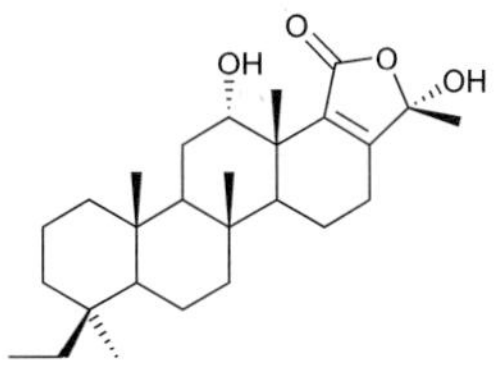

244

245

lamellose [132]. A species of *Phyllospongia* collected in Indonesia yielded the scalarane sesterterpenoid **245** [133].

7
Conclusion

Along with the development of new analytical instruments and techniques over the past 30 years, a variety of furanosesterterpenoids have been isolated and characterized from marine sponges. Further chemical and biological studies on these furanosesterterpenoids should contribute to a deeper understanding of their roles in nature. Also, intensive studies involving the comprehensive evaluation of these molecules may lead to the creation of a new field in bioscience. This review has pointed out the continued finding of new furanosesterterpenoids from marine sponges.

Acknowledgements The authors gratefully acknowledge the grant from the Knowledge Innovation Program (KZCX2-YW-216) and the Hundred Talents Project of the Chinese Academy of Sciences, and the support of the K.C. Wong Education Foundation, Hong Kong.

References

1. Faulkner DJ (2002) Nat Prod Rep 19:1
2. Blunt JW, Copp BR, Munro MHG, Northcote PT, Prinsep MR (2006) Nat Prod Rep 23:26
3. Hanson JR (1996) Nat Prod Rep 13:529
4. Hanson JR (1992) Nat Prod Rep 9:481
5. Hanson JR (1986) Nat Prod Rep 3:123
6. Liu Y, Zhang S, Abreu P (2006) Nat Prod Rep 23:630
7. Rifai S, Fassouane A, Kijjoa A, Van Soest R (2004) Mar Drugs 2:147
8. Guo YW, Trivellone E (2000) Chin Chem Lett 11:327
9. Fontana A, Albarella L, Scognamiglio G, Uriz M, Cimino G (1996) J Nat Prod 59:869
10. Bando T, Shishido K (1996) Chem Commun, p 1357
11. Garrido L, Zubía E, Ortega MJ, Salvá J (1997) J Nat Prod 60:794
12. Capon RJ, Jenkins A, Rooney F, Ghisalberti EL (2001) J Nat Prod 64:638

13. Lenis LA, Nunez L, Jimenez C, Riguera R (1996) Nat Prod Lett 8:15
14. Liu Y, Mansoor TA, Hong J, Lee C-O, Sim CJ, Im KS, Kim ND, Jung JH (2003) J Nat Prod 66:1451
15. Fürstner A, Gastner T, Rust J (1999) Synlett, p 29
16. Chill L, Rudi A, Aknin M, Loya S, Hizi A, Kashman Y (2004) Tetrahedron 60:10619
17. Rochfort SJ, Atkin D, Hobbs L, Capon RJ (1996) J Nat Prod 59:1024
18. Liu Y, Hong J, Lee C-O, Im KS, Kim ND, Choi JS, Jung JH (2002) J Nat Prod 65:1307
19. Shen Y-C, Lo K-L, Lin Y-C, Khalil AT, Kuo Y-H, Shih P-S (2006) Tetrahedron Lett 47:4007
20. Shin J, Rho JR, Seo Y, Lee HS, Cho KW, Sim CJ (2001) Tetrahedron Lett 42:3005
21. Issa HH, Tanaka J, Higa T (2003) J Nat Prod 66:251
22. Schmitz WD, Messerschmidt NB, Romo D (1998) J Org Chem 63:2058
23. Choi K, Hong J, Lee C-O, Kim D, Sim CJ, Im KS, Jung JH (2004) J Nat Prod 67:1186
24. Liu Y, Bae BH, Alam N, Hong J, Sim CJ, Lee C, Im KS, Jung JH (2001) J Nat Prod 64:1301
25. Choi HJ, Choi YH, Yee SB, Im E, Jung JH, Kim ND (2005) Mol Carcinog 44:162
26. Martínez A, Duque C, Sato N, Fujimoto Y (1997) Chem Pharm Bull 45:181
27. Höller U, König GM, Wright AD (1997) J Nat Prod 60:832
28. Takabe K, Hashimoto H, Sugimoto H, Nomoto M, Yoda H (2004) Tetrahedron Asymmetry 15:909
29. Jiang YH, Ryu S-H, Ahn E-Y, You S, Lee B-J, Jung JH, Kim D-K (2004) Nat Prod Sci 10:272
30. McPhail K, Davies-Coleman MT, Coetzee P (1998) J Nat Prod 61:961
31. Noda Y, Hashimoto H, Norizuki T (2001) Heterocycles 55:1839
32. De Rosa S, De Giulio A, Crispino A, Iodice C, Tommonaro G (1997) Nat Prod Lett 10:267
33. Bourguet-Kondracki M-L, Debitus C, Guyot M (1996) J Chem Res Synop 192
34. Tsoukatou M, Siapi H, Vagias C, Roussis V (2003) J Nat Prod 66:444
35. Elkhayat E, Edrada R, Ebel R, Wray V, Van Soest R, Wiryowidagdo S, Mohamed MH, Müller WEG, Proksch P (2004) J Nat Prod 67:1809
36. Charan RD, McKee TC, Boyd MR (2002) J Nat Prod 65:492
37. De Rosa S, Carbonelli S (2006) Tetrahedron 62:2845
38. Sayed KAE, Mayer AMS, Kelly M, Hamann MT (1999) J Org Chem 64:9258
39. De Rosa S, Milone A, De Giulio A, Crispino A, Iodice C (1996) Nat Prod Lett 8:245
40. De Rosa S, De Giulio A, Crispino A, Iodice C, Tommonaro G (1997) Nat Prod Lett 10:7
41. Martínez A, Duque C, Fujimoto Y (1997) Lipids 32:565
42. Coombs J, Lattmann E, Hoffmann HMR (1998) Synthesis 1367
43. Pommier A, Kocienski PJ (1997) Chem Commun, p 1139
44. De Rosa M, Soriente A, Sodano G, Scettri A (2000) Tetrahedron 56:2095
45. Kocienski PJ, Pommier A, Stepanenko V, Jarowicki K (2003) J Org Chem 68:4008
46. Soriente A, De Rosa M, Apicella A, Scettri A, Sodano G (1999) Tetrahedron Asymmetry 10:4481
47. Soriente A, Crispino A, De Rosa M, De Rosa S, Scettri A, Scognamiglio G, Villano R, Sodano G (2000) Eur J Org Chem 947
48. D'Acquisto F, Lanzotti V, Carnuccio R (2000) Biochem J 346:793
49. Kirsch G, König GM, Wright AD, Kaminsky R (2000) J Nat Prod 63:825
50. Tsuda M, Endo T, Mikami Y, Fromont J, Kobayashi J (2002) J Nat Prod 65:1507
51. Namikoshi M, Suzuki S, Meguro S, Nagai H, Koike Y, Kitazawa A, Kobayashi H, Oda T, Yamada J (2004) Fish Sci 70:152

52. Wonganuchitmeta S-N, Yuenyongsawad S, Keawpradub N, Plubrukarn A (2004) J Nat Prod 67:1767
53. Carotenuto A, Fattorusso E, Lanzotti V, Magno S, Carnuccio R, D'Acquisto F (1997) Tetrahedron 53:7305
54. Zhou G-X, Molinski TF (2006) J Asian Nat Prod Res 8:15
55. Coll JC, Kearns PS, Rideout JA, Hooper J (1997) J Nat Prod 60:1178
56. Uenishi J, Kawahama R, Yonemitsu O (1997) J Org Chem 62:1691
57. Uenishi J, Kawahama R, Imakoga T, Yonemitsu O (1998) Chem Pharm Bull 46:1090
58. Fontana A, Fakhr I, Mollo E, Cimino G (1999) Tetrahedron Asymmetry 10:3869
59. Müller EL, Faulkner DJ (1997) Tetrahedron 53:5373
60. Fu X, Ferreira ML, Schmitz FJ, Kelly M (1999) J Nat Prod 62:1190
61. Makarieva TN, Rho J-R, Lee H-S, Santalova EA, Stonik V, Shin J (2003) J Nat Prod 66:1010
62. Yang S-W, Chan T-M, Pomponi SA, Gonsiorek W, Chen G, Wright AE, Hipkin W, Patel M, Gullo V, Pramanik B, Zavodny P, Chu M (2003) J Antibiot 56:783
63. Phuwapraisirisan P, Matsunaga S, Van Soest RWM, Fusetani N (2004) Tetrahedron Lett 45:2125
64. Cheung AK, Snapper ML (2002) J Am Chem Soc 124:11584
65. Schmidt EW, Faulkner DJ (1996) Tetrahedron Lett 37:3951
66. De Rosa S, Crispino A, De Giulio A, Iodice C, Benrezzouk R, Terencio MC, Ferrándiz ML, Alcaraz MJ, Payá M (1998) J Nat Prod 61:931
67. Charan RD, McKee TC, Boyd MR (2001) J Nat Prod 64:661
68. De Rosa S, Crispino A, De Giulio A, Iodice C, Amodeo P, Tancredi T (1999) J Nat Prod 62:1316
69. Demeke D, Forsyth CJ (2003) Org Lett 5:991
70. Gunasekera SP, McCarthy PJ, Kelly-Borges M, Lobkovsky E, Clardy J (1996) J Am Chem Soc 118:8759
71. Corey EJ, Roberts BE (1997) J Am Chem Soc 119:12425
72. Boukouvalas J, Cheng Y-X, Robichaud J (1998) J Org Chem 63:228
73. Magnuson SR, Sepp-Lorenzino L, Rosen N, Danishefsky SJ (1998) J Am Chem Soc 120:1615
74. Shimazawa R, Suzuki T, Dodo K, Shirai R (2004) Bioorg Med Chem Lett 14:3291
75. Brohm D, Philippe N, Metzger S, Bhargava A, Muller O, Lieb F, Waldmann H (2002) J Am Chem Soc 124:13171
76. Demeke D, Forsyth CJ (2002) Tetrahedron 58:6531
77. Jung ME, Nishimura N (2001) Org Lett 3:2113
78. Miyaoka H, Kajiwara Y, Yamada Y (2000) Tetrahedron Lett 41:911
79. Takahashi M, Dodo K, Hashimoto Y, Shirai R (2000) Tetrahedron Lett 41:2111
80. Piers E, Caillé S, Chen G (2000) Org Lett 2:2483
81. Demeke D, Forsyth CJ (2000) Org Lett 2:3177
82. Fontana A, Ciavatta ML, Cimino G (1998) J Org Chem 63:2845
83. Marcos IS, Pedrero AB, Sexmero MJ, Diez D, Basabe P, Hernandez FA, Broughton HB, Urones JG (2002) Synlett, p 105
84. Miyaoka H, Yamanishi M, Kajiwara Y, Yamada Y (2003) J Org Chem 68:3476
85. Marcos IS, Pedrero AB, Sexmero MJ, Diez D, Basabe P, García N, Moro RF, Broughton HB, Mollinedo F, Urones JG (2003) J Org Chem 68:7496
86. Miyaoka H, Yamanishi M, Mitome H (2006) Chem Pharm Bull 54:268
87. Basabe P, Delgado S, Marcos IS, Diez D, Diego A, De Roman M, Urones JG (2005) J Org Chem 70:9480
88. Hansen TV, Stenstrom Y (2001) Tetrahedron Asymmetry 12:1407

89. Spinella A, Caruso T, Martino M, Sessa C (2001) Synlett, p 1971
90. Hata T, Tanaka K, Katsumura S (1999) Tetrahedron Lett 40:1731
91. Furuichi N, Hata T, Soetjipto H, Kato M, Katsumura S (2001) Tetrahedron 57:8425
92. De Rosa S, Crispino A, De Giulio A, Iodice C, Tommonaro G (1997) J Nat Prod 60: 844
93. Kimura J, Ishizuka E, Nakao Y, Yoshida WY, Scheuer PJ, Kelly-Borges M (1998) J Nat Prod 61:248
94. Carotenuto A, Fattorusso E, Lanzotti V, Magno S, Carnuccio R, Iuvone T (1998) Comp Biochem Physiol 119C:119
95. Randazzo A, Debitus C, Minale L, Pastor PG, Alcaraz MJ, Payá M, Gomez-Paloma L (1998) J Nat Prod 61:571
96. Capasso A, Casapullo A, Randazzo A, Gomez-Paloma L (2003) Life Sci 73:611
97. Busserolles J, Payá M, D'Auria MV, Gomez-Paloma L, Alcaraz MJ (2005) Biochem Pharmacol 69:1433
98. Monti MC, Casapullo A, Riccio R, Gomez-Paloma L (2004) Bioorg Med Chem 12:1467
99. De Marino S, Iorizzi M, Zollo F, Debitus C, Menou J-L, Ospina LF, Alcaraz MJ, Payá M (2000) J Nat Prod 63:322
100. Buchanan MS, Edser A, King G, Whitmore J, Quinn RJ (2001) J Nat Prod 64:300
101. Lunardi I, Santiago GMP, Imamura PM (2002) Tetrahedron Lett 43:3609
102. Basabe P, Diego A, Díez D, Marcos IS, Urones JG (2000) Synlett, p 1807
103. Qiu Y, Deng Z, Pei Y, Fu H, Li J, Proksch P, Lin W (2004) J Nat Prod 67:921
104. Paloma LG, Randazzo A, Minale L, Debitus C, Roussakis C (1997) Tetrahedron 53:10451
105. Lal AR, Cambie RC, Rickard CEF, Bergquist PR (1994) Tetrahedron Lett 35:2603
106. Cambie RC, Lal AR, Rickard CEF (1996) Acta Crystallogr C 52:709
107. Carotenuto A, Fattorusso E, Lanzotti V, Magno S, Mayol L (1996) Liebigs Ann Chem p 77
108. Ledroit V, Debitus C, Ausseil F, Raux R, Menou J-L, Hill BT (2004) Pharm Biol 42:454
109. Cambie RC, Rickard CEF, Rutledge PS, Yang X-S (1999) Acta Crystallogr C 55:112
110. Lu Q, Faulkner DJ (1997) J Nat Prod 60:195
111. Rueda A, Zubía E, Ortega MJ, Carballo JL, Salvá J (1997) J Org Chem 62:1481
112. Cambie RC, Rutledge PS, Yang X-S, Bergquist PR (1998) J Nat Prod 61:1416
113. Tsuchiya N, Sato A, Hata T, Sato N, Sasagawa K, Kobayashi T (1998) J Nat Prod 61:468
114. Pettit GR, Cichacz ZA, Tan R, Hoard MS, Melody N, Pettit RK (1998) J Nat Prod 61:13
115. Pettit GR, Tan R, Melody N, Cichacz ZA, Herald DL, Hoard MS, Pettit RK, Chapuis J-C (1998) Bioorg Med Chem Lett 8:2093
116. Rueda A, Zubía E, Ortega MJ, Carballo JL, Salvá J (1998) J Nat Prod 61:258
117. Kimura J, Hyosu M (1999) Chem Lett 61
118. Miyaoka H, Nishijima S, Mitome H, Yamada Y (2000) J Nat Prod 63:1369
119. Youssef DTA, Yamaki RK, Kelly M, Scheuer PJ (2002) J Nat Prod 65:2
120. Tsukamoto S, Miura S, Van Soest RWM, Ohta T (2003) J Nat Prod 66:438
121. Pettit GR, Tan R, Cichacz ZA (2005) J Nat Prod 68:1253
122. Youssef DTA, Shaala LA, Emara S (2005) J Nat Prod 68:1782
123. Hernandez-Guerrero CJ, Zubia E, Ortega MJ, Luis Carballo J (2006) Tetrahedron 62:5392
124. Yu Z-G, Bi K-S, Guo Y-W (2005) Helv Chim Acta 88:1004
125. Stessman CC, Ebel R, Corvino AJ, Crews P (2002) J Nat Prod 65:1183

126. Nam S-J, Ko H, Shin M, Ham J, Chin J, Kim Y, Kim H, Shin K, Choi H, Kang H (2006) Bioorg Med Chem Lett 16:5398
127. Aoki S, Higuchi K, Isozumi N, Matsui K, Miyamoto Y, Itoh N, Tanaka K, Kobayashi M (2001) Biochem Biophys Res Commun 282:426
128. Wan Y-Q, Su J-Y, Zeng L-M, Wang M-Y (1996) Gaodeng Xuexiao Huaxue Xuebao 17:1747
129. Jaspars M, Jackson E, Lobkovsky E, Clardy J, Diaz MC, Crews P (1997) J Nat Prod 60:556
130. Jiménez JI, Yoshida WY, Scheuer PJ, Lobkovsky E, Clardy J, Kelly M (2000) J Org Chem 65:6837
131. Jiménez JI, Yoshida WY, Scheuer PJ, Kelly M (2000) J Nat Prod 63:1388
132. Chang LC, Otero-Quintero S, Nicholas GM, Bewley CA (2001) Tetrahedron 57:5731
133. Roy MC, Tanaka J, De Voogd N, Higa T (2002) J Nat Prod 65:1838

Top Heterocycl Chem (2007) 11: 259–281
DOI 10.1007/7081_2007_080
© Springer-Verlag Berlin Heidelberg
Published online: 4 July 2007

Natural Sulfated Polysaccharides for the Prevention and Control of Viral Infections

Carlos A. Pujol[1] (✉) · María J. Carlucci[1] · María C. Matulewicz[2] ·
Elsa B. Damonte[1]

[1]Departamento de Química Biológica, Facultad de Ciencias Exactas y Naturales,
Universidad de Buenos Aires, Pabellon 2, Ciudad Universitaria, 1428 Buenos Aires,
Argentina
capujol@qb.fcen.uba.ar

[2]Departamento de Química Orgánica, CIHIDECAR-CONICET,
Facultad de Ciencias Exactas y Naturales, Universidad de Buenos Aires, Pabellon 2,
Ciudad Universitaria, 1428 Buenos Aires, Argentina

Abstract The inhibitory action of polyanionic substances on virus replication was reported more than 50 years ago. Seaweeds, marine invertebrates, and higher plants represent abundant sources of novel compounds of proved antiviral activity. Natural sulfated polysaccharides (SPs) are potent in vitro inhibitors of a wide variety of enveloped viruses, such as herpes simplex virus (HSV) types 1 and 2, human immunodeficiency virus (HIV), human cytomegalovirus (HCMV), dengue virus (DENV), respiratory syncytial virus (RSV), and influenza A virus. Several polysulfate compounds have the potential to inhibit virus replication by blocking the virion binding to the host cell. In contrast, their in vivo efficacy in animal and human systemic infections has undesirable drawbacks, such as poor absorption, toxic side effects, inability to reach target tissues, and anticoagulant properties. At the present time, SPs have been tested in clinical trials as topical microbicides to prevent sexually transmitted diseases caused by diverse pathogens including viruses, bacteria, fungi, and parasites. The resistance to antiviral agents that arises during drug treatment is one of the reasons for the continuous search for new compounds. In this respect, SPs are considered suitable tools to prevent viral infections in humans and to be used as a new strategy for antiviral chemotherapy.

Keywords Antiviral activity · Chemical structure · Natural sulfated polysaccharides · Polyanions · Seaweeds

Abbreviations

AIDS	Acquired immunodeficiency syndrome
ARI	Acute respiratory infection
CC_{50}	50% Cytotoxic concentration
DENV	Dengue virus
DS	Dextran sulfate
GAG	Glycosaminoglycan
HA	Hemagglutinin
HAART	Highly active antiretroviral therapy
HCMV	Human cytomegalovirus
HCV	Hepatitis C virus
HIV	Human immunodeficiency virus
HN	Hemagglutinin neuraminidase
HPV	Human papillomavirus
HS	Heparan sulfate
HSV	Herpes simplex virus
HVEM	Herpesvirus entry mediator
IC_{50}	50% Inhibitory concentration
JEV	Japanese encephalitis virus
MVEV	Murray Valley encephalitis virus
PIV	Parainfluenza virus
RSV	Respiratory syncytial virus
SI	Selectivity index
SP	Sulfated polysaccharide
SX	Sulfated xylan
TBEV	Tick-borne encephalitis virus
TK	Thymidine kinase
WHO	World Health Organization
WNV	West Nile virus
YFV	Yellow fever virus

1
Introduction

There are many classes of chemical compounds with putative antiviral effects. One such class is known broadly as sulfated polysaccharides (SPs). They include sulfated homopolysaccharides, sulfated heteropolysaccharides, sulfo-glycolipids, carrageenans, and fucoidans. SPs are believed to be of potential therapeutic importance because they can mimic sugar-rich molecules known as glycosaminoglycans (GAGs) present in cell membranes. Examples of GAGs which are important in mammalian physiology are heparan sulfate (HS), dermatan sulfate, and chondroitin sulfate. HS receptors on cell surfaces are important in many physiological and pathological processes and are essential points for viral entry in susceptible cells. It has been postulated that SPs may compete for binding sites normally occupied by GAGs and thus inhibit these processes.

Polysaccharides of biological origin (from yeasts, algae, marine inverte-brates, bacteria, higher plants, and fungi) constitute a class of natural products with low mammalian toxicity that are currently regarded as hav-ing many biological properties, such as anticoagulant, antithrombotic, anti-inflammatory, antitumoral, contraceptive, and antiviral activities. They have diverse functions in their tissues of origin, presenting complex and often het-erogeneous structures, and therefore they are not easy subjects for structural determination [1].

Many studies have been conducted to investigate the in vitro antiviral activity of various SPs. Studies have generally concentrated on synthetic dex-tran sulfates (DSs), pentosan sulfates, clinically used heparins, and seaweed-derived carrageenans. Some reviews reported that sulfated homopolysaccha-rides are more potent than sulfated heteropolysaccharides [2,3]. In general, polysaccharides exhibiting antiviral potential are highly sulfated [4].

The marine environment provides a rich source of chemical diversity for the screening and identification of new compounds with desirable antivi-ral properties [5,6]. Of interest in this context are polysaccharides produced by some species of algae. These compounds have shown promising activ-ity against a variety of animal enveloped viruses, such as herpes simplex virus (HSV) types 1 and 2, human immunodeficiency virus (HIV), human cytomegalovirus (HCMV), dengue virus (DENV), respiratory syncytial virus (RSV), human papillomavirus (HPV), and influenza A virus [3]. Previous studies showed that polysulfates have the potential to inhibit virus replication by shielding off the positively charged sites of the viral envelope glycopro-tein which are necessary for virus attachment to cell surface HS, a primary binding site, before more specific binding occurs to the cell receptors [7,8]. This general mechanism also explains the broad antiviral activity of polysul-fates against enveloped viruses. Variations in the viral envelope glycoprotein region may result in differences in the susceptibility of different enveloped viruses to this type of compound [3]. On the contrary, other studies showed that these polysaccharides did not interfere with virus attachment or penetra-tion, but they did prevent viral protein synthesis [9,10].

In view of the dramatic situation of the global HIV/AIDS epidemic, and a possible spread of avian influenza and other viral diseases, the search for potent antiviral agents is urgent. Effective antiviral therapeutics are not available, and the presently approved therapy for HIV (HAART) has been recognized to be toxic, unable to eradicate the causative virus, and to induce severe drug resistance [11]. In this situation, more attention should be paid to the search for antiviral agents present in natural products. Marine algae are one of the richest sources of bioactive compounds, and need to be thoroughly investigated [12].

2
Structural Characteristics

2.1
Sulfated Polysaccharides from Red Seaweeds

The major matrix-phase polysaccharides are sulfated galactans. These galactans consist mainly of linear chains of alternating 3-linked β-D-galactopyranosyl and 4-linked α-galactopyranosyl units, which are classified either as carrageenans if the 4-linked residue is in the D configuration or agarans if the 4-linked residue is in the L configuration. A substantial part of the 4-linked residues may exist in the form of a 3,6-anhydro derivative.

2.1.1
Carrageenans

Carrageenans are only seldom pyruvylated or methoxylated. Variations in their structure originate mainly from their sulfation pattern and/or the appearance of 3,6-anhydrogalactose [13–15]. Taking into account the sulfation of the 3-linked unit, carrageenans may be grouped into two main families, each of them including different idealized structures. In the κ family, the 3-linked unit is sulfated in the 4-position and comprises four natural idealized structures which are designated by the Greek letters μ-, ν-, κ-, and ι-carrageenan (Fig. 1). Sulfation on the 2-position gives rise to the λ family, λ- and θ-carrageenan being the corresponding natural idealized structures (Fig. 2). Recently, it has been reported [16] that *Callophyllis hombroniana* biosynthesizes mainly θ-carrageenan; in *C. variegata*, θ-carrageenan was also found, but together with a novel carrageenan backbone constituted by β-D-galactose 2-sulfate linked to α-D-galactose 2,3,6-trisulfate and β-D-galactose 2,4-disulfate linked to 3,6-anhydro-α-D-galactose 2-sulfate as dominant repeating units (Fig. 3) [17]. Native polysaccharides obtained from various seaweeds are constituted mainly by one of these repeating units, but more often they are "hybrids" containing the same molecule regions formed by the "ideal" sequences shown above [13–15].

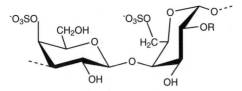

R = H μ -carrageenan

R = SO₃⁻ ν-carrageenan

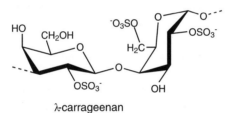

R = H κ-carrageenan

R = SO₃⁻: ι-carrageenan

Fig. 1 Carrageenans of the κ family

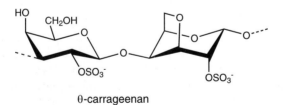

λ-carrageenan

θ-carrageenan

Fig. 2 Carrageenans of the λ family

2.1.2
Agarans

Agarans are typically low in sulfate ester substitution, but those from numerous sources are rich in methyl ether or pyruvate acetal substitution; in addition, glycosyl substitution (galactosyl, 4-O-methyl-L-galactosyl and/or xylo-

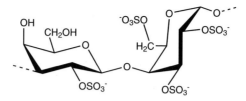

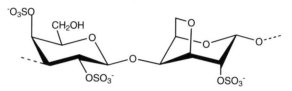

Fig. 3 Novel repeating units in the carrageenans from *Callophyllis variegata*

syl) has also been reported. Due to the complexity and diversity of the substitution pattern, it has not been possible to classify agarans into "ideal structures". Actually, only the term agarose has a strict chemical sense (Fig. 4), whereas other polysaccharides of the group are usually termed according to the algal species from which they were isolated [13–15]. Figure 4 also shows the basic structure of the agaran backbone sulfated on the 6-position of the α-L-galactopyranosyl residue (porphyran).

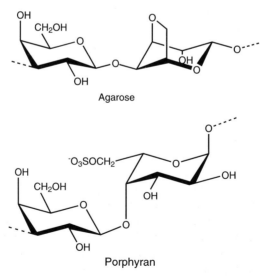

Fig. 4 Agarose and porphyran repeating units

2.1.3
DL-Hybrid Galactans

As has been mentioned above, in carrageenans the 4-linked residues are in the D configuration, while in agarans they belong to the L series. However, in several red seaweeds, galactans which do not fit into this classification were found. These polysaccharides, called DL-hybrid galactans, contain 4-linked units in both D and L configuration. Even though there is a lot of circumstantial evidence which suggests the real presence of these hybrids, no absolute proof of their existence has ever been obtained. Actually, the absolute proof would be the isolation of oligosaccharide fragments derived from the junction region of the agaran and carrageenan moieties [14, 15, 18]. Figure 5 shows a possible agaran and carrageenan domain in DL-hybrid galactans.

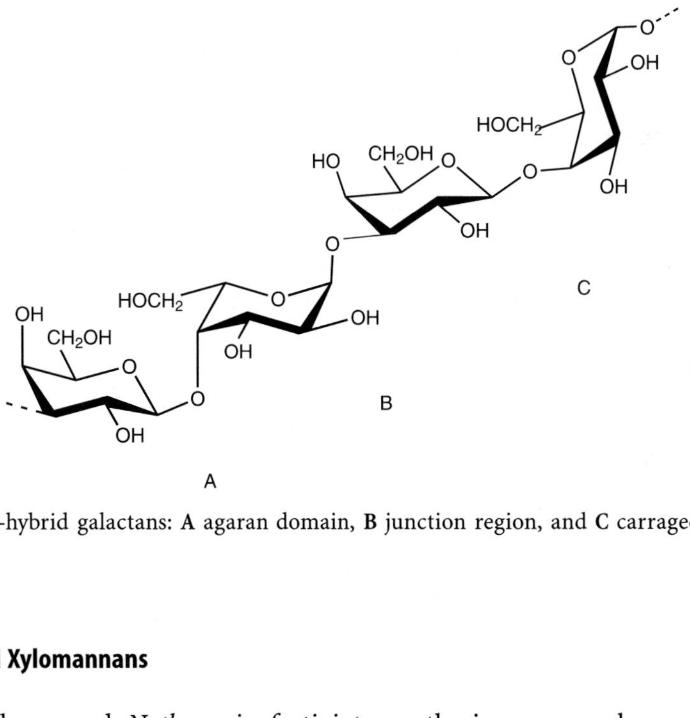

Fig. 5 DL-hybrid galactans: **A** agaran domain, **B** junction region, and **C** carrageenan domain

2.1.4
Sulfated Xylomannans

The red seaweed *Nothogenia fastigiata* synthesizes a complex system of polysaccharides comprising neutral xylans of the β-D-(1→3), β-D-(1→4) "mixed linkage" type, sulfated xylogalactans of the agaran type, and a family of α-(1→3)-linked D-mannans 2- and 6-sulfated and having single stubs of β-(1→2)-linked D-xylose [19–21]. Figure 6 shows the general substitution pattern of these xylomannans. Similar mannan backbones, but sulfated in the 4- and 6-positions, were detected in *Nemalion vermiculare* [13, 22] and *Liagora valida* [13, 23].

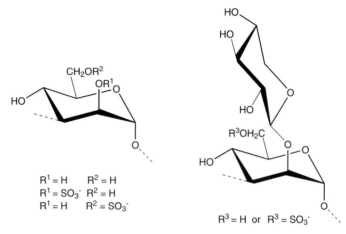

$R^1 = H \quad R^2 = H$
$R^1 = SO_3^- \quad R^2 = H$
$R^1 = H \quad R^2 = SO_3^-$

$R^3 = H$ or $R^3 = SO_3^-$

Fig. 6 Mannans from *Nothogenia fastigiata*

2.2
Polysaccharides from Brown Seaweeds

Brown seaweeds are known to produce different polysaccharides, namely alginates, laminarans, and fucoidans. Fucoidans always contain essentially L-fucose and sulfate, together with minor amounts of D-xylose, D-galactose, D-mannose, and D-glucuronic acid [24]. In spite of the many studies attempting to determine the fine structure of the fucoidans, only a few examples of regularity were found. The differences in the structural details are not due to the heterogeneity of the samples but to extreme compositional and structural dispersion, much larger than that normally found in plant polysaccharides [15, 24, 25].

When fucoidans of *Adenocystis utricularis* and *Sargassum stenophyllum* were precipitated with cetrimide and the insoluble complexes were subjected to fractional solubilization in solutions of increasing sodium chloride concentration, it was observed that fractions with high percentages of uronic acid and low sulfate content were dissolved at sodium chloride concentrations lower than 2.0–2.5 M: these fractions contained major amounts of L-fucose, D-xylose, and D-galactose, together with mannose and glucose. At higher sodium chloride concentrations, fractions with high sulfate content and comprising mainly fucose and galactose were obtained. Thus, fucoidans may be grouped into two sets: uronofucoidans and galactofucans [24, 25].

Structural analysis of the galactofucans from *Adenocystis utricularis* showed that the fucan constituent was mainly composed of a 3-linked α-L-fucopyranosyl backbone, mostly sulfated at C-4, and branched at C-2 with nonsulfated fucopyranosyl and fucofuranosyl units and 2-sulfated fucopyranosyl residues. The galactan moiety was more heterogeneous, with galactopyranose units linked on C-3 and C-6 and sulfation mostly on C-4 [24].

When Bilan et al. [26] fractionated the crude fucoidan of *Fucus evanescens* by ion-exchange chromatography on DEAE-Sephacel, using aqueous sodium chloride of increasing sodium chloride concentration as eluent, a fraction which was essentially a homofucan sulfate was isolated. This homofucan was shown to contain a linear backbone of alternating 3- and 4-linked-α-L-fucopyranose 2-sulfate residues. Also, some additional sulfation was observed on C-4 of the 3-linked residues, whereas a part of the remaining hydroxyl groups was randomly acetylated.

2.3
Other Polysaccharides

A broad range of polysaccharides have emerged as an important class of bioactive molecules which occur naturally in a great variety of plants and microorganisms. Recently, it has been reported that sulfated cellulose could be used to prevent and treat HPV; however, there is no clear information about the positions of sulfation [27, 28] or if cellulose is persulfated.

Previous studies revealed that the sodium spirulan isolated from *Spirulina platensis* consists of 3-linked rhamnopyranosyl, 2-linked 3-*O*-methylrhamnopyranosyl, and 3,4-linked hexuronopyranosyl residues; the first unit was sulfated on C-2 or disulfated on C-2 and C-4, and the second one was sulfated on C-4. Recently, Lee et al. [29] evaluated the effect of partial desulfation and oversulfation of this polysaccharide against HSV-1 and HSV-2 (see below).

3
Spectrum of Antiviral Activity

Natural SPs are effective in inhibiting a wide range of enveloped viruses, whereas nonenveloped viruses are not significantly susceptible to these compounds. The degree of inhibitory activity varies with the compound and the virus. The current status of antiviral studies performed with the human pathogenic viruses more susceptible to these polysulfates is presented in this section.

3.1
Herpesviruses

HSV was one of the first viruses to be reported as susceptible to the antiviral action of SPs more than 50 years ago, when the antiherpetic effect of synthetic and natural polyanionic substances was reported [30, 31]. From these initial studies, and based on the need to find new antiviral strategies to combat herpetic infections, the more prevalent human pathogens of the family *Herpesviridae*, such as HSV type 1 (HSV-1), HSV type 2 (HSV-2), and HCMV, have turned out to be one of the main candidate groups for evaluation of polysulfates [32]. In fact, these herpesviruses represent viral agents highly susceptible to the effect of SPs. This field of antiviral research came into the focus of interest of many recently published studies evaluating the inhibitory effect against herpesviruses of natural polysaccharides isolated from algae, cyanobacteria, and plants with diverse structural characteristics.

Among algal derivatives, the SPs extracted from red seaweeds represent the more extensively analyzed type of polyanionic virus inhibitors. In this group, the highly effective antiherpesvirus compounds include: xylomannans from *Nothogenia fastigiata* [21]; carrageenans from *Gigartina skottsbergii* [33], *Gymnogongrus griffithsiae* [34], *Meristiella gelidium* [35], and *Stenogramme interrupta* [36]; galactans from *Bostrychia montagnei* [37], *Gymnogongrus torulosus* [38], *Cryptonemia crenulata* [34], *Callophyllis variegata* [17], and *Schizymenia binderi* [39]; and agarans from *Acantophora spicifera* [40].

Although much less studied in their chemical properties and biological activities than the red seaweeds, the brown and green algae have also been reported as providers of antiviral SPs. In general, sulfated fucans and fucoidans are the main antiviral polysaccharidic components extracted from American and Asian located brown algae of the species *Leathessia difformis* [41], *Adenocystis utricularis* [24], *Stoechospermum marginatum* [42], *Undaria pinnatifida* [43, 44], *Sargassum horneri* [45], and *Sargassum patens* [46, 47]. Some SPs from the green algae *Enteromorpha compressa*, *Monostroma nitidum*, *Caulerpa brachypus*, *Caulerpa okamurai*, *Caulerpa scapelliformis*, *Caulerpa racemosa*, *Chaetomorpha crassa*, *Chaetomorpha spiralis*, *Codium adhaerens*, *Codium fragile*, and *Codium latum*, and four synthetic sulfated xylans (SXs) prepared from the β-1,3-xylan of *Caulerpa brachypus*, showed potent anti-HSV-1 activities with 50% inhibitory concentrations (IC_{50}) of 0.38–8.5 $\mu g\,ml^{-1}$, while having low cytotoxicities. Anti-HSV-1 activities of SXs were dependent on their degrees of sulfation. Some polysaccharides obtained from *Caulerpa brachypus* and *Codium latum* showed strong anti-HSV-1 activities even when added to the medium 8 h postinfection [10]. The compounds mainly containing glucose, xylose, galactose, and arabinose are promising antiherpetic agents [10, 48, 49].

Microorganisms are also an interesting alternative source of SPs with antiviral activity against herpesvirus. In particular, calcium spirulan and intracellular and extracellular spirulan-like substances with pronounced and selective antiviral activity against HSV-1 and HCMV were isolated from cyanobacteria (previously named blue-green algae), such as *Spirulina platensis* [50] and *Arthrospira platensis* [51], whereas marine *Pseudomonas* species produced anti-HSV-1 active extracellular GAGs and SPs [52]. Interestingly, an acidic polysaccharide named nostoflan, isolated from the terrestrial cyanobacterium *Nostoc flagelliforme*, was also found to be active against HSV-1 and HCMV [53]. Nostoflan might be mainly composed of two types of sugar sequence and exhibited a selectivity index of 13 000, as high as those corresponding to the most effective SPs. Also naviculan, a polysaccharide isolated from the diatom *Navicula directa*, showed antiviral activity against HSV-1 and HSV-2 [54].

Polysaccharides active against herpesvirus were also obtained from plant extracts like *Prunella vulgaris* [55, 56] and *Cedrela tubiflora* [57]. In an interesting study performed by Liu et al. [58], they reported the anti-HSV activity of a neutral polysaccharide isolated from a traditional Chinese medicinal herb, *Polygonatum cyrtonema* Hua, as well as the hydrolyzed fragment derived from the polymer. The authors demonstrated that oligosaccharides with a degree of polymerization of 3–5 without branches retained antiviral activity. Similarly, a low molecular weight substance named PI-88, which is a mixture of highly sulfated mannose-containing di- to hexasaccharides, inhibited HSV infection of cells and cell-to-cell spread of HSV-1 and HSV-2 [59]. Compared to a relatively large heparin polysaccharide, PI-88 demonstrated weaker inhibition of HSV infectivity but more efficient reduction of cell-to-cell spread of HSV. A tetrasaccharide fraction of PI-88 was the minimum fragment necessary to inhibit HSV-1 infectivity, while a trisaccharide was sufficient to reduce cell-to-cell spread. These small oligosaccharides, of natural or synthetic origin, represent a therapeutically valuable alternative to be assayed in systemic in vivo model infections.

From all this wide spectrum of polysulfates, the red seaweed-derived polysaccharides represent the most potent and selective antiviral agents able to block HSV replication in cell culture at concentrations as low as $0.1-1 \mu g \, ml^{-1}$ without causing cell toxicity at concentrations up to $1-5 \, mg \, ml^{-1}$. Thus, the selectivity index (SI), i.e., the relationship between the cytotoxic dose (measured as 50% cytotoxic concentration (CC_{50}), concentration required to reduce cell viability by 50%) and the effective antiviral dose (determined as IC_{50}, concentration required to reduce virus cytopathic effect by 50%), is in the order 1000–50 000, values not easy to afford for any compound. Furthermore, carrageenans and galactans are effective inhibitors of herpesviruses independently of the antiviral assay, either cytopathic effect, plaque number, virus yield, or antigen expression reduction tests, and, more importantly, their effectiveness is not significantly affected by the input multiplicity of infection. For example, a substantial increase in virus inoculum, approximately 1000–10 000 times, required for the virus yield experiments in comparison with plaque reduction assays did not greatly alter the antiviral effectiveness of diverse types of carrageenans [33], in contrast to the results obtained with other antiviral drugs. An additional hallmark is the fact that these selective antiviral polysulfates can be obtained from the sea algae, and thus they can be prepared and made available in large quantities at low cost. At this point it must be noticed that both homogeneous purified compounds and crude polysaccharide extracts are highly active and selective anti-HSV agents.

Mode of action studies are consistent with the initial virus attachment to the host cell receptor as the main target of SPs isolated from natural sources or of commercial origin [33, 38, 60]. In the antiviral assays performed in vitro, this class of compounds was effective only when added simultaneously with virus or immediately after infection, and usually full inhibitory activity was achieved only when the compound was present during the virus adsorption period. For HSV, the initial step in the multiplication cycle is the binding of the glycoprotein gC, and in some cases the glycoprotein gB, to cell surface GAGs, preferentially HS chains, which are found in the form of proteoglycans on cell surfaces or in the extracellular matrix of all mammalian organs and tissues [61–63]. After attachment to HS, the viral glycoprotein gD binds to a second set of receptors or coreceptors required for entry and termed herpesvirus entry mediators (HVEMs) [64]. The interaction gD–HVEMs triggers the fusion of the virion envelope with the cell membrane to allow virus penetration to the cell cytoplasm. The coreceptors are variable according to the cell type, but always the initial binding to HS allows concentration of the virus on the cell surface and hence facilitates its subsequent binding to the coreceptor and the efficiency of the infection. Thus, the SP inhibitors of herpesviruses interfere with the interaction of gC with HS by occupying the sites in the virion envelope necessary for attachment of virions to HS [65].

The interaction of SP with the external glycoprotein may not only prevent the attachment of virus to the cell receptor blocking adsorption, but also the possibility of a direct inactivating effect on virion infectivity can be considered. This virucidal activity is measured by pretreatment of the virus suspensions with the polysulfates in a cell-free system, and then the mixtures are diluted and incubated with the cell cultures to determine the remaining virus infectivity. In most cases, virion pretreatment with SP did not produce loss of infectivity. Occasionally, virucidal activity was detected for some compounds but at concentrations highly exceeding the IC_{50}, and a noticeable exception of polysulfate with virucidal activity against HSV was the λ-carrageenan from *Gigartina skottsbergii* [33]. It may be assumed that virucidal polysulfates bind with high affinity to the virion, leading to the formation of a very stable virion-compound complex without chance of reversibility when the complex is added to the cell. The irreversible and virucidal action of the λ-carrageenan against HSV-2 seems to be responsible for the protective effect on vaginal infection in a murine model [66]. Besides, despite the lack of virucidal activity, the partially cyclized μ/ν-carrageenan from the same alga protected mice against intraperitoneal infection with HSV-1 [67]. None of these carrageenans exhibited significant levels of cytotoxicity or anticoagulant activity [68]. Zeitlin et al. tested a range of antiviral substances for their possible effectiveness as vaginal microbicides against genital herpes in mice, and found that carrageenan and fucoidan are good candidates for further development [69].

The serial passage in cell culture of HSV-1 in the presence of increasing concentrations of carrageenans has shown a very slow induction of drug resistance. In contrast to the behavior of other antiviral drugs such as acyclovir, which select very rapidly for highly resistant mutants, several passages were required to obtain HSV variants with a reduced susceptibility to the SP [70]. However, the isolation of viral resistant mutants is very slow but consistent, indicating the requirement of a specific interaction between virus and polysulfates. In addition, polysulfates were inhibitory against thymidine kinase (TK) acyclovir-resistant HSV-1 mutants [33, 34]. The lack of cross-resistance to nucleoside analogues is not surprising, given the differential target of both types of inhibitors. But this finding opens an interesting alternative for the tentative use of SPs in combination with nucleoside analogues against HSV infections, particularly in immunocompromised patients who require prolonged antiviral treatment and are prone to select for drug-resistant strains.

3.2
Retroviruses

HIV infection in humans is now pandemic. As of January 2006, the Joint United Nations Programme on HIV/AIDS (UNAIDS) and the World Health Organization (WHO) estimate that AIDS has killed more than 25 million people since it was first recognized on December 1, 1981, making it one of the most destructive pandemics in recorded history. Antiretroviral treatment reduces both the mortality and the morbidity of HIV infection, but routine access to antiretroviral medication is not available in all countries. An alternative therapy to circumvent this problem is the use of polyanionic substances, which demonstrated a number of promising features as potential anti-HIV drug candidates. In this respect, various SPs (e.g., heparin, DS, dextrin sulfate, cyclodextrin sulfate, curdlan sulfate, pentosan polysulfate, mannan sulfate, sulfoevernan, and fucoidan) and derivatives thereof (e.g., O-acylated heparin, polyacetal polysulfate, polyvinyl alcohol sulfate, and modified cyclodextrin sulfates) have been found to inhibit HIV replication in vitro at concentrations that are up to 10 000-fold lower than the cytotoxic concentration [2, 71].

These compounds are targeted at the interaction between the viral envelope glycoprotein gp120 and the CD4 receptor. They not only inhibit HIV-1-induced cytopathogenicity and HIV-1 antigen expression, but also inhibit the activity of purified reverse transcriptase and RNase H, which are essential for retrovirus replication [72]. SPs may act synergistically with other anti-HIV drugs (e.g., azidothymidine (AZT)). They are known to lead very slowly to virus drug-resistance development and they show activity against HIV mutants that have become resistant to reverse transcriptase inhibitors, such as AZT, tetrahydroimidazo[4,5,1-*jk*]-1,4-benzodiazepin-2(1*H*)-thione (TIBO), and others [3].

In the search for new sources of anti-HIV compounds it was found that the SP named naviculan, isolated from *Navicula directa*, had a marked inhibitory effect on cell–cell fusion between CD4-expressing and HIV gp160-expressing cells [54]. The gametic and tetrasporic reproductive stages from the Mediterranean red alga *Asparagopsis armata* contain sulfated galactans that inhibited HIV replication in vitro at 10 and 8 μg ml^{-1}, respectively, as measured by HIV-induced syncytium formation as well as reverse transcriptase activity in cell-free culture supernatant. The maximum antiviral effect involved the presence of the polysaccharides after or during infection but not before infection [4]. In addition, a synthetic polysaccharide prepared by sulfation of astragalus polysaccharide, obtained from *Astragalus membranaceus* used as a Mongolian herbal medicine, showed high anti-HIV activity and low cytotoxicity when assayed in vitro [73].

Many results show that several sulfonated polysaccharides inhibit both X4 and R5 viruses in vitro. HIV-1 isolates from newly infected individuals are predominantly M-tropic and utilize CCR5 (R5), while T-tropic isolates that use CXCR4 (X4) evolve later in the course of the disease. Given that these sulfonated polysaccharides are negatively charged molecules, it is believed that their mechanism of action is to bind to the positively charged region of the viral envelope [74, 75].

A number of reports have suggested that sulfated compounds, such as DS, bind the V3 loop of X4 viruses more readily than they bind the loop of R5 viruses [76]. Consistent with this is the demonstration that DS fails to inhibit R5 strains in vitro [75]. Because the envelope of R5 viruses is less positively charged than that of X4 viruses [77, 78], there is controversy over whether sulfated polymers will be effective in blocking R5 viruses. Ability to block R5 viruses is critical for a microbicide, because even if both phenotypes are transmitted together, it is the R5 viruses that are amplified during the initial infection [79, 80].

3.3
Flaviviruses

Flavivirus is a genus of the family Flaviviridae composed of nearly 70 arthropod-borne viruses that cause important human diseases, such as yellow fever virus (YFV), DENV, West Nile virus (WNV), and Japanese encephalitis virus (JEV). They cause a variety of diseases including fever, encephalitis, and hemorrhagic fever. In particular, DENV has reemerged in recent years as an increasingly important public health threat affecting more than 100 countries worldwide, with nearly 50 million infections each year and over 2.5 billion people at risk [81].

Flaviviruses are included among the enveloped viruses recently reported as dependent on cell surface HS to efficiently initiate cell infection. An involvement of HS during attachment and entry through its binding to the virion envelope glycoprotein E was initially demonstrated for DENV [82] and then extended to YFV [83], tick-borne encephalitis virus (TBEV) [84], and Murray Valley encephalitis virus (MVEV) [85], as well as to hepatitis C virus (HCV), a member of the *Hepacivirus* genus of Flaviviridae [86].

In accordance with the finding of highly sulfated HS as a putative primary receptor for these viruses, the effectiveness of heparin to prevent infection of cultured cells by DENV was demonstrated [82, 83, 87], whereas diverse algal SPs were also assayed and found able to block DENV infection. The effective compounds included galactans [17, 38] and carrageenans [35, 88] extracted from red seaweeds collected from South America. The IC_{50} values of these natural SPs against DENV-2 infection of monkey and human cell lines were in the range $0.1 - 1\ \mu g\ ml^{-1}$. Given their lack of toxicity at concentrations as high as $1\ mg\ ml^{-1}$, these SPs can be considered very selective antidengue agents able to inhibit the in vitro replication of DENV-2 at concentrations that were up to 10 000-fold lower than the cytotoxic concentrations. A noticeable property exhibited by the carrageenans and galactans derived from the seaweed *Cryptonemia crenulata* was the dependence of their antiviral activity on the DENV serotype and the host cell [88]. This virus presents four serotypes which cocirculate worldwide. The serotypes 2 and 3 are very susceptible to the inhibitory activity of the polysulfates but, unfortunately, the effectiveness of these compounds against DENV-1 and DENV-4 is very low or negligible. A similar situation has been described for heparin [89]. With respect to the host cell, the polysaccharides were inhibitors of DENV-2 and DENV-3 infection in mammalian cells but were inactive in mosquito cells [88].

Surprisingly, mechanistic studies demonstrated that carrageenans and galactans act not only by preventing DENV-2 adsorption to the host cell, but also by blocking the complete process of internalization and release of viral genome into the cytoplasm. In fact, the DL-galactan hybrid C2S-3 was found to be an inhibitor of HSV attachment without any subsequent effect in the HSV multiplication cycle, but the same compound could block both events of attachment and internalization in DENV-2 infection [88, 90]. Probably, the dissimilar action against DENV and HSV may be ascribed to differences in the internalization process between both viruses.

Interestingly, the antiviral activity of HS mimetics against flaviviruses was also demonstrated in vivo. Two sulfated galactomannans extracted from seeds of *Mimosa scabrella* and *Leucaena leucocephala*, named BRS and LLS, respectively, protected mice against intraperitoneal infection with YFV [91]. More recently, the oligosaccharide PI-88 was assayed in vitro and in murine models of flavivirus infection. This low molecular weight compound did not show in vitro effectiveness but ameliorated disease in JEV and DENV-2 infected mice, providing a note of caution about the predictive accuracy of in vitro assays for the in vivo therapeutic activity [92].

Structure–activity relationship studies carried out with different heparin-like polyanions, including small polyanions, GAGs, and persulfated GAGs, have demonstrated the need for a minimum chain size equivalent to the heparin decasaccharide, as well as a high charge density and structural flexibility for optimal interaction between the polyanion and the E protein of DENV [93]. The heparin-derived decasaccharide was similar in potency to heparin with IC_{50} values of 0.3 μg ml^{-1}. Taking account of this information may be helpful for the design of new DENV binding/entry inhibitors with maximum antiviral effectiveness.

3.4
Respiratory Viruses

The main causative agents of acute respiratory infections (ARI) in infants and children are mostly thought to be viruses. When ARIs affect immunocompromised patients or the elderly, the mortality rates are significantly higher than in immunocompetent individuals. Many types of viruses cause ARI. Among them, influenza viruses A and B and RSV are very harmful because of the severity of illness after infection and their high communicability in the human population [94]. Influenza, commonly known as flu, is an infectious disease of birds and mammals caused by an RNA virus of the family Orthomyxoviridae. In humans, common symptoms of influenza infection are fever, sore throat, muscle pains, severe headache, coughing, weakness, and fatigue. Typically, influenza is transmitted from infected mammals through the air by coughs or sneezes, creating aerosols containing the virus, and from infected birds through their feces. Influenza can also be transmitted by saliva, nasal secretions, feces, and blood.

At the beginning of the infection, the virus particles are attracted to the cell membrane electrostatically and then bind to a specific receptor via the hemagglutinin (HA), hemagglutinin neuraminidase (HN), or G proteins of virions. SPs, which are negatively charged, are thought to disturb nonspecifically the approach and binding of the virions to receptors. Following this mode of action, DS inhibits the replication of influenza A virus and RSV but not the replication of influenza B virus, measles virus, or parainfluenza viruses type 3 (PIV-3) [95].

Fusion experiments with different influenza subtypes (H1N1 and H3N2) demonstrated that DS (8 and 500 kDa) and pentosan polysulfate strongly inhibit the fusion activity as well as the in vitro replication of influenza virus [96]. The good correlation of antifusion and antireplication effects strongly suggests that the antiviral properties of polymeric anions might be based on their antifusion activity.

As regards the antiviral compounds derived from microorganisms, the polysaccharides nostoflan and naviculan proved to be active against influenza A virus [54, 97]. Chemically synthesized oversulfated derivatives of extracellular GAG and SP, produced by a marine *Pseudomonas*, prepared by dicyclohexylcarbodiimide-mediated reaction for both polysaccharides, showed antiviral activities against influenza virus type A but not against type B [52].

A natural SP OKU-40 was extracted from the marine microalga *Dinoflagellata* and was found to inhibit the replication of HIV, RSV, influenza A and B viruses, measles virus, and parainfluenza viruses type 2 (PIV-2). However, it did not inhibit the replication of mumps virus or PIV-3 [98]. The action of negatively charged polysaccharides is not merely one of nonspecific inhibition of the binding of an enveloped virus to receptors. In fact, OKU-40 did not inhibit the binding of HIV or influenza A virus to the cell membrane, but it did inhibit the fusion of the membranes of HIV-infected MOLT-4 cells to those of uninfected cells and the fusion of the influenza A virus envelope to uninfected MDCK cells [99].

3.5
Papillomavirus

Sexually transmitted HPV infections are very common. Although most HPV infections do not cause noticeable symptoms, persistent infection with some genital HPV types can lead to cervical cancer or other anal/genital cancers [100]. Another subset of HPV types can cause genital warts. Recent studies have suggested that condoms are not highly effective in preventing HPV infection. Although a new vaccine called Gardasil became available, it will not protect against all genital HPV types and is too expensive for use in the developing world [101]. Inexpensive HPV-inhibitory compounds (known as topical microbicides) might be useful for blocking the spread of HPV. Comparison of a variety of compounds revealed that carrageenan is an extremely potent inhibitor for a broad range of sexually transmitted HPVs. Although carrageenan can inhibit HSV and some strains of HIV in vitro, genital HPVs are about 1000-fold more susceptible, with IC_{50} values in the low $ng\,ml^{-1}$ range. Carrageenan acts primarily by preventing the binding of HPV virions to cells. This finding is consistent with the fact that carrageenan resembles HS, an HPV cell-attachment factor. However, carrageenan is three orders of magnitude more potent than heparin, a form of cell-free HS that has been regarded as a highly effective model HPV inhibitor.

Since carrageenan might have utility as a topical microbicide for preventing the sexual transmission of HPVs, it is important to consider the fact that human vaginal pH is typically below 4.5 [102]. Experiments performed in culture medium buffered to pH 4.5 or 5.0 with lactic or acetic acid, respectively, showed that ι-carrageenan remained effective for blocking infectivity under acidic conditions [103]. Carrageenan can also block HPV infection through a second, postattachment HS-independent effect. Half of the infectious titer remained susceptible to inhibition by ι-carrageenan for up to 12 h after initial binding to cells [103].

Although carrageenan was highly effective for neutralizing different genital HPV types in vitro, it was substantially less potent against several papillomavirus types tropic for nongenital skin. Since common genital HPVs occupy a single genus, and nongenital papillomavirus types are phylogenetically distant from the genital types [104], it is tempting to speculate that all HPVs tropic for the genital mucosa would be comparably susceptible to inhibition by carrageenan. However, the possibility that some genital HPVs might exhibit natural resistance to inhibition by carrageenan would be an important factor to consider in the design of clinical efficacy trials. More clinical trials are still needed to determine whether carrageenan-based products are effective as topical microbicides against genital HPVs.

4
Structural Requirements for Antiviral Activity

The structural requirements for antiviral activity are the following:

- The molecular weight: It is known that the antiviral activity of SPs increases with the molecular weight, tending to level off after ~ 100 kDa, and that the highest activity is in the range of 10–100 kDa [15].
- The anionic groups: Most of the seaweed polysaccharides with antiviral activity carry only sulfate as their anionic group. Nevertheless, in the case of fucoidans both sulfate groups and uronic acids are present. Sulfated seaweed polysaccharides with degrees of sulfation lower than 20–22% usually do not show activity, unless other structural factors compensate the sulfate deficiency.
- The sulfate distribution: a low degree of sulfation does not eliminate the possibility of highly charged zones in the polysaccharide backbone.
- The shape of the carbohydrate chain: The interaction between the sulfate groups of the polysaccharide and the positive charges of the amino acids in the HSV-1 gC protein would be increased if the polysaccharide could not only approach the virus surface but also adapt to it. Examination of the secondary structures of carrageenans, agarans, DL-hybrid galactans, xylomannans, and fucans suggests that all of them are random coils at room temperature but can adopt ordered forms at compelling conditions.
- The influence of the hydrophobic sites: In fucoidans the interaction of the methyl groups with the hydrophobic pocket of HSV-1 gC seems, in addition to the electrostatic forces, to be decisive in the binding of the polysaccharide to the viral protein.

Those polysaccharides with a sulfate pattern structurally resembling the negative charge distribution of the binding sites of HS will competitively inhibit viral binding. For example, in λ-carrageenan sulfation on C-2 of the 3-linked residue and on C-6 of the 4-linked residue mimics the binding sites of HS. In θ-carrageenan the location of the sulfate groups is also adequate. The polysaccharides from *Callophyllis variegata*, which showed a potent antiviral activity, fulfill the features of adequate molecular weight and sulfate arrangement described above [17]. But not only ionic interactions are important; the contribution of hydrophobic interactions to the binding should also be considered [15].

The α-(1→3)-linked D-mannans of *Nothogenia fastigiata* sulfated on C-2 and C-6 also mimic the minimal binding domain of HS. However, the lower antiviral activity of the fractions with D-xylose side chains suggests that branching precludes binding to the gC viral protein [21].

5
Antiviral Effectiveness in Humans

Polysulfates suffer from a number of drawbacks which seem to argue against their potential usefulness in humans. These are short plasma half-life (approximately 1–2 h), rapid degradation in the gut and in plasma, and a poor ability to penetrate and target infected tissues and cells [8]. However, high oral bioavailability can be obtained by appropriate chemical modifications, as shown for the modified β-cyclodextrin sulfates (mCDS11 and mCDS71). Sulfated polymers are also notorious for their anticoagulant activity, but as has been demonstrated with periodate-treated heparin and O-acylated heparin, this problem can be overcome by appropriate chemical modifications [71]. Furthermore, several natural SPs with structural characteristics differing from those of heparin have a potent antiviral action without significant anticoagulant properties [3].

Taking into account the above considerations, the SPs were suggested to be ideal microbicides for topical use [105]. A number of potential candidate microbicides have been shown to inhibit virus attachment, fusion, and entry into host target cells for sexually transmitted infections. These include cellulose sulfate, poly(styrene 4-sulfonate), polystyrene sulfonate, polymethylene-hydroquinone sulfonate, naphthalene sulfonate polymer, and a carrageenan derived from seaweed (Carraguard) that may prevent viral entry. Several candidate compounds have already progressed to various stages of clinical trials [106]. Currently, some of them have been formulated and are in phase II/III clinical trials.

In addition, the majority of microbicide trials are being planned to take place in Sub-Saharan Africa where the majority of infections occur through R5 strains of HIV with a high prevalence of clade C viruses [107]. Currently, PC-815, a novel combination microbicide containing carrageenan and the nonnucleoside reverse transcriptase inhibitor MIV-150, is being tested in a phase I clinical trial [108].

6
Conclusions

In the last few decades, the discovery of SPs from natural sources with potent antiviral activities has increased significantly, but their clinical application against human viral infections is still far from being satisfactory. The therapeutic perspectives of SPs will probably improve with an adequate formulation in a clinically useful vehicle. The development of new drug delivery systems, such as encapsulation in liposomes or nanoparticles, is a strategy currently gaining attention to improve the in vivo effectiveness and reduce the adverse effects of polysulfates. The potential of these natural compounds to prevent a wide spectrum of severe viral diseases warrants further investigation to ameliorate their administration in systemic virus infections.

References

1. Mulloy B (2005) An Acad Bras Cienc 77:651
2. Schaeffer DJ, Krylov VS (2000) Ecotoxicol Environ Saf 45:208
3. Witvrouw M, DeClerq E (1997) Gen Pharmacol 29:497
4. Haslin C, Lahaye M, Pellegrini M, Chermann JC (2001) Planta Med 67:301
5. Kirk R, Gustafson KR, Oku N, Milanowski DJ (2004) Curr Med Chem Anti-Infective Agents 3:233
6. Tziveleka LA, Vagias C, Roussis V (2003) Curr Top Med Chem 3:1512
7. McClure M, Moore J, Blanco D, Scotting P, Cook G, Keyner R, Weber J, Davies D, Weiss R (1992) AIDS Res Hum Retroviruses 8:19
8. Lüscher-Mattli M (2000) Antivir Chem Chemother 11:249
9. Biesert L, Adamski M, Zimmer G, Suhartono H, Fuchs J, Unkelbach U, Mehlhorn R, Hideg K, Milbradt R, Rubsamen-Waigmann H (1990) Med Microbiol Immunol Berl 179:307
10. Lee JB, Hayashi K, Maeda M, Hayashi T (2004) Planta Med 70:813
11. Luescher-Mattli M (2003) Curr Med Chem Anti-Infective Agents 2:219
12. Smit AJ (2004) J Appl Phycol 16:245
13. Usov AI (1992) Food Hydrocolloids 6:9
14. Stortz CA, Cerezo AS (2000) Curr Top Phytochem 4:121
15. Damonte EB, Matulewicz MC, Cerezo AS (2004) Curr Med Chem 11:2399
16. Falshaw R, Furneaux RH, Stevenson DE (2005) Carbohydr Res 340:1149
17. Rodriguez MC, Merino ER, Pujol CA, Damonte EB, Cerezo AS, Matulewicz MC (2005) Carbohydr Res 340:2742
18. Takano R, Hayashi K, Hara S (1997) Recent Res Dev Phytochem 1:195
19. Matulewicz MC, Cerezo AS (1987) Carbohydr Polym 7:121

20. Kolender AA, Matulewicz MC, Cerezo AS (1995) Carbohydr Res 273:179
21. Kolender AA, Pujol CA, Damonte EB, Matulewicz MC, Cerezo AS (1997) Carbohydr Res 304:53
22. Usov AI, Adamyants KS, Yarotskii SV, Anoshina AA (1975) Zh Obshch Khim 45:916
23. Usov AI, Dobkina IM (1988) Bioorg Khim 14:642
24. Ponce NMA, Pujol CA, Damonte EB, Flores ML, Stortz CA (2003) Carbohydr Res 338:153
25. Duarte MER, Cardoso MA, Noseda MD, Cerezo AS (2001) Carbohydr Res 333:281
26. Bilan MI, Grachev AA, Ustuzhanina NE, Shashkov AS, Nifantiev NE, Usov AI (2002) Carbohydr Res 337:719
27. Christensen ND, Reed CA, Culp TD, Hermonat PL, Howett MK, Anderson RA, Zaneveld LJD (2001) Antimicrob Agents Chemother 45:3427
28. Cheshenko N, Keller MJ, MasCasullo V, Jarvis GA, Cheng H, John M, Li J-H, Hogarty K, Anderson RA, Waller DP, Zaneveld LJD, Profy AT, Klotman ME, Herold BC (2004) Antimicrob Agents Chemother 48:2025
29. Lee JB, Hayashi T, Hayashi K, Sankawa U, Maeda M (1998) J Nat Prod 61:1101
30. Nahmias AJ, Kibrick S, Bernfeld P (1964) Proc Soc Exp Biol Med 115:993
31. Takemoto KK, Fabisch P (1964) Proc Soc Exp Biol Med 116:140
32. Witvrouw M, Desmyter J, De Clercq E (1994) Antivir Chem Chemother 5:345
33. Carlucci MJ, Ciancia M, Matulewicz MC, Cerezo AS, Damonte EB (1999) Antivir Res 43:93
34. Talarico LB, Zibetti RGM, Faria PCS, Scolaro LA, Duarte MER, Noseda MD, Pujol CA, Damonte EB (2004) Int J Biol Macromol 34:63
35. Tischer PCSF, Talarico LB, Noseda MD, Guimaraes SMPB, Damonte EB, Duarte MER (2006) Carbohydr Polym 63:459
36. Cáceres PJ, Carlucci MJ, Damonte EB, Matsuhiro B, Zúñiga EA (2000) Phytochemistry 53:81
37. Duarte MER, Noseda DG, Noseda MD, Tulio S, Pujol CA, Damonte EB (2001) Phytomedicine 8:53
38. Pujol CA, Estévez JM, Carlucci MJ, Ciancia M, Cerezo AS, Damonte EB (2002) Antivir Chem Chemother 13:83
39. Matsuhiro B, Conte AF, Damonte EB, Kolender AA, Matulewicz MC, Mejías EG, Pujol CA, Zúñiga EA (2005) Carbohydr Res 340:2392
40. Duarte MER, Cauduro JP, Noseda MD, Noseda DG, Goncalves AG, Pujol CA, Damonte EB, Cerezo AS (2004) Carbohydr Res 339:335
41. Feldman SC, Reynaldi S, Stortz CA, Cerezo AS, Damont EB (1999) Phytomedicine 6:335
42. Adhikari U, Mateu CG, Chattopadhyay K, Pujol CA, Damonte EB, Ray B (2006) Phytochemistry 67:2474
43. Thompson KD, Dragar C (2004) Phytother Res 18:551
44. Lee JB, Hayashi K, Hashimoto M, Nakano T, Hayashi T (2004) Chem Pharm Bull 52:1091
45. Preeprame S, Hayashi K, Lee JB, Sankawa U, Hayashi T (2001) Chem Pharm Bull 49:484
46. Zhu W, Chiu LCM, Ooi VEC, Chan PKS, Anj PO Jr (2004) Int J Antimicrob Agents 24:81
47. Zhu W, Chiu LCM, Ooi VEC, Chan PKS, Ang PO Jr (2006) Phytomedicine 13:695
48. Jassim SAA, Naji MA (2003) J Appl Microbiol 95:412
49. Ghosh P, Adhikari U, Ghosal PK, Pujol CA, Carlucci MJ, Damonte EB, Ray B (2004) Phytochemistry 65:3151

50. Lee BJ, Srisomporn P, Hayashi K, Tanaka T, Sankawa U, Hayashi T (2001) Chem Pharm Bull 49:108
51. Rechter S, Konig T, Auerochs S, Thulke S, Walter H, Domenburg H, Walter C, Marschall M (2006) Antiviral Res 72:197
52. Matsuda M, Shigeta S, Okutani K (1999) Mar Biotechnol 1:68
53. Kanekiyo K, Lee JB, Hayashi K, Takenaka H, Hayakawa Y, Endo S, Hayashi T (2005) J Nat Prod 68:1037
54. Lee JB, Hayashi K, Hirata M, Kuroda E, Suzuki E, Kubo Y, Hayashi T (2006) Biol Pharm Bull 29:2135
55. Xu HX, Lee SH, Lee SF, White RL, Blay J (1999) Antiviral Res 44:43
56. Chiu LC, Zhu W, Ooi VE (2004) J Ethnopharmacol 93:63
57. Craig MI, Benencia F, Coulombié FC (2001) Fitoterapia 72:113
58. Liu F, Liu Y, Meng Y, Yang M, He K (2004) Antiviral Res 63:183
59. Nyberg K, Ekblad M, Bergstrom T, Freeman C, Parish CR, Ferro V, Trybala E (2004) Antiviral Res 63:15
60. Neyts J, Snoeck R, Schols D, Balzarini J, Esko JD, Van Schepdael A, De Clercq E (1992) Virology 189:48
61. Herold B, WuDunn D, Soltys N, Spear P (1991) J Virol 65:1090
62. Herold BC, Visalli R, Susmarksi N, Brandt C, Spear PG (1994) J Gen Virol 75:1211
63. Shieh M, WuDunn D, Montgomery R, Esko J, Spear P (1992) J Cell Biol 116:1273
64. Spear PG (1993) Semin Virol 4:167
65. Mardberg K, Trybala E, Glorioso JC, Bergstrom T (2001) J Gen Virol 82:1941
66. Carlucci MJ, Scolaro LA, Noseda MD, Cerezo AS, Damonte EB (2004) Antiviral Res 64:137
67. Pujol CA, Scolaro LA, Ciancia M, Matulewicz MC, Cerezo AS, Damonte EB (2006) Planta Med 72:121
68. Carlucci MJ, Pujol CA, Ciancia M, Noseda MD, Matulewicz MC, Damonte EB, Cerezo AS (1997) Int J Biol Macromol 20:97
69. Zeitlin L, Whaley KJ, Hegarty TA, Moench TR, Cone RA (1997) Contraception 56:329
70. Carlucci MJ, Scolaro LA, Damonte EB (2002) J Med Virol 68:92
71. De Clercq E (1995) Clin Microbiol Rev 8:200
72. De Clercq E (2000) Med Res Rev 20:323
73. Liu GG, Borjihan G, Baigude H, Nakashima H, Uryu T (2003) Polym Adv Technol 14:471
74. Jagodzinski PP, Wiaderkiewicz R, Kurzawski G, Kloczewiak M, Nakashima H, Hyjek E, Yamamoto N, Uryu T, Kaneko Y, Posner MR (1994) Virology 202:735
75. Lynch G, Low L, Li S, Sloane A, Adams S, Parish C, Kemp B, Cunningham AL (1994) J Leukoc Biol 56:266
76. Moulard M, Lortat-Jacob H, Mondor I, Roca G, Wyatt R, Sodroski J, Zhao L, Olson W, Kwong PD, Sattentau QJ (2000) J Virol 74:1948
77. Nabatov AA, Pollakis G, Linnemann T, Kliphius A, Chalaby MS, Paxton WA (2004) J Virol 78:524
78. Bartolini B, Di Caro A, Cavallaro RA, Liverani L, Mascellani G, La Rosa G, Marianelli C, Muscillo M, Benedetto A, Cellai L (2003) Antiviral Res 58:139
79. Neurath AR, Strick N, Jiang S, Li YY, Debnath AK (2002) BMC Infect Dis 2:6
80. Scordi-Bello IA, Mosoian A, He C, Chen Y, Cheng Y, Jarvis GA, Keller MJ, Hogarty K, Waller DP, Profy AT, Herold BC, Klotman ME (2005) Antimicrob Agents Chemother 49:3607
81. Gubler DJ (2002) Trends Microbiol 10:100

82. Chen Y, Maguire T, Hileman RE, Fromm JR, Esko JD, Linhardt RJ, Marks RM (1997) Nat Med 3:866
83. Germi R, Crance JM, Garin D, Guimet J, Lortat-Jacob H, Ruigrok RWH, Zarski JP, Drouet E (2002) Virology 292:162
84. Kroschewski H, Allison SL, Heinz FX, Mandl CW (2003) Virology 308:92
85. Lee E, Lobigs M (2000) J Virol 74:8867
86. Cribier B, Schmitt C, Kirn A, Stoll-Keller F (1998) Arch Virol 143:375
87. Hung SL, Lee PL, Chen HW, Chen LK, Kao CL, King CC (1999) Virology 257:156
88. Talarico LB, Pujol CA, Zibetti RGM, Faría PCS, Noseda MD, Duarte MER, Damonte EB (2005) Antiviral Res 66:103
89. Lin YL, Lei HY, Lin YS, Yeh TM, Chen SH, Liu HS (2002) Antiviral Res 56:93
90. Talarico LB, Zibetti RGM, Faría PCS, Scolaro LA, Duarte MER, Noseda MD, Pujol CA, Damonte EB (2004) Int J Biol Macromol 34:63
91. Ono L, Wollinger W, Rocco IM, Coimbra TLM, Gorin PAJ, Sierakowski MR (2003) Antiviral Res 60:201
92. Lee E, Pavy M, Young N, Freeman C, Lobigs M (2006) Antiviral Res 69:31
93. Marks RM, Lu H, Sundaresan R, Toida T, Suzuki A, Imanari T, Hernáiz MJ, Linhardt RJ (2001) J Med Chem 44:2178
94. Shigeta S (1996) Antivir Chem Chemother 9:93
95. Hosoya M, Balzarini J, Shigeta S, De Clercq E (1991) Antimicrob Agents Chemother 36:2515
96. Luescher-Mattli M, Glueck R, Kempf CH, Zanoni-Grassi M (1993) Arch Virol 130:317
97. Kanekiyo K, Lee JB, Hayashi K, Takenaka H, Hayakawa Y, Endo S, Hayashi T (2005) J Nat Prod 68:1037
98. Hasui M, Matsuda M, Okutani K, Shigeta S (1995) Int J Biol Macromol 17:293
99. Hashimoto K, Kodama E, Mori S, Watanabe J, Baba M, Okutani K, Matsuda M, Shigeta S (1996) Antivir Chem Chemother 7:189
100. Schiffman M, Kjaer SK (2003) J Natl Cancer Inst Monogr 31:14
101. Mao C, Koutsky LA, Ault KA, Wheeler CM, Brown DR, Wiley DJ, Alvarez FB, Bautista OM, Jansen KU, Barr E (2006) Obstet Gynecol 107:18
102. Howett MK, Kuhl JP (2005) Curr Pharm Des 11:3731
103. Buck CB, Thompson CD, Roberts JN, Muller M, Lowy DR, Schiller JT (2006) PLoS Pathog 2:e69
104. De Villiers EM, Fauquet C, Broker TR, Bernard HU, Zur Hausen H (2004) Virology 324:17
105. Pearce-Pratt R, Phillips DM (1996) Biol Reprod 54:173
106. Witvrouw M, Fikkert V, Pluymers W, Matthews B, Mardel K, Schols D, Raff J, Debyser Z, De Clercq E, Holan G, Pannecouque C (2000) Mol Pharmacol 58:1100
107. Essex M (1999) Adv Virus Res 53:71
108. Fernandez-Romero JA, Thorn M, Turville SG, Titchen K, Sudol K, Li J, Miller T, Robbiani M, Maguirre R, Buckheit RW, Hartman TL, Phillips DM (2007) Sex Transm Dis 34:9

Top Heterocycl Chem (2007) 11: 283–301
DOI 10.1007/7081_2007_062
© Springer-Verlag Berlin Heidelberg
Published online: 2 June 2007

4-Hydroxy Coumarine: a Versatile Reagent for the Synthesis of Heterocyclic and Vanillin Ether Coumarins with Biological Activities

Naceur Hamdi[1,2] (✉) · Mustapha Saoud[3,4] · Antonio Romerosa[5]

[1] Borj Cedria Higher Institute of Sciences and Technology of Environment,
Ecopark of Borj Cedria Touristic road of Soliman, Tunisia
hamdi_naceur@yahoo.fr

[2] Personal address:
maison no. J99, DIAR BENMAHMUD Elakba, 2011 Denden, Tunisia

[3] Departamento de química Física, Bioquímica y química Inorgánica,
Cañada de san Urbano, s/n, 04120 Almería, Spain

[4] Personal address:
Campoverde 23, 4°3, 04008 Almería, Spain

[5] Institut de Chimie de Rennes, UMR 6509 CNRS – Université de Rennes,
campus de Beaulieu, 35042 Rennes, France

Abstract We report here the synthesis of newly coumarinic derivatives by using 4-hydroxy coumarine as starting material. These newly compounds were screened in vitro for their antimicrobial and antifungal activities. The structures of the synthesized compounds were proved by IR, ^{1}H NMR, ^{13}C NMR and mass-spectral.

1
Introduction

The varied biological activity of coumarins fused with a benzopyrano pyrimidines ring system [1–6] has continued to stimulate a great deal of interest in the development of new methodologies for the synthesis of multi-substituted [1]benzopyranopyrimidines.

Recently, 4-chloro-3-formylcoumarin and amino derivatives have been used in the synthesis of 2-functionalized [1]benzopyrano[4,3-b]pyrrol-4(1H)-one derivatives via the Fischer–Fink reaction [7]. Although this new synthesis requires neither protection nor harsh conditions, there are still several drawbacks with this synthetic route. First, the competing Knorr-type transformation might be observed if the amino derivatives possessed a low electrophilic group or very reactive alkylketone. Second, since the 3-formyl group on the coumarin is highly susceptible to nucleophilic attack, a second attack of an amino derivative may occur, complicating the products. Third, only 1,2-disubstituted, but not 1,2,3-trisubstituted [1]benzopyrano[4,3-b]pyrrol-4(1H)-one derivatives, can be prepared by this synthesis. Thus, an efficient and protection-free synthesis of multifunctionalized [1]benzopyrano[4,3-b]pyrrol-4(1H)-one derivatives from inexpensive materials under mild conditions remains to be discovered. Therefore, it would be useful to create a review of the data, making it available for ready and easy reference.

2
Synthetic Ring Systems

Several derivatives of the pyran or of fused pyran ring systems are endowed with different types of biological activities. It has been reported that pyran derivatives exhibit antimicrobial activity [8], growth stimulating effects [9], antifungal and plant growth regulation effects [10],antitumor activity [11], central nervous system (CNS) activity [12, 13] and hypotensive effect [14]. Moreover, pyran derivatives are well known for antihistaminic activity [15], platelet antiaggregating activity and local anaesthetic activity [16–18], antiallergenic effect [19], antidepressant effect [20] and as antiproliferation agents [21, 22]. With this in mind and in continuation of our previous work [23–29] on the synthesis of newly fused 4-H-pyran using enaminonitriles as a starting material, we report the synthesis of a variety of new benzopyrano pyrimidines along with their antimicrobial activity.

This synthesis involves Michael cycloaddition reaction of the readily available 4-hydroxycoumarine 1 with α cyanocrotonitrile 2 in ethanolic piperidine to afforded 2-amino-3-cyano-4-methyl-4H, 5H-pyrano-benzopyran-5-one (3). Treatment of 3 with acetic anhydride for 0.5 h and/or 3 h under reflux afforded N-acetyl and [1]benzopyrano[3',4':5,6]pyrano(2,3-d) pyrimidine-6,8-dione derivatives (4) and (5a), respectively. Also, interaction of 3 with benzoyl chloride or formic acid gave the corresponding pyrimidine derivatives (5b,c) while its treatment with formamide afforded the aminopyrimidine derivative (6).

On the basis of spectral data, structure 5B was excluded. Structure 5A was established on the basis of IR spectrum, which showed an absorption band at 1665 (5a), 1659 (5b), 1790 cm^{-1} (5c) characteristic to (CO). The ^{1}H NMR spectrum which revealed a broad single at ∂ 7.47 ppm (5a) and at 7.33 ppm (5c) characterised for NH proton. Structures 3–6 were established by spectral data and analogy with our previous work (Scheme 1).

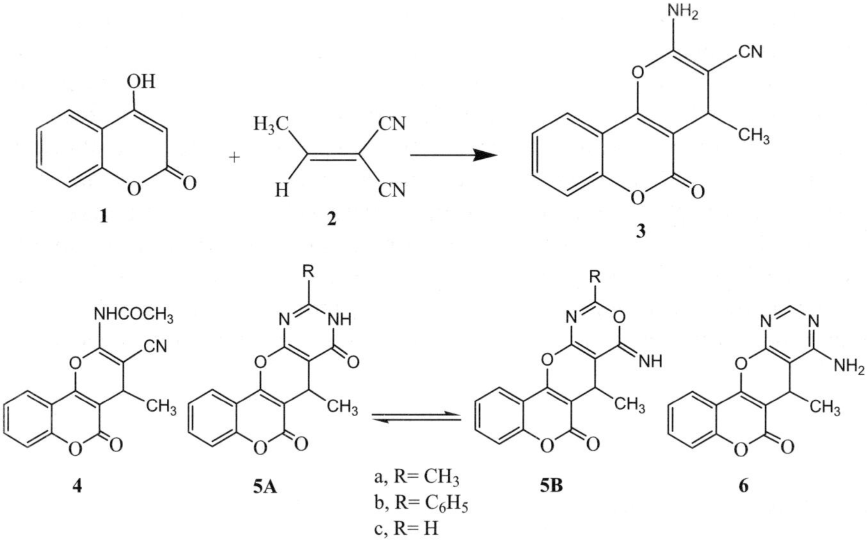

Scheme 1 Preparation of compounds **3**, **4**, **5A**, **5B** and **6**

The enaminonitrile (**3**) proved to be a useful key intermediate in the synthesis of a variety of new pyranopyrimidine, pyranotriazolopyrimdine and pyranopyrimidotriazine derivatives. Thus, treatment of **3** with triethylorthoformate – Ac$_2$O – gave the corresponding ethoxymethyleneamino derivatives (**7**). Hydrazinolysis of **7** in ethanol at room temperature yielded 9-amino-8,9-dihydro-8-imino-7-methyl-6H,7H-[1]benzopyrano[3′,4′:5,6]-[3,2-d]pyrimidie-6-one (**8a**).

Aminolysis of **7** with aliphatic primary amines gave the corresponding pyranopyrimidine-6-one derivatives (**8b–d**), while dimethylamine gave the dimethylaminomethyleneamino derivative (**9**). Ammonolysis of **7** in

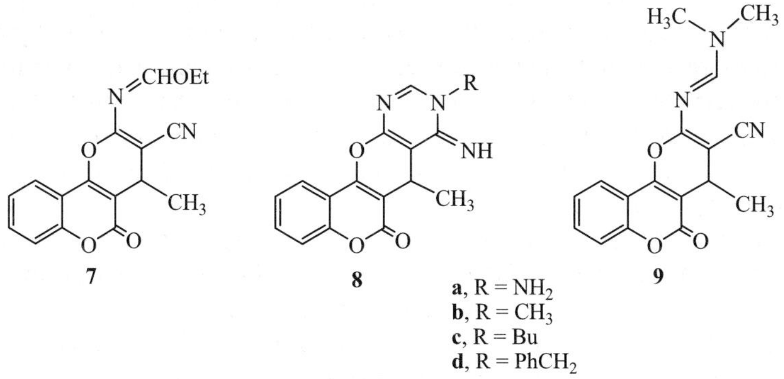

Scheme 2 Preparation of compounds **7**, **8** and **9**

methanol gave 8-amino-7-methyl-6H, 7H-[1]benzopyrano[3′,4′ : 5,6]pyrano-[2,3-d]pyrimidine-6-one (6) (Scheme 2).

Treatment of 8a with some carboxylic acid chlorides gave the corresponding 14-methyl-2-substituted 13H,14H-[1]Benzopyrano[3′,4′ : 5,6]pyrano[3,2-e] [1,2,3]triazolo-[1,5-c]pyrimidine-13-one (10b–e), while cyclocondensation of 8a with ethylcyanoacetate and diethyloxalate gave the corresponding 2-cyanomethyl and diethyloxalate gave the corresponding 2-cyanomethyl and 2-ethoxycarbonyl derivatives (10f,g), respectively. Also, (8a) was reacted with triethyl orthoformate, affording the corresponding (1,2,4) triazolo-[1,5-c]pyrimidine-13-one derivative (10a). Structures 7–10 were established by spectral data and analogy with our previous work (Scheme 3).

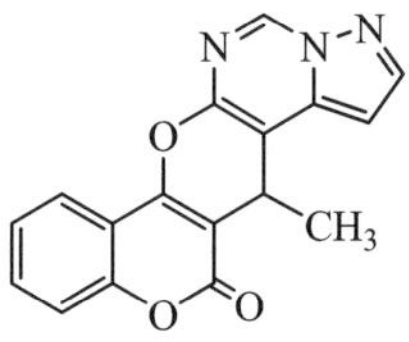

10 (a-g)

a, R = H   e, R = C_6H_4Cl-p
b, R = CH_3 f, R = CH_2CN
c, R = CH_2Cl g, R = CO_2Et
d, R = Ph

Scheme 3 Structure of compounds 10

Instead of the anticipated formation of the triazolopyrimidine derivative 13 [30–33], the reaction of 8a with ethylchloroformate, through nucleophilic displacement followed by spontaneous hydrolysis of the ester intermediate 11, led to the corresponding carbamic acid derivative 12 (Scheme 4).

Interaction of 8a with ethylchloroacetate in methanolic sodium methoxide leads to cyclocondensation with elimination of EtOH and HCl, affording the triazine-3,14–dione derivative (14), while 8a was reacted with alcoholic CS_2-KOH to give 14-methyl-2,3-dihydro-13-oxo-2H,13H,14H-[1]Benzopyrano [3′,4′ : 5′,6′]pyrano[3,2-e]-[1,2,4]triazolopyrimidine-2-thione (15) (Scheme 5).

Finally, treatment of 8a with aromatic aldehydes gave the pyrimidine derivatives 16a–c instead of the expected triazolopyrimidine derivatives such as (10d,e) [34–36].

Derivatives of 2-aminothiazolines are important pharmacological compounds and precursors in the synthesis of medications [37], such as the antibiotic sulfathiazole and the anthelmintic thiabedazole. Moreover, recent research indicates that they are also inhibitors of enzymes, such as kinurenin 3-hydroxylase [38]. On the other hand, derivatives of 4-hydroxy-chromen-2-one are known as anti-coagulants and antitumor compounds [39–41]. The 2-aminothiazolines are obtained by means of the Hantzsch reaction [42–44], which is the reaction of α-haloketones with thioureas.

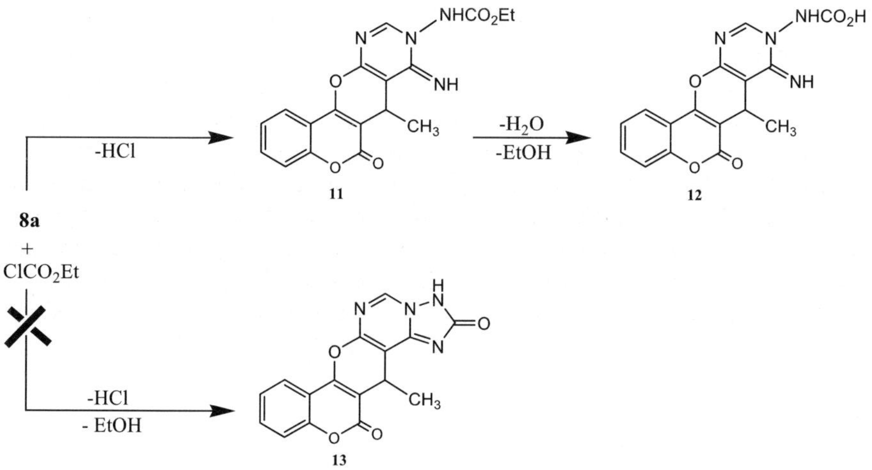

Scheme 4 Reaction of **8a** with ethylchloroformate

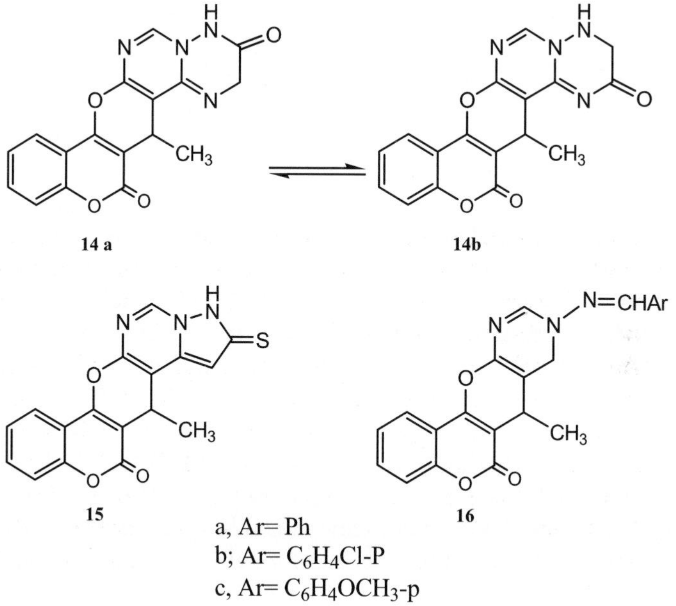

a, Ar= Ph
b; Ar= C$_6$H$_4$Cl-P
c, Ar= C$_6$H$_4$OCH$_3$-p

Scheme 5 Interaction of **8a** with ethylchloroacetate

For these reasons, in the present work, the reaction of 3-(2-bromoacetyl)-4-hydroxychromen-2-one (**18**) with thioureas, thioacetamides and ammonium dithiocarbamate have been investigated. Compound **18** was synthesized from 3-acetyl-4-hydroxychromen-2-one (**17**) and used as a suitable synthone for further reactions. The 4-hydroxycoumarin nucleus is very susceptible to electrophilic substitution [45, 46] and the preparation of **18** using bromine is difficult and isnt regiospecific.

Thus, 3-acetyl-4-hydroxychromen-2-one (**17**) reacts with bromine in a conventional manner (bromine/acetic acid) to give substitution products at the aromatic nucleus as the major product [47, 48]. For example, 3-acetyltropolone and 4-acetyltropolone were reacted with bromine to afford the corresponding substitution products at the tropolone nucleus as the main products [49, 50]. For this reason, **17** was treated with phenyltrimethylammonium tribromide [51–53] (Scheme 6). The reaction was carried out at room temperature using tetrahydrofuran as the solvent. The structure of **18** was determined on the basis of spectral data and elemental analysis.

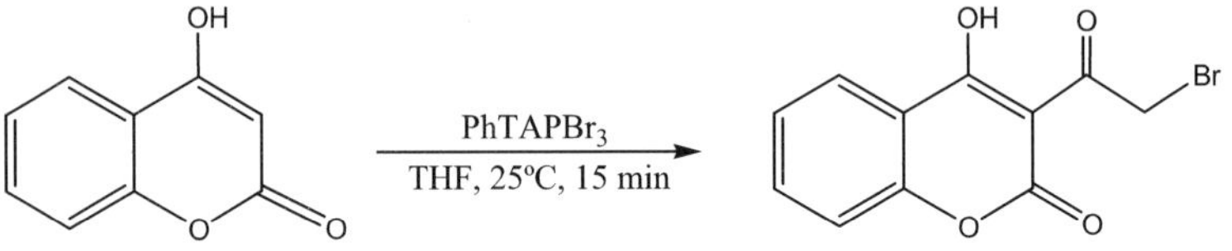

Scheme 6 Preparation of 3-(2-bromoacetyl)-4-hydroxychromen-2-one **18**

The compound 3-(2-Bromoacetyl)-4-hydroxychromen-2-one (**18**) reacts with thiourea to afford 3-(2-amino-thiazol-4-yl)-4-hydroxychromen-2-one hydrobromide (**19a**). This reaction was carried out in boiling ethanol for 30 min. Compound **19a** also gave positive coloration with iron(III) chloride solution (Scheme 7). In the reaction of compound **18** with 1-methylthiourea under identical experimental conditions as above, 4-hydroxy-3-(2-methylaminothiazol-4-yl)chromen-2-one (**19b**) in 67% yield was obtained. In a similar manner, **18** reacted with three arylthiourea derivatives affording the corresponding 3-(2-arylthiazol-4-yl)chromen-2-ones (**19c–e**) in different yields (70, 63 and 74%, respectively). The derivatives formed in these reactions were identified as 4-hydroxy-3-(2-phenylaminothiazol-4-yl)chromen-2-one(**19c**), 4-hydroxy-3-(2-*p*-tolylaminothiazol-4-yl)chromen-2-one (**19d**) and 4-hydroxy-3-2-(4-methoxyphenylamino)thiazol-4-yl)chromen-2-one (**19e**). The identities of these compounds were established by spectral data and elemental analysis.

The reaction of **18** with thioacetamide and thiobenzamide gave 4-hydroxy-3-(2-methylthiazol-4-yl)chromen-2-one (**20a**) and 4-hydroxy-3-(2-phenylthiazol-4-yl)chromen-2-one (**20b**). The derivatives **20a** were isolated in 72 and 74% yield, respectively (Scheme 8).

Finally, compound **18** was also reacted with ammonium dithiocarbamate to give 4-hydroxy-3-(2-mercaptothiazol-4-yl)chromen-2-one (**21**) (Scheme 9).

Treatment of 2-amino-4-(4′-chlorophenyl)-3-cyano-4*H*,5*H*-pyrano[3,2-*c*][1] benzopyran-5-one (**22**) [54, 55] with triethyl orthoformate in acetic an-

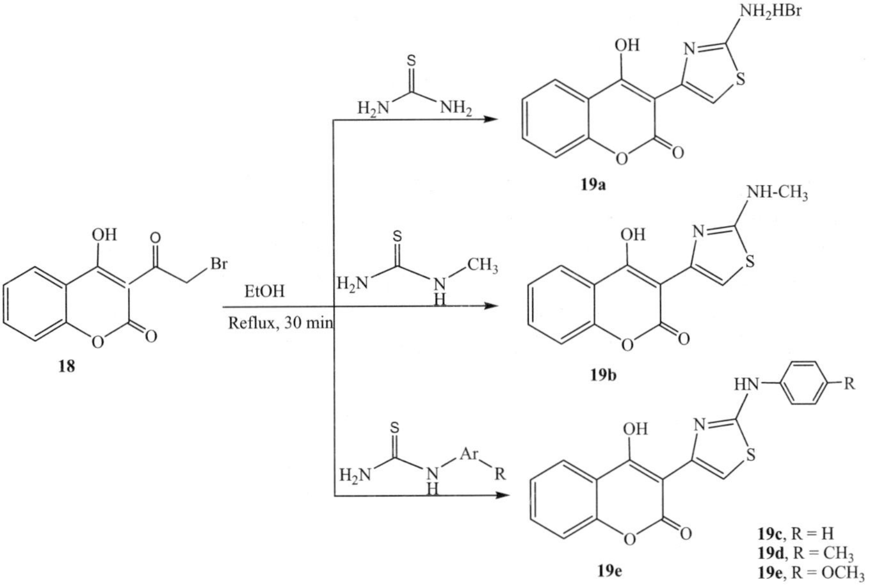

Scheme 7 Reaction of compound **18** with thioureas

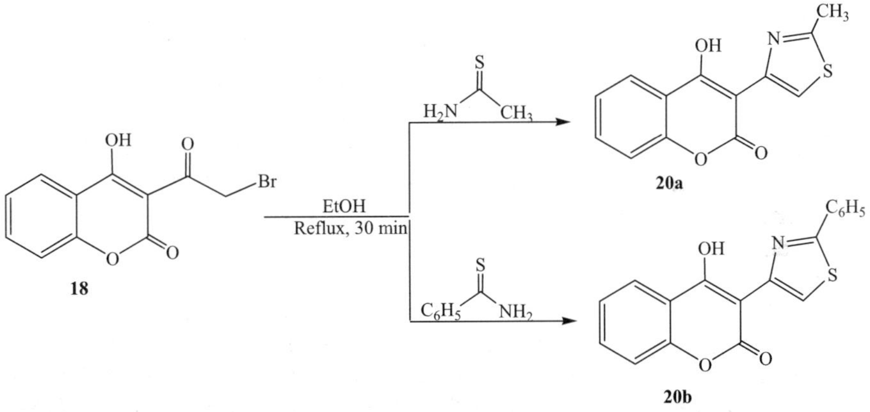

Scheme 8 Reaction of compound **18** with thioacetamide and thiobenzamide

hydride at reflux afforded 4-(4′-chlorophenyl)-3-cyano-2-ethoxymethylene-amino-4H,5H-pyrano[3,2-c][1]benzo-pyran-5-one (**23**). Hydrazinolysis of the latter in ethanol at room temperature yielded 9-amino-7-(4′-chloro-phenyl)-8,9-dihydro-8-imino-6H,7H-[1]benzopyrano[3′,4′ : 5,6]-pyrano[2,3-d]pyrimidine-6-one (**24**) (Scheme 10).

Refluxing compound **24** with triethyl orthoformate afforded the [1,2,4]tri-azolo[1,5-c]pyrimidine **25a**, while with acetyl chloride or chloroacetyl chlo-

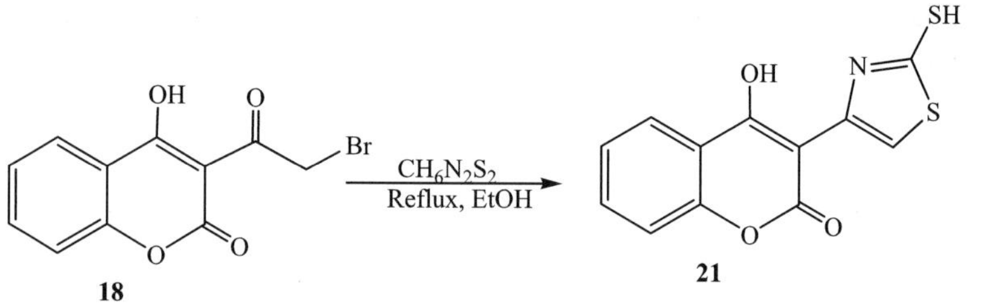

Scheme 9 Reaction of compound **18** with ammonium dithiocarbamate

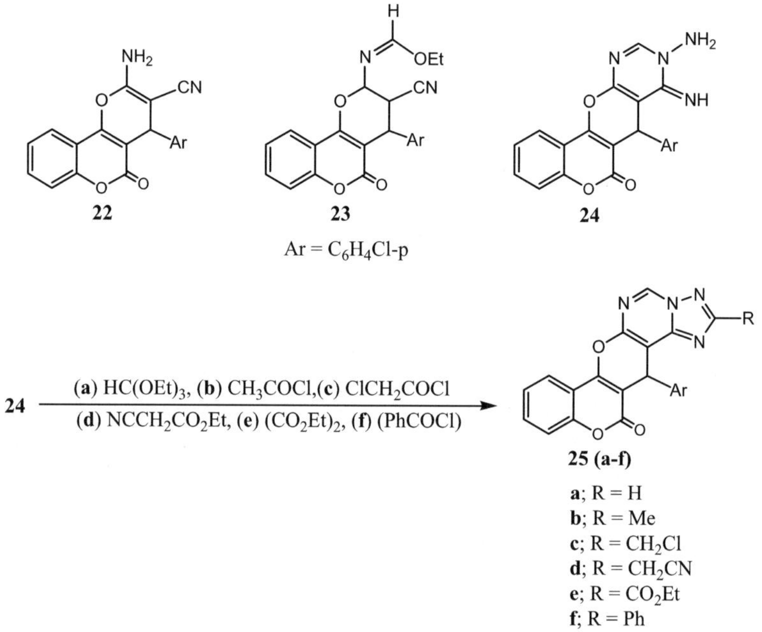

Ar = C₆H₄Cl-p

Scheme 10 Preparation of compounds **22–25**

ride compounds, **25b** and **25c** were formed, respectively. Reaction of **24** with ethyl cyanoacetate and diethyl oxalate afforded the heterocycles **25d** and **25e**, respectively, while with benzoyl chloride, the 2-phenyl derivative **25f** was obtained (Scheme 1). Structure **25** was established by spectral data and analogy with our previous work [56, 57]. When **24** was treated with methyl chloroformate for 30 min, the methoxycarbonyl derivative **26a** was formed, while heating of **24** with methyl chloroformate under reflux for 6 h afforded [1,2,4]triazolo[1,5-*c*]pyrimidine **26b** via elimination of methanol from **26a**. The structure of **26b** was supported by an independent synthesis from **26a** and ethanol under reflux for 5 h. When **24** was treated with ethyl chloroformate, an intermediate bis(ethoxycarbonyl) derivative was formed, which eliminated ethanol to furnish the ester **26d**. Treatment of **24** with carbon disulfide in alcoholic potassium hydroxide solution gave the 2-thione derivative **26c** (Scheme 11).

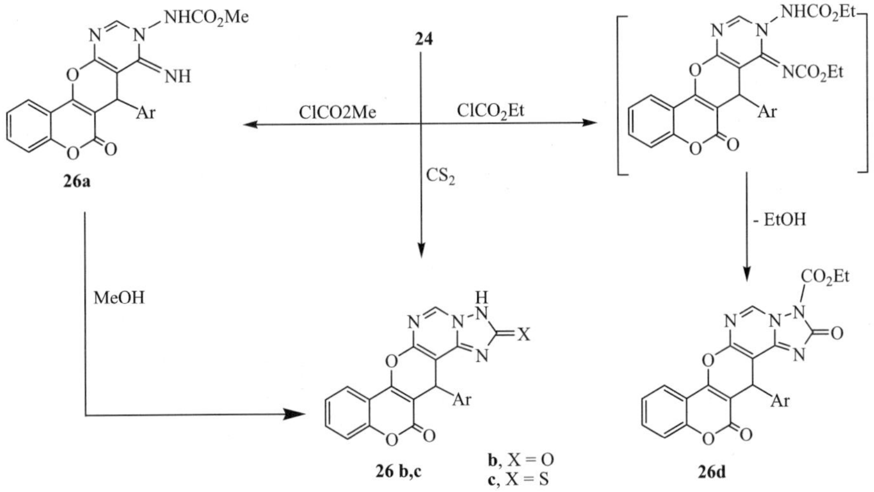

Scheme 11 Reactivity of compound **24**

Interaction of **24** with ethyl chloroacetate in methanolic sodium methoxide afforded the triazin-3,14-dione derivative **27**. The alternate structure **28** was excluded on the basis of spectral data [58, 59]. Based on the reaction conditions, (sodium methoxide) structure **27** is thought to result from the initial formation of a sodium salt on the less basic imino nitrogen atom, which cyclizes into **27** with elimination of NaCl and EtOH (Scheme 12). The IR spectrum of **27** showed a characteristic $C = O$ absorption band at $1650 \, cm^{-1}$,

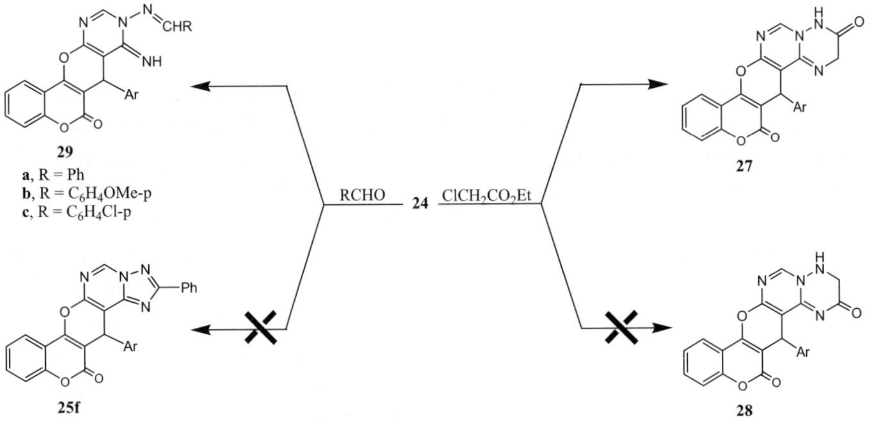

Scheme 12 Interaction of **24** with ethyl chloroacetate and aromatic aldehydes

whereas if structure **28** were correct, one would expect an absorption band for the carbonyl band at a higher frequency than that observed for **27**. The ^{1}H-NMR spectrums revealed a singlet at δ 4.95 ppm, characteristic for the methylene proton.

The reaction of **24** with aromatic aldehydes gave pyrimidines **29** (Scheme 12) instead of the anticipated formation of triazolopyrimidines such as **25f**. The proposed structures of **29** were also supported by the spectral data.

2.1
Antibacterial Activity

Newly synthesized compounds **22**, **23**, **25c–e**, **26d** and **29e** were screened in vitro for their antimicrobial activities against Gram positive bacteria *Staphylococcus aureus* (NCTC-7447), *Bacillus cereus* (ATCC-14579) and Gram negative bacteria *Serratia marcesens* (IMRU-70) and *Proteus merabitis* (NTCC-289) using the paper disk diffusion method for the antibiotic sensitivity technique [60]. The tested compounds were dissolved in *N,N*-dimethylformamide (DMF) to obtain a 1 mg/mL solution. The inhibition zones of microbial growth produced by different compounds were measured in millimeters at the end of an incubation period of 48 h at 28 °C. DMF alone showed no inhibition zone.

An ampicillin standard (25 mg) was used as a reference to evaluate the potency of the tested compounds. The results are illustrated in Table 1.

Table 1 Antibacterial activity of some compounds

Comp.	*Staphylococcus Aureus* (NCTC-7447)	*Bacillus cereus* (ATCC-14579)	*Serratia Marcesens* (IMRU-70)	*Proteus Merabitis* (NTCC-289)
22	23	19	18	20
23	22	24	24	23
25c	23	22	23	22
25d	22	19	24	22
25e	23	22	25	23
26d	24	22	21	23
29e	18	20	24	21
Ampicillin	26	25	26	27

2.2
Antifungal Activity

Newly synthesized compounds **22**, **23**, **25c–e**, **26d** and **29e** were screened for their antifungal activity against two species of fungi, *Aspergillus ochraceus*

Wilhelm (AUCC-230) and *Penicillium chrysogenum Thom* (AUCC-530) using the paper disk diffusion method [60]. The tested compounds were dissolved in DMF to get a 1 mg/mL solution. The inhibition zones were measured in 10^{-3} m at the end of an incubation period of 48 h at 28 °C. A standard of mycostatin (30 mg) was used as a reference and the results are shown in Table 2.

Table 2 Antifungal activity of some compounds

Comp.	*Aspergillas ochraceus* *Wilheim* (AUCC-230)	*Penicillium chrysogenum* *Thom* (AUCC-530)
22	14	16
23	19	16
25c	19	19
25d	19	13
25e	19	20
26d	15	17
29e	19	18
Mycostatin	22	24

The synthesis of heterocycles fused with a chromone moiety has attracted much attention because of their pharmacological importance [61]. 2-Alkyl-/aryl-amino-3-formylchromone **30** has been utilized recently in the synthesis of several heterocycles [62]. 3-(Arylaminomethylene)chroman-2,4-dione **31** (R' = aryl) has been transformed into tricoumarols and coumarinoquinolines [63]. Considering 3-(alkyl-/aryl-aminomethylene)chroman-2,4-dione **31** to be a very good precursor of heterocycles fused at the 3,4-position of 1-benzopyran, we became interested in synthesizing 3-(alkylaminomethylene) chroman-2,4-diones **31** (R' = alkyl), which are expected to be more reactive than **31** (R' = aryl). β-Alkylamino-α,β-unsaturated ketones have been utilized in the synthesis of different heterocycles [64]. Our earlier method for the synthesis of **31** (R' = aryl) using K-10 montmorillonite and aromatic amines [65] with **32** is difficult to accomplish with aliphatic amines. A report on the thermal rearrangement of nitrone **33e** to **34e** (70%) and **34e** (25%) gave us an impetus to utilize this rearrangement as a route to our target system **31**(R' = alkyl). Recently, we reported a one-pot synthesis of **34** (R' = alkyl or aryl) from **32** and some differences in the reactivity of N-alkyl- and N-aryl nitrones **33** towards hydrolysis reactions [66].

In continuation of our studies on the reactivities of nitrones **33**, we report herein the solvent directed rearrangement of nitrones **33** to **34** and/or **31** and conversion of **5** (R' = alkyl) to coumarino [3,4-d]-isoxazole. Nitrones **33** (R' = alkyl) were prepared by reaction of **32** with nitroalkanes **35** and Zn in the presence of HOAc in EtOH under an inert atmosphere. Nitrones **33** Were heated under reflux in different solvents for varying times to obtain **34** and/or **31** (Table 3).

It was observed that polar solvents facilitated the formation of **34**, whereas nonpolar solvents allow the formation of both **34** and **31**. The rearrangement of **33b** to **34b** takes place in 7 h when heated under reflux in methanol, but needs only 2 h in ethanol (entries 1, 2 and 3), which indicates that higher temperatures facilitate this rearrangement. However, the same transformation is incomplete even after heating under reflux for 41 h in benzene (entry 4). Thus, the polarity of the solvent also has some effect on the rearrangement. To check the necessity for a protic solvent, the same transformation was also

Table 3 Compounds **34** and **31** prepared from nitrones **33** by heating in different solvents

Entry	Nitrones	R	R′	Medium	Time/h	% Yield of 34 [a]	Yield of 31	E:Z of 31	Mp of 31 (°C) (E+Z)
1	33a	H	Et	MeOH	7	98	–	–	–
2	33b	Me	Et	MeOH	7	95	–	–	–
3	33b	Me	Et	EtOH	2	98	–	–	–
4	33b	Me	Et	C$_6$H$_6$	41 [b]	30	40	2 : 5	188–90
5	33b	Me	Et	CH$_3$CN	2	90	–	–	–
6	33b	Me	Et	Acetone	30	80	–	–	–
7	33a	H	Et	MeOH/TsOH	5	95	–	–	–
8	33b	Me	Et	Toluene	14	20	60	1 : 3	186–88
9	33b	Me	Et	Xylene	2	10	70	2 : 5	188–90
10	33b	Me	Et	Benzene/TsOH	1	90	–	–	–
11	33b	Me	Et	AcOH [c]	4.5	90	–	–	–
12	33a	H	Me	Xylene	2	10	70	2 : 5	184–86
13	33c	H	Me	Xylene	2	20	65	1 : 2	194–96
14	33b	Me	Me	Xylene	2	15	72	1 : 2	192–94
15	33f	H	Ar [d]	Toluene	6	15	70	5 : 2	194–97
16	33g	Me	Ar [d]	Toluene	7	17	70	1 : 2	186–90
17	33g	Me	Ar [d]	Xylene	2	10	85	1 : 2	188–90
18	33e	H	Ph	MeOH	14	–	–	–	No reaction
19	33f	H	Ar [d]	MeOH	20	–	–	–	No reaction
20	33g	Me	Ar [d]	MeOH	20	–	–	–	No reaction
21	33g	Me	Ar [d]	EtOH	4	90	–	–	–
22	33g	Me	Ar [d]	Benzene/TsOH	4	60	–	–	–
23	33g	Me	Ar [d]	CH$_3$CN	2	90	–	–	–
24	33g	Me	Ar [d]	AcOH [c]	10	95	–	–	–

[a] All compounds have the same mp and mmp with authentic samples
[b] 10% unreacted starting material was recovered
[c] Reactions were carried out at room temperature with stirring
[d] Ar stands for C$_6$H$_4$Me-p

carried out in dipolar aprotic solvents. The rearrangement was complete in 2 h when heated at reflux in acetonitrile (entry 5) and for 30 h in acetone (entry 6). By comparing the rearrangements in MeOH and acetone, it is clear that a protic solvent has little effect on this rearrangement, however, the outcome of the rearrangement depends on the polarity of the solvent and also on the temperature. On heating in benzene (entry 4), **33** gave a mixture of **34b** and **31b**. Like compound **31** (R′ = aryl), 3;5 compound **31** (R′ = alkyl) showed a single spot on TLC, however, ^{1}H NMR measurements showed a diastereomeric mixture (E and Z) [67]. The higher deshielding effect on the β-H when cis to the ester function in an α,β-unsaturated ester compared to an α,β-unsaturated ketone [68] helped to distinguish the E and Z isomers. From these observations, it is presumed that although the transformation of **33** to **34** is guided by both the polarity of the solvent and heating, the transformation of **33** to **31** is only thermally controlled. Based on this, the above rearrangement was carried out in toluene and xylene (entries 8 and 9), where the reaction times were 14 h and 2 h, respectively. The yield of **31** was also found to increase by changing the solvent from toluene to xylene. Interestingly, the rearrangement of **33b** in benzene was complete in 1 h by adding a trace of p-toluene sulfonic acid. Compound **34b** was the only product (entry 10). The addition of TsOH to a methanolic solution of **33a** also enhanced the reaction rate (entry 7). The rearrangement of **33b** to **34b** can also be accomplished by stirring **33b** in AcOH at room temperature (entry 11). The N-aryl nitrones **33e–g** were also heated under reflux in various solvents (entries 15–23). The results are similar to those using the N-alkyl nitrones **3a–d**. In most cases the aryl nitrones needed a longer rearrangement time than the alkyl nitrones. The aryl nitrones are less susceptible to rearrangement in comparison to alkyl nitrones. Aryl nitrones (**33e–g**) failed to rearrange when heated under reflux in methanol for 20 h (entries 18–20), though **33g** readily rearranged to **34g** in good to excellent yields when heated under reflux in ethanol, acetonitrile or benzene-TsOH (entries 21–23). As with alkyl nitrones, aryl nitrone **33g** also underwent rearrangement to **34g** when stirred in AcOH at room temperature (entry 24). Considering the mechanism for the formation of **34** and **31** from **33** (Scheme 1), it is observed that during the formation of **34**, the pyran ring opens to form **35** (Scheme 13, path a) followed by a 1,5-H-shift. This route is facilitated by the polarity of the solvent. In contrast, formation of **32** requires a tandem electrocyclic ring closure and a 1,5-H shift (Scheme 1, path b), both of which are thermally allowed processes. The above experiments enabled us to synthesize **34** (R′ = alkyl) or **31** (R′ = alkyl) selectively from **33** (R′ =alkyl). Compound **31** (R′ = alkyl), having a balkylamino-α,β-unsaturated ketone moiety, produced coumarino[3,4-d] isoxazole **36** [69] in quantitative yield when heated with hydroxylamine hydrochloride in ethanol at reflux for 2 h. Similar treatment of **31** (R′ = alkyl) with phenylhydrazine hydrochloride produced the hydrazone derivative **37** (Scheme 14).

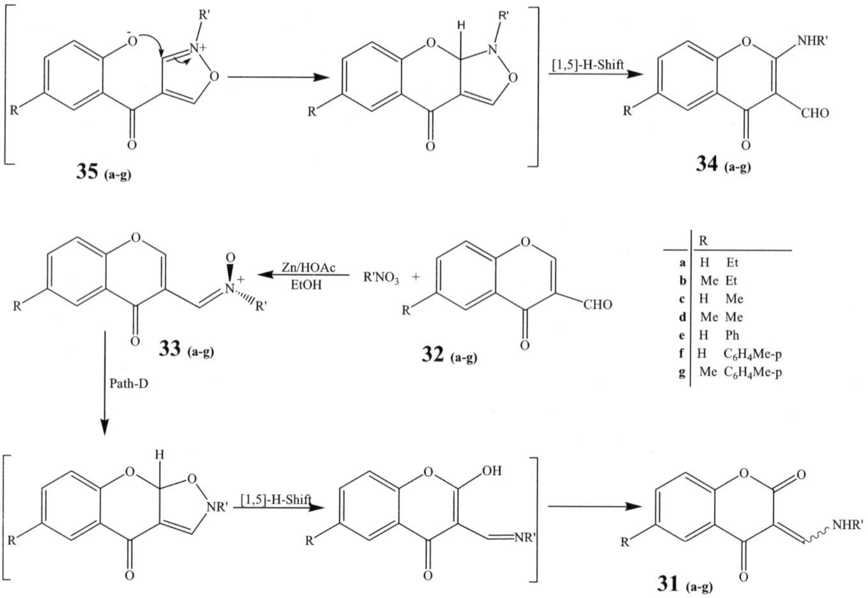

Scheme 13 Preparation of compounds **31–35**

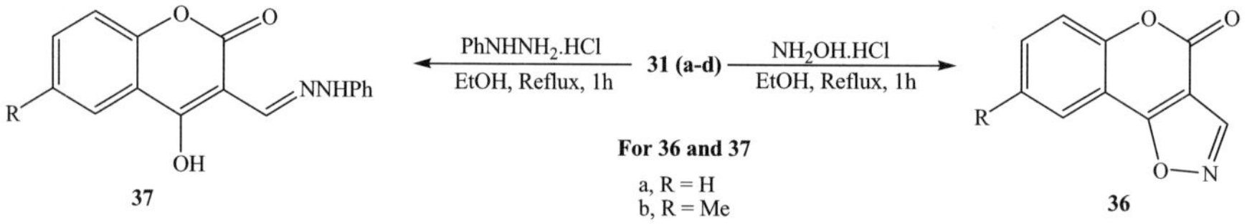

Scheme 14 Preparation of compounds **36, 37**

4-aryloxy methyl and heteroaryl coumarin derivatives are known for their anti-inflammatory activity [70–72]. Previously, we have reported that vanallinyl ethers of 4-bromomethyl coumarins can exhibit good anti-inflammatory activity and low acute toxicity [73]. As an extension of our interest for the search of new heterocyclic moieties as potent anti-inflammatory agents devoid of side effects such as ulcerogenic activity, we have synthesized a series of o-alkylated coumarin derivatives. We aimed to synthesize a series of heterocyclic moieties like benzofuran, chromone and 4-hydroxy coumarin, which were linked to coumarin moiety at 4th position by ether linkage. These coumarinyl ethers were investigated for analgesic and anti-inflammatory activity.

According to our previous reports [74, 76], the anti-inflammatory activity of ethers of 4-bromomethyl coumarins can be significantly modified by using different aryl systems. Here we have modified the aryl moieties to heterocyclic moieties like chromone, benzofuran and coumarin ring, which were synthesized for further investigation. 4-(Bromomethyl) coumarins **38** were synthesized by the Pechmann cyclization of phenols with 4-bromoethyl acetoacetate [77]. The 4-bromomethyl coumarin reacted with 2,4-dihydroxyacetophenone gave a corresponding 4-(4'-acetyl-3'-hydroxy-phenoxy methyl) coumarin **39a–d**. The reaction of the **39a–d**, which underwent Kostanecki synthesis [78] using sodium acetate and acetic anhydride, gave the corresponding chromonyl ethers **40a–d**. The benzofuranyl ethers **41a–d** were obtained by the treatment of phenacyl bromide with corresponding 4-(4'-acetyl-3'-hydroxy-phenoxy methyl)-coumarin [79] (Scheme 14). The 4-hydroxy coumarinyl ethers were synthesized by the reaction of **39a–d** with diethyl carbonate and sodium using the Boyd method of synthesis [80]. The compound **42a** was also prepared by another route using 4,7-dihydroxy coumarin with 4-bromomethylcoumarin in the presence of K_2CO_3 in dry acetone under stirring conditions (Scheme 15).

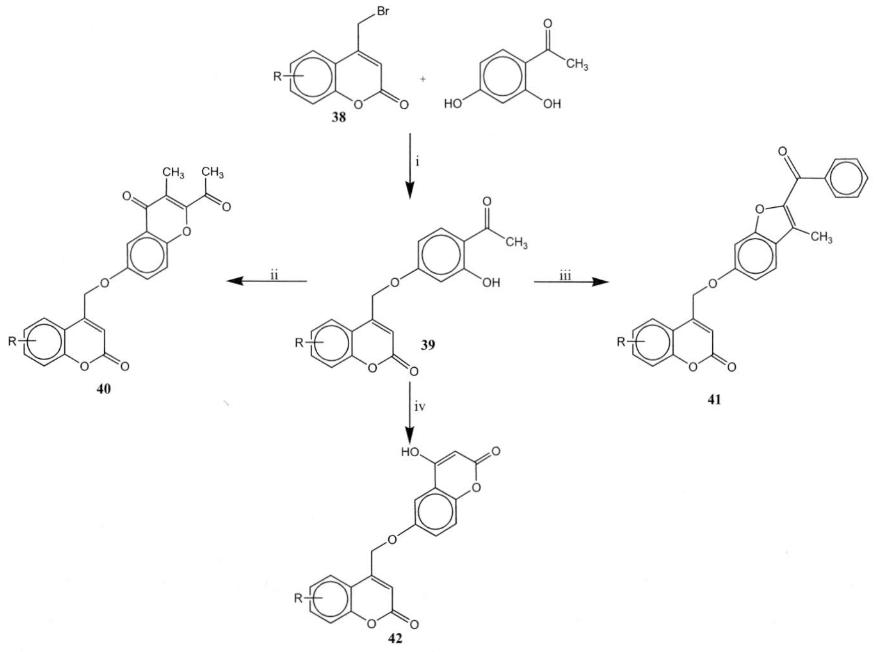

Scheme 15 Preparation of compounds **39–42**

The required 4,7-dihydroxy coumarins were prepared from resorcinol using known methods [81]. Postulated structures of the newly synthesized compounds **39a–d**, **40a–d**, **41a–d** and **42a–d** were in agreement with their IR, ^{1}H NMR spectral and elemental analysis data. In the IR spectrum of compound **39a**, (R = 6-CH$_3$) exhibited prominent bands around 1709, 1648 and 3042 cm^{-1} due to carbonyl lactone of coumarin, carbonyl of acetophenone and OH stretching vibrations. The lower stretching of OH and carbonyl of acetophenone were observed due to the intramolecular hydrogen bonding between OH and COCH$_3$. In the ^{1}H NMR spectrum, the two sharp singlets at d 2.49 and 2.67 are due to 6-CH$_3$ and COCH$_3$ protons, respectively. The OH proton was observed at the down field region at d 12.87.

The compound **40a**, **41a** and **42a** (R = 6-CH$_3$) showed the absence of OH stretching in IR at 3000–3400 cm^{-1} and absence in ^{1}H NMR at δ 12–13. This means that the free OH present in the 4th position has undergone cyclization to respectively afford chromone and benzofuran. However, the compound **42** showed presence of OH in the IR spectrum at 3444 cm^{-1}, which is due to the presence of 4-hydroxy moiety. The absence of peak in the down field region at δ 12–13 and presence of an additional peak is observed at δ 5.50 due the OH of 4-hydroxycoumarin in the NMR spectrum. The compound **41a** (R = 6-CH$_3$) showed three peak the IR region at 1720, 1648 and 1633 Cm^{-1} assigned as lactone carbonyl of coumrin, lactone carbonyl of chromone, and carbonyl of COCH$_3$. The NMR spectra showed the presence of three singlets at δ 2.33, 2.45 and 2.54 are due to the methyl of chromone, coumrin and COCH$_3$ respectively indicating the formation of 3-acylated product. It was worthy to note that the addition of equimolar quantity of acetic anhydride the reaction does not precede instead the addition of excess amount of the acetic anhydride forms the acylated chromone. The IR spectrum of compound **6a** showed the two stretching bands at 1720 and 1702 Cm^{-1} due to the carbonyl stretching of two coumarins moities. The OH stretching is observed as broad band at 3444 Cm^{-1}. The NMR spectrum showed the presence of two C$_3$–H of coumarin at 6.06 and 6.66, the OH proton observed as singlet at δ 5.50.

3
Conclusion

In conclusion, we have synthesized **4** (R$'$ = alkyl/aryl) in excellent yield compared to earlier reports by modifying the solvent for the rearrangement of **33**. A synthetic route to previously unreported 3-(alkylaminomethylene)chroman-2,4-diones **31** (R$'$ = alkyl) with moderate yields has been revealed and those compounds have been shown to be the synthetic equivalents of the versatile substrate 3-formyl-4-hydroxycoumarin.

Furthermore, eight 3-(thiazol-4-yl)-4-hydroxychromen-2-one derivatives were prepared in good yields in the reaction of 3-(2-bromoacetyl)-4-hydroxychromen-2-one with thioureas, thioacetamide, thiobenzamide and ammonium dithiocarbamate. The obtained coumarin derivatives can be used as potentially bioactive compounds and as precursors in the synthesis of medications.

A new series of heterocyclic ethers were synthesized. The analgesic and anti-inflammatory activities of these compounds were comparable with the drugs standardly used. The benzofuranyl ethers of coumarins were found to be the most active amongst all of the compounds and chloro and methoxy substitution in coumarin rings increased this activity.

References

1. Colotta V, Cecchi L, Melani F, Filacchioni G, Martini C, Giannaccini G, Lucacchini A (1990) J Med Chem 33:2646–2651
2. Colotta V, Cecchi L, Melani F, Filacchioni G, Martini C, Gelli S, Lucacchini A (1991) Il Farmaco 46:1139–1154
3. Colotta V, Cecchi L, Melani F, Filacchioni G, Martini C, Gelli S, Lucacchini A (1991) J Farm Sci 80:276–279
4. Singh P, Ojha TN, Sharma RC, Tiwari S (1994) Indian J Chem B 32B:555–561
5. Branca Q, Jakob-Rotne R, Ketler R, Roever S, Scalone M (1995) Chimia 49:381–385
6. Alberola A, Alvaro R, Ortega AG, Sadaba ML, Sanudo MC (1999) Tetrahedron 55:13211–13224
7. Alberola A, Calvo L, Ortega AG, Encabo AP, Sanudo MC (2001) Síntesis pp 1941–1948
8. Hantzsch A (1888) Ann Chem 249:1
9. Wiley RH, England DC, Behr IC (1951) Org React 6:367
10. Sammes PG (1990) Sulfonamides and Sulfones, Comprehensive Medicinal Chemistry, Vol. 2. Pergamon, Oxford, pp 255–270
11. Rover S, Cesura MA, Huguenin P, Szente A (1997) J Med Chem 40:4378
12. Wattenberg LW, Low LKT, Fladmoe AV (1979) Cancer Res 39:1651
13. Willette RE, Soine TO (1961) J Pharm Sci 51:149
14. Dean FM (1963) Naturally Occurring Oxygen Ring Compounds. Butterworth, London, pp 176–220
15. Ian C-Y, Jin Zh-T, Yin B-Zh (1989) J Heterocyclic Chem 26:601
16. Iwataki I (1972) Bull Chem Soc Japan 45:3218
17. Nozoe T (1959) In: Ginsburg D (ed) Non-benzenoid Aromatic Compounds. Interescience Publishers Inc., New York, pp 339–463
18. Dou HJM, Vernin G, Metzger J (1969) J Heterocyclic Chem 6:575
19. Baule M, Vivaldi R, Poite JC, Dou HJM, Vernin G, Metzger J (1972) Bull Soc Chim France p 2679
20. Li Z-H, Jin Z-T, Yin B-Z, Nozoe T (1987) J Heterocyclic Chem 24:779
21. Takase K, Kasai K, Shimisu K, Nozoe T (1971) Bull Chem Soc Japan 44:2466
22. Grakauskas V (1970) J Org Chem 35:723
23. Grakauskas V (1969) J Org Chem 34:3825
24. Jackson WD, Polaya JB (1951) Aust J Sci 13:149
25. Walker HA, Wilson S, Atkins EC, Garrett HE, Richardson AR (1951) J Pharmacol Exp Ther 101:368
26. Lewenstein MJ (1954) US Patent 106 683
27. Lewenstein MJ (1954) Chem Abstr 48:13175
28. Lepetil G, S. P. A. (1974) DE Patent 2 424 670
29. Lepetil G, S. P. A. (1975) Chem Abstr 83:20628
30. Hardtmann GE, Kathawala FG (1977) US Patent 4 053 600
31. Hardtmann GE, Kathawala FG (1978) Chem Abstr 88:22970
32. Kathawala FG (1974) US Patent 3 850 932
33. Kathawala FG (1975) Chem Abstr 82:140175

34. Clark RL, Pessolano AA, Shen TY (1977) ZA Patent 7 603 163
35. Clark RL, Pessolano AA, Shen TY (1977) Chem Abstr 86:P189951a
36. El-Agrody AM (1994) J Chem Res (S) p 280
37. El-Agrody AM, Hassan SM (1995) J Chem Res (S) p 100
38. El-Agrody AM, Emam HA, El-Hakim MH, Abd El-latif MS, Fakery AH (1997) J Chem Res (S) 320(M):2039
39. El-Agrody AM, Abd El-latif MS, Fakery AH, Bedair AH (2000) J Chem Res (S) p 26
40. Shaker RM (1996) Pharmazie 51:148
41. Elliott AJ, Guzik H, Soler JR (1982) J Heterocycl Chem 19:1437
42. Aly AS, Fathy NM, Swelam SA, Abdel-Megeid FME (1995) Egypt J Pharm Sci 36:177
43. Hewitt W, Vincent S (1989) Theory and Application of Microbiological Assay. Academic Press Inc, New York
44. Cremer A (1980) Antibiotic Sensitivity and Assay Tests, 4th edn. Butterworths, London, p 521
45. Darbarwar M, Sundaramurthy V (1982) Synthesis pp 337–388
46. Ghosh CK, Bandyopadhyay C, Maiti J (1987) Heterocycles 26:1623–1656
47. Ghosh CK (1990) J Indian Chem Soc 67:5–15
48. Ghosh CK (1991) J Indian Chem Soc 68:21–28
49. Singh G, Singh R, Girdhar NK, Ishar MPS (2002) Tetrahedron 58:2471–2480
50. Singh G, Singh G, Ishar MPS (2003) Synlett pp 256–258
51. Bandyopadhyay C, Sur KR, Patra R, Banerjee S (2003) J Chem Res (S) pp 459–460
52. Bandyopadhyay C, Sur KR, Patra R, Banerjee S (2003) J Chem Res (M) pp 847–856
53. Bandyopadhyay C, Sur KR, Patra R, Sen A (2000) Tetrahedron 56:3583–3587
54. Smith EM, Doll RJ (1984) EU Patent 102 046
55. Smith EM, Doll RJ (1984) Chem Abstr 101:7052
56. Chantegrel B, Nadi AI, Gelin S (1984) J Org Chem 49:4419–4424
57. Bellassoued-Fargeau MC, Maitte P (1986) J Heterocycl Chem 23:1753–1756
58. Alberola A, Gonzalez-Ortega A, Sadaba ML, Sanudo MC (1998) J Chem Soc Perkin Trans 1, pp 4061–4065
59. Ishar MPS, Kumar K, Singh R (1998) Tetrahedron Lett 39:6547–6550
60. Ghosh T, Patra R, Bandyopadhyay C (2004) J Chem Res (S) pp 47–49
61. Pascual C, Meier J, Simon W (1966) Helv Chim Acta 49:164–168
62. Moorty SR, Sundaramurthy V, Subba Rao NV (1973) Indian J Chem 11:854–856
63. Ghosh T, Bandyopadhyay C (2004) Tetrahedron Lett 45:6169–6172
64. Kulkarni MV, Pujar BG, Patil VD (1983) Studies on coumarins II. Arch Pharm (Weinheim) 316(1):15
65. Kulkarni MV, Patil VD, Biradar VN, Nanjappa S (1981) Synthesis and biological properties of some 3-heterocyclic substituted coumarins. Arch Pharm (Weinheim) 34(5):435
66. Buckle DR, Outred DJ, Ross JW, Smith H, Smith RJ, Spicer BA, Gasson BC (1979) Aryloxyalkyloxy- and aralkyloxy-4-hydroxy-3-nitrocoumarins which inhibit histamine release in the rat and also antagonize the effects of a slow reacting substance of anaphylaxis. J Med Chem 22(2):158
67. Ghate M, Kulkarni MV, Shobha R, Kattimani SY (2003) Synthesis of vanillin ethers from 4-(bromomethyl) coumarins as anti-inflammatory agents. Eur J Med Chem 38:297
68. Shastri LA, Ghate M, Kulkarni MV (2004) Dual fluorescence and biological evaluation of paracetamol ethers from 4-bromomethylcoumarins. Ind J Chem 43B(11):2416
69. Kusanur R, Ghate M, Kulkarni MV (2004) Synthesis and biological activities of some substituted 4-{4-(1,5-diphenyl-1H-pyrazol-3-yl)phenoxymethyl} coumarins. Indian J Het Chem 13:201

70. Burger A, Ullyot GE (1947) Analgesic studies. b-ethyl and b-isopropylamine derivatives of pyridine and thiazole. J Org Chem 12:342
71. Kostanecki SV, Rozycki A (1901) Formation of chromones or coumarins by acylation of o-hydroxyaryl ketones. Ber 34:102
72. Rao PS, Reddy KV, Ashok D (2000) A facile synthesis of 7-(substituted aryl)-2-benzoyl-3-methyl-5H-furo[3,2-g] [1]benzopyran-5-ones and their antifeedant activity. Indian J Chem 39B:557
73. Boyd J, Robertson A, Whalley W-B (1948) J Chem Soc p 174
74. Pandey G, Muralikrishna C, Bhalerao UT (1989) Mushroom tyrosinase catalysed synthesis of coumestans, bebzofuranderivatives and related heterocyclic compounds. Tetrahedron 45:6867
75. Shah VR, Bose JL, Shah RC (1960) New synthesis of 4-hydroxy coumarins. J Org Chem 25:677
76. Winter CA, Risley EA, Nuss GW (1962) Carrageenin-induced edema in hind paw of rat as the assay for antiinflammatory drugs. Proc Soc Exp Biol 3:544
77. Koster R, Anderson M, De Beer EJ (1959) Acetic acid for analgesic screening. Fed Proc 18:412
78. Litchfield JT, Wilkoxon FJ (1949) A simplified method of evaluating dose-effect experiments. Pharmacol Exp Ther 96:99
79. Ghate M et al. (2005) Eur J Med Chem 40:882–887
80. Liao Y-X et al. (2003) Tetrahedron Lett 44:1599–1602
81. Egan D, Kennedy RO, Moran E, Cox D, Prosser E (1999) Drug Metab Rev 22:503–529

Top Heterocycl Chem (2007) 11: 303–323
DOI 10.1007/7081_2007_072
© Springer-Verlag Berlin Heidelberg
Published online: 1 August 2007

Antiviral and Antimicrobial Evaluation of Some Heterocyclic Compounds from Turkish Plants

Ilkay Orhan[1] (✉) · Berrin Özcelik[2] · Bilge Şener[1]

[1]Department of Pharmacognosy, Gazi University, 06330 Ankara, Turkey
iorhan@gazi.edu.tr

[2]Department of Pharmaceutical Microbiology, Faculty of Pharmacy, Gazi University, 06330 Ankara, Turkey

Abstract Antibiotic resistance has become a problem since the discovery of antibiotics. Not long after the introduction of penicillin, *Staphylococcus aureus*, which can be also transmitted to humans via milk and milk products, began developing penicillin-resistant strains. Therefore, one approach that has been used for the discovery of new antimicrobial agents from natural sources is based on the evaluation of traditional plant extracts. Natural products have played a pivotal role in antibiotic drug innovation and include aminoglycosides, cephalosporins, macrolides, cycloserine, novobiocin, and lipoproteins. However, only a few antiviral agents are available on the market. To this purpose, we have screened a great number of herbal extracts along with some pure natural substances and obtained interesting findings. This chapter covers the results of our rigorous search for new antiviral and antimicrobial alternative compounds from a number of Turkish plants.

Keywords Antimicrobial · Antiviral · Cytotoxicity · Natural compound · Plant extract

Abbreviations

AIDS	Acquired immunodeficiency syndrome
AMP	Ampicillin
ATCC	American type of culture collection
CPE	Cytopathogenic effect
FLU	Flucanozole
GC-MS	Gas chromatography–mass spectrometry
HIV	Human immunodeficiency virus
HSV	*Herpes simplex virus* type-1
KET	Ketocanozole
MAL	*N*-(Methylsuccinyl)-anthranoyllycoctonine

MDBK Madin–Darby bovine kidney
MIC   Minimum inhibitory concentration
MNTC Maximum non-toxic concentration
MRSA Methycillin-resistant *Staphylococcus aureus*
OFX   Oflaxocin
PI-3 *Parainfluenza* type-3
SAR   Structure–activity relationship
TNF   Tumor necrosis factor
WHO   World Health Organization

1
Introduction

Since their discovery during the twentieth century, antimicrobial agents (antibiotics and related medicinal drugs) have substantially reduced the threat posed by infectious diseases. On the other hand, antibiotic resistance can be defined as the ability of any microorganism to withstand the effects of an antibiotic, which is a specific type of drug resistance and evolves naturally via natural selection through random mutation [1, 2]. The antibiotic action is an environmental pressure; those bacteria that have a mutation allowing them to survive will live on to reproduce. They will then pass this trait to their offspring, which will be a fully resistant generation [3, 4]. Without doubt, antibiotic resistance is a global issue. Because of cross-continental travel of both humans and goods, antibiotic-resistant bacteria are spread from one country to another. Therefore, antibiotic resistance has been called one of the world's most pressing public health problems.

Much evidence supports the view that the total consumption of antimicrobials is the critical factor in selecting resistance. Paradoxically, underuse through lack of access, inadequate dosing, poor adherence, and substandard antimicrobials may play as an important role as overuse [5, 6]. Since misuse of antibiotics jeopardizes the usefulness of essential drugs, decreasing inappropriate antibiotic use is the ideal way to control resistance [7]. Natural selection of penicillin-resistant strains in a bacterium known as *Staphylococcus aureus* began soon after penicillin was introduced in the 1940s. Recently more hospital-acquired infections are becoming resistant to the most powerful antibiotics available, such as vancomycin, which emerged in the United States in 2002, presenting physicians and patients with a serious problem [8]. A number of cases of community-associated methicillin-resistant *S. aureus* (MRSA) have also been reported, including cases in patients without established risk factors [9].

Modern medical science and practice have something of an armory of effective tools, ranging from antiseptics and anesthetics to vaccines and antibiotics. However, one field has been weak in finding drugs to deal with viral infections, although viral diseases have affected humans many centuries. The emergence of antivirals is the product of a greatly expanded knowledge of the genetic and molecular function of organisms, allowing biomedical researchers to understand the structure and function of viruses.

On the other hand, herbal medicine is the oldest form of healthcare known to mankind and has been used by all cultures throughout history. The World Health Organization (WHO) estimates that 4 billion people, 80% of the world population, presently use plants for some aspect of primary health care. WHO also notes that of 119 plant-derived pharmaceutical medicines, about 74% are used in modern medicine in ways that correlate directly with their traditional uses as plant medicines by native cultures.

In our ongoing study on the plant extracts and their pure compounds from Turkish medicinal plants, we have so far screened a number of plant extracts and pure components for their antimicrobial and antiviral activities. In this chapter, we intend to cover the recent results obtained from our antimicrobial and antiviral studies on heterocyclic compounds isolated from several Turkish medicinal plants and marine organisms.

2
Terpene-Type Compounds

Although biodiversity in terrestrial environments is extraordinary, oceans covering more than 70% of our planet represent a greater diversity of life. A small number of marine plants, animals, and microbes have already yielded over 12 000 novel compounds [10]. Among them, a great number of substances of marine origin have been reported to possess antimicrobial activity such as loloatins A–D, myticins A and B, psammaplin A, etc., which have been covered in several excellent reviews [10–12].

In our previous study, ten terpene-type compounds, which we isolated from the marine sponges *Ircinia spinulosa* and *Spongia officinalis* of the Aegean Sea collection, were assayed against the American-type culture collection (ATCC) strains of *Bacillus subtilis* (6633), *S. aureus* (6536), *Pseudomonas aeruginosa* (9027), and *Escherichia coli* (8739) by the disk diffusion method [13]. The results were evaluated by comparing the inhibition zones of the compounds, namely, furospinulosin-1, furospongin-1, 2-(hexaprenylmethyl)-2-methylchromenol, heptaprenyl-*p*-quinol, 12-*epi*-deoxoscalarin, 1,4,44-trihydroxy-2-octa-prenylbenzene, demethylfurospongin-4, 4-hydroxy-3-octaprenylbenzoic acid, 4-hydroxy-3-tetraprenylacetic acid, and 11-β-acetoxy-12-en-16-one. We found that furospinulosin-1, furospongin-1, 2-(hexaprenylmethyl)-2-methylchromenol, and heptaprenyl-*p*-quinol did not exhibit any inhibition against those bacterium strains, whereas 12-*epi*-deoxoscalarin (**1**) exerted a weak activity against *B. subtilis* and *S. aureus*, causing inhibition zones of 12 and 11 mm in diameter, respectively (Fig. 1). Demethylfurospongin-4 (**2**), 4-hydroxy-3-tetraprenylacetic acid (**3**), and 11-β-acetoxy-12-en-16-one (**4**) were also active against Gram-positive bacteria (Fig. 1). Among them, 4-hydroxy-3-tetraprenylacetic acid (**3**) was the most effective causing 20 mm of inhibition zone (Fig. 1). None of the compounds tested had an ability to inhibit *E. coli*. Owing to scarcity of the compounds, only one concentration (0.5 mg/disk) could be tested. In conclusion, we suggested that this assay should be better with higher yield, if repeated.

In a similar study reported previously, 23 hydroquinone and quinone derivatives from the sponge *Ircinia spinulosa* were tested for their antibacterial activity, and relevant structure–activity relationships (SAR) were established [14]. As a consequence, SAR studies indicated that the optimum length of side chain in the compounds for antibacterial activity should be 5–15 carbon atoms, which is in accordance with our most effective compound, 4-hydroxy-3-tetraprenylacetic acid (**3**). Besides, it was reported that long-chain alcohols exert higher antimicrobial activity compared to the corresponding acids and aldehydes [15]. Similarly, a relationship between antibacterial activity and the structure of aliphatic alcohols was described, which suggested that maximum activity against *S. aureus* might be depend on the number of carbon atoms in hydrophobic chain. This should be less than 12, but as close to 12 as possible [16]. However, according to the study of Inoue et al., this finding did not support the anti-*Stapylococcus* effect of the aliphatic terpene alcohols, farnesol, nerolidol, and plaunotol, which may result from configuration of functional groups and double bonds that affected activity [17]. Therefore, in that study, it was concluded that anti-*Stapylococcus* activity depends not only on aliphatic side chains, but also on the configurations of functional groups and double bonds. Possibly, the difference observed in the antibacterial effect of our compounds might be due to this reason as well.

2.1
Flavonoids

Flavonoids are a group of polyphenolic compounds ubiquitous in many plants, in which they occur as the free forms, glycosides, as well as as methylated derivatives. In addition to their diverse biological activities, there is an increasing interest in flavonoids due to their anti-infective properties [18]. For instance, the flavonoids quercetin and kaempferol, as well as the flavonoid glycosides rutin and isoquercitrin, were reported to have antibacterial and antifungal activities [19]. Quercetin and kaempferol are known to be the most common flavonols present in many plants, and occur in different glycosidic forms. In many studies, they or their various glycosides have been proved to possess antimicrobial activity or, in other words, the antimicrobial activity of plant extracts (e.g., *Rubus ulmifolius*, *Combretum erythrophyllum*, *Morus alba*, *Trollius chinensis*, and propolis) has been attributed to quercetin and kaempferol [20–25].

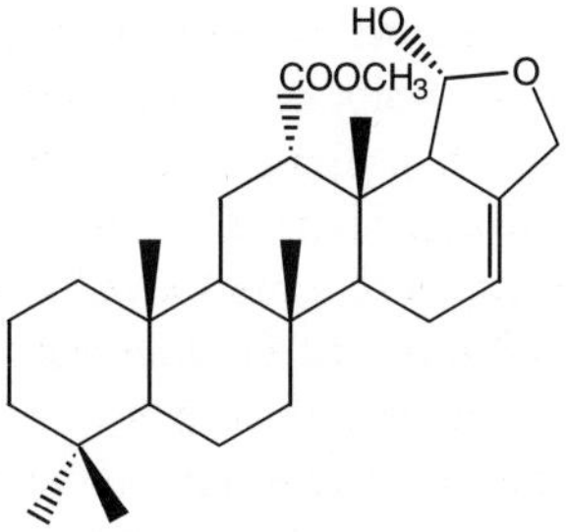

(1)

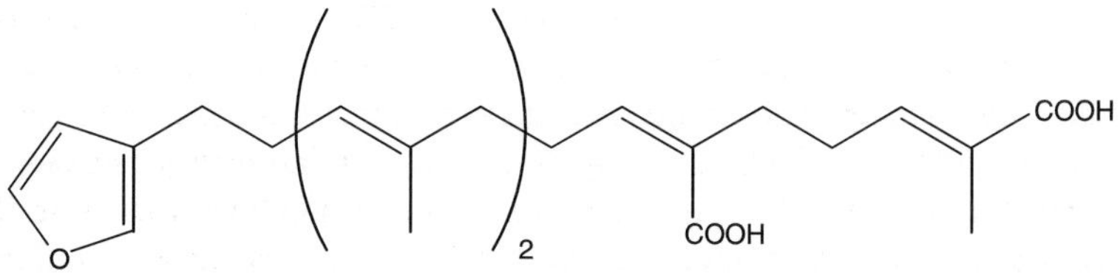

(2)

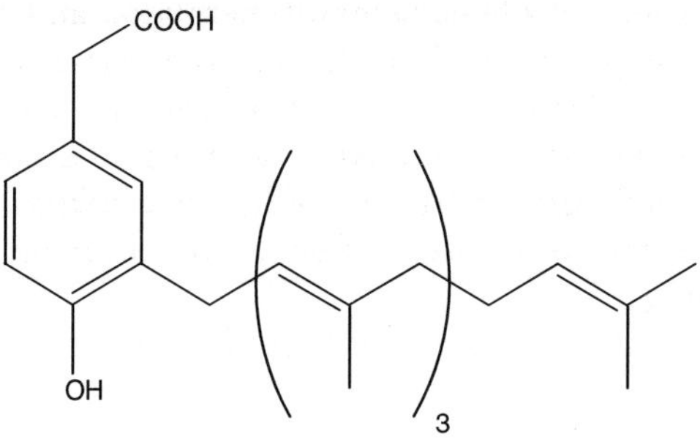

(3)

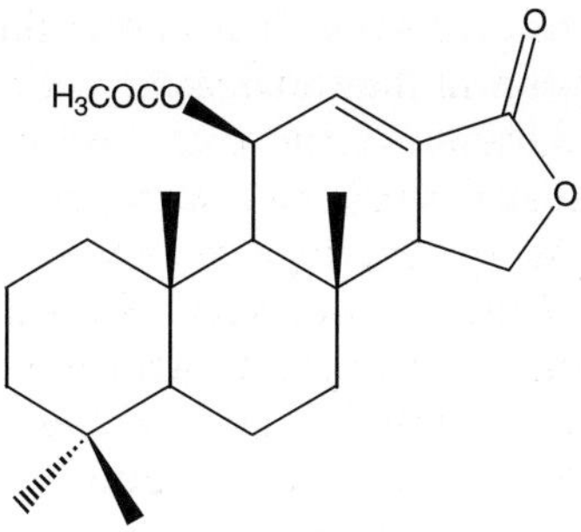

(4)

Fig. 1 Chemical structures of the antibacterial terpenes of marine origin

In one of our recent studies, we examined antimicrobial and antiviral activities of four flavonoid derivatives, namely, scandenone (5), tiliroside (6), quercetin-3,7-O-α-L-dirhamnoside (7), and kaempferol 3,7-O-α-L-dirhamnoside (8) as shown in Fig. 2. These were tested against *E. coli*, *P. aeruginosa*, *Proteus mirabilis*, *Klebsiella pneumoniae*, *Acinetobacter baumannii*, *S. aureus*, *B. subtilis*, and *Enterococcus faecalis*, as well as the fungus *Candida albicans* by the microdilution method [26]. In addition, both DNA virus *Herpes simplex* (HSV) and RNA virus *Parainfluenza* (PI-3) were employed for antiviral assessment of the compounds using Madin–Darby bovine kidney (MDBK) and Vero cell lines.

All four compounds were found to be most active against *S. aureus* and *E. faecalis*, with a minimum inhibitory concentration (MIC) of 0.5 μg/mL, followed by *E. coli* (2 μg/mL), *K. pneumoniae* (4 μg/mL), *A. baumannii*, and *B. subtilis* (8 μg/mL). *P. mirabilis* and *P. aeruginosa* were the most resistant bacteria against the compounds (16 and 32 μg/mL, respectively). Notably, antibacterial activity of the compounds was as potent as ampicillin (AMP) and oflaxocin (OFX) towards *S. aureus* and *E. faecalis*. These compounds also possessed quite remarkable antifungal activity against *C. albicans*, as much as ketocanozole (KET) (1 μg/mL).

As shown in Table 1, none of the compounds had the ability to inhibit HSV, while only quercetin-3,7-O-α-L-dirhamnoside had inhibitory activity against PI-3 in the range 8–32 μg/mL of maximum and minimum cytopathogenic effect (CPE) inhibitory concentrations, respectively. The inhibitory concentration range of this compound is on a vast scale, which resembles that of oseltamivir (32 to < 0.25 μg/mL). Besides, its maximum non-toxic concentration (MNTC) (64 μg/mL) was observed to be better than oseltamivir (32 μg/mL).

One of the undisputed functions of flavonoids and related polyphenols is their role in protecting plants against microbial invasion, which accumulate phytoalexins in response to microbial attack plants. Moreover, it is evident that a structure–activity relationship exists between the various flavonoids and their antimicrobial activity in most cases. A large number of antimicrobial flavonoids have been reviewed brilliantly and their SARs discussed [27–32]. The majority of antifungal flavonoids have been observed to have either isoflavonoid, flavan, or flavanone structures such as maackiain, mucronulatol, luteolin 7-(2″-sulfatoglucoside), etc., which is consistent with our data on flavonoids. Accordingly, the presence of a phenolic group in the flavonoid structure suggests that it is necessary for antimicrobial activity. Interestingly, increasing the number of hydroxyl, methoxyl, or glycosyl substituents resulted in the steady loss of antifungal effect of the flavonoids [33]. In the review of Bylka et al. [34], it was suggested that the antibacterial effect towards Gram-negative bacteria is higher with flavones, while flavonoids containing two or three hydroxy groups in rings A and B are more active in inhibition of Gram-positive bacteria. However, in our study, all four flavonoid derivatives,

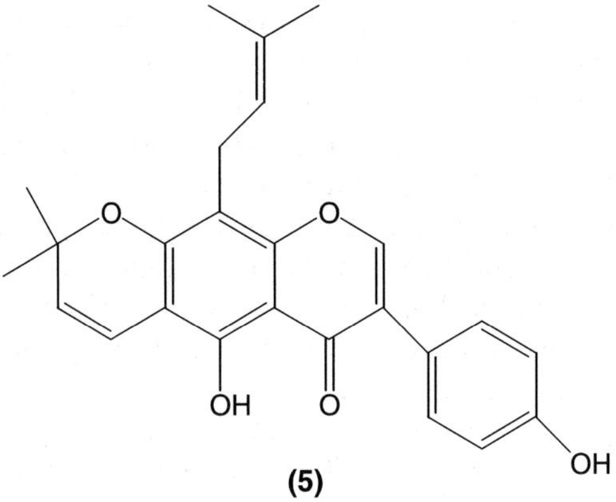

(5)

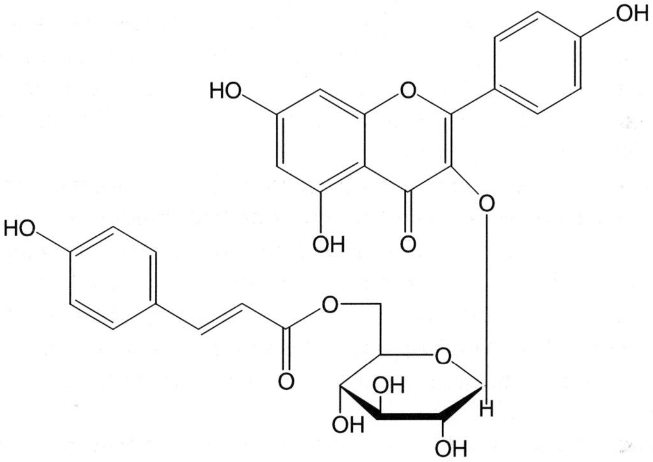

(6)

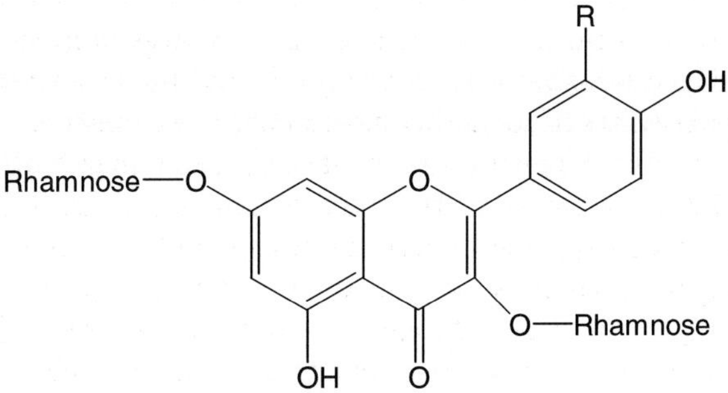

(7) R=OH
(8) R=H

Fig. 2 Chemical structures of the flavonoid derivatives with antimicrobial activity

Table 1 Antiviral activity and cytotoxicity of the flavonoid derivatives screened

Compounds	MDBK cells			Vero cells		
	MNTC	CPE inhibitory conc. vs HSV		MNTC	CPE inhibitory conc. vs PI-3	
		Max.	Min.		Max.	Min.
	(μg/mL)	(μg/mL)	(μg/mL)	(μg/mL)	(μg/mL)	(μg/mL)
Scandenone	64	–	–	64	–	–
Tiliroside	64	–	–	64	–	–
Quercetin-3,7-*O*-α-L-dirhamnoside	64	–	–	64	32	8
Kaempferol-3,7-*O*-α-L-dirhamnoside	64	–	–	64	–	–
Acyclovir	16	16	< 0.25	–	–	–
Oseltamivir	–	–	–	32	32	< 0.25

consisting of one prenylated isoflavone and three flavonol glycosides, exhibited an equal strength of antibacterial and antifungal activities, independent of their structural substitutions.

Recently, flavonoids have been investigated from the viewpoint of their antiviral effect, particularly against the human immunodeficiency virus (HIV), the causative agent of acquired immonodeficiency syndrome (AIDS). Among them, quercetin has been shown to be effective against divergent virus types by many researchers, which supports our data on quercetin-3,7-*O*-α-L-dirhamnoside [26]. In one of the earliest studies, oral application of quercetin in mice was found to display a protective effect towards intraperitoneal encephalomyocarditis, *Mengo*M,L and *Mengo*M virus infections, but not against intracerebral challenge with *Mengo*M virus. It was not virucidal and did not interfere with *Mengo* virus replication in L cells [35]. The potentiative interaction of quercetin with murine α/β interferon in mice against *Mengo* virus infection [36] was also proved. Moreover, quercetin was reported to greatly enhance the antiviral effect of tumor necrosis factor (TNF) that produces a dose-dependent inhibition of vesicular *stomatitis* virus, *Encephalomyocarditis* virus, and HSV type-1 replication in WISH cells [37]. In another study, the effect of different substituents of quercetin and luteolin on the ability to inhibit the HSV type-1 replication in RK-13 cells was studied [38].

In conclusion, our results demonstrated that scandenone, tiliroside, quercetin-3,7-*O*-α-L-dirhamnoside, and kaempferol-3,7-*O*-α-L-dirhamnoside possessed severe antibacterial and antifungal activities, whereas only quercetin-3,7-*O*-α-L-dirhamnoside exerted noticeable anti-PI-3 activity.

2.2
Alkaloids

2.2.1
Diterpene Alkaloids

The genus *Consolida*, *Aconitum*, and *Delphinium* (Ranunculaceae) are well-known to be rich in diterpene alkaloids, which possess a diverse range of biological activities. These plants have also been the cause of poisonings, which primarily occur in cattle as well as human beings, due to toxicity of their alkaloids. In one of our recent studies, five diterpenoid-derivative alkaloids, lycoctonine (9), 18-*O*-methyllycoctonine (10), delcosine (11), 14-acetyldelcosine (12), and 14-acetylbrowniine (13) (as shown in Fig. 3) were screened for their antibacterial, antifungal, and antiviral activities [39].

Once more, HSV and PI-3 were employed for antiviral assessment of the compounds using MDBK and Vero cell lines. Their MNTC and CPE values were determined using acyclovir and oseltamivir as the references. Besides antibacterial and antifungal activities, the alkaloids were tested against *E. coli*, *P. aeruginosa*, *P. mirabilis*, *K. pneumoniae*, *A. baumannii*, *S. aureus*, and *B. subtilis*, as well as the fungus *C. albicans* by the microdilution method as compared to the references AMP, OFX, and KET.

The results pointed out that these alkaloids possessed the highest antibacterial activity against *K. pneumoniae* and *A. baumannii* at 8 μg/mL concentration (Table 2), whereas they were moderately active to the rest of the bacteria. However, all the alkaloids tested were highly effective against *C. albicans* in a comparable manner to KET in the antifungal screening. Conversely, a selective inhibition was observed towards PI-3 virus by these alkaloids, while they were entirely unsuccessful on inhibition of HSV (Table 3).

PI-3 inhibitory activity of the alkaloids was fairly analogous to that of oseltamivir, ranging between 8–32 μg/mL as minimum and maximum inhibitory concentrations for the CPE. Our results showed that the alkaloids possessed rather high antifungal activity against *C. albicans* and a compelling antibacterial effect only against *K. pneumoniae* and *A. baumannii*, while they exerted a strong inhibition against PI-3.

Even though much is already known about the toxicity of diterpene alkaloids that contribute to the toxicity of *Consolida*, *Delphinium*, and *Aconitium* species, no antiviral study has been so far reported on this type of alkaloids. Therefore, no SAR studies have been encountered by us on the antiviral or antimicrobial activities of these alkaloids. However, a quantitative SAR analysis performed on a number of diterpene alkaloids isolated from an *Aconitum* sp. indicated that biological activity of these alkaloids may be related to their toxicity rather than to a specific pharmacological action [40]. In a current study on 43 norditerpenoid alkaloids from *Consolida*, *Delphinium*, and *Aconitum* species against several tumor cell lines, lycoctonine and browniine were

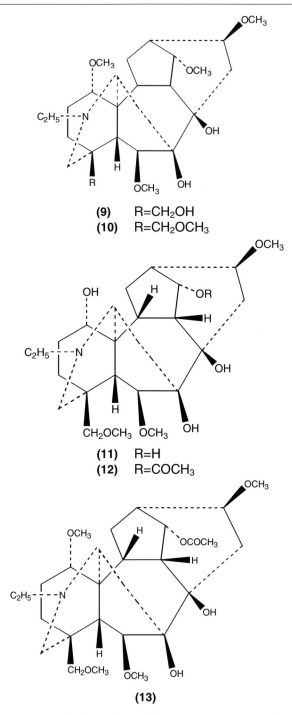

Fig. 3 Chemical structures of the diterpene alkaloids with antimicrobial activity

Table 2 Antimicrobial activity of the diterpene alkaloids

Alkaloids	Escherichia coli	Pseudomonas aeruginosa	Proteus mirabilis	Klebsiella pneumoniae	Acinetobacter baumannii	Staphylococcus aureus	Bacillus subtilis	Candida albicans
Lycoctonine	32	64	32	8	8	64	128	coli4
18-O-Methyllycoctonine	32	64	32	8	8	64	128	4
Delcosine	32	64	32	8	8	64	128	4
14-Acetyldelcosine	32	64	32	8	8	64	128	4
14-Acetylbrowniine	32	64	32	8	8	64	128	4
AMP	2	–	2	2	2	< 0.12	0.12	–
OFX	0.12	1	< 0.12	0.12	0.12	0.5	0.5	–
KET	–	–	–	–	–	–	–	2

Table 3 Antiviral activity and cytotoxicity of the diterpene alkaloids

Alkaloids	MDBK cells			Vero cells		
	MNTC	CPE inhibitory conc. vs HSV		MNTC	CPE inhibitory conc. vs PI-3	
		Max.	Min.		Max.	Min.
	(μg/mL)	(μg/mL)	(μg/mL)	(μg/mL)	(μg/mL)	(μg/mL)
Lycoctonine	64	–	–	32	32	8
18-O-methyllycoctonine	64	–	–	64	32	1
Delcosine	64	–	–	64	32	1
14-Acetyldelcosine	64	–	–	64	32	1
14-Acetylbrowniine	64	–	–	64	32	1
Acyclovir	16	16	< 0.25	–	–	–
Oseltamivir	–	–	–	32	32	< 0.25

found to be active among those screened [41]. In contrast to this data, ly-coctonine and 14-acetyl derivative of browniine in our study showed a lesser amount of cytotoxicity on MDBK and Vero cell lines at 64 μg/mL. Gonzales-Coloma et al. studied antifeedant activity and toxicity of some diterpene alkaloids (15-acetylcardiopetamine, cardiopetamine along with its amino alcohol, the β, γ unsaturated ketone, and the acetylated ketone derivatives) from *Delphinium* sp. on the insects *Spodoptera littoralis* and *Leptinotarsa decemlineata* [42]. Results of the study showed that the C13 and C15 hydroxy substituents are essential features of the active molecule, while the C11 benzoate group enhanced the biological effect on both insect species, where all of our alkaloids lacked those two substituents.

In a taxonomic study done in 2002, lycoctonine-type alkaloids isolated from three *Delphinium* species were classified into three groups according to the degree of their toxicity: N-(methylsuccinyl)-anthranoyllycoctonine (MAL)-type with high toxicity, lycoctonine-type with moderate toxicity, and 7,8-methylene-dioxylycoctonine (MDL)-type with low toxicity [43]. In that paper, it was reported that the moiety attached to C14 is quite important for the toxicity of these alkaloids, which is also in accordance with our present data. Furthermore, other functionalities on these molecules are also notable in terms of toxicity. It was noticed that the tertiary nitrogen, anthranilic acid substitution, and C1 moiety affect the toxicity degree within those alkaloids. For instance, when the methylsuccinyl group is removed from MAL (which then converts to lycoctonine), lycoctonine becomes 93 times less toxic.

Briefly, our report was the first on antiviral, antibacterial, and antifungal activities of lycoctonine, 18-O-methyllycoctonine, delcosine, 14-acetyldelcosine, and 14-acetylbrowniine. Furthermore, our data also suggest that all of the diterpene alkaloids are worthy of being evaluated for their antimicrobial and antiviral activities for future-promising results.

2.2.2
Isoquinoline Alkaloids

In our recent work, we focused on investigation of antiviral activity of 33 isoquinoline alkaloids and seven derivatives of them, which are classified as protopine-type [protopine (**14**) and β-allocryptopine (**15**)], benzylisoquinoline-type [(+)-reticuline (**16**) and (+)-norjuziphine], benzophenantridine-type [sanguinarine (**17**), norsanguinarine, and chelidimerine (**18**)], spirobenzylisoquinoline-type [fumarophycine (**19**), fumarophycine acetate, (–)-corpaine, (±)-sibiricine (**20**), sibiricine acetate, (±)-dihydrosibiricine, (+)-fumariline, (–)-dihydrofumariline, (+)-parfumine, parfumine acetate, dihydroparfumine acetate], phtalideisoquinoline-type [α-hydrastine (**21**), (+)-bicuculline, (–)-bicuculline, and (–)-adlumidine], aporphine-type [(+)-bulbocapnine (**22**) and (+)-isoboldine], protoberberine-type [berberine (**23**), (–)-stylopine, (–)-canadine, (–)-sinactine, (–)-ophiocarpine (**24**), ophiocarpine-*N*-oxide, corydalmine, palmatine, (±)-corydalidzine, dehydrocorydaline, and dehydrocavidine], cularine-type [(+)-cularicine (**25**), oxocularine, oxosarcocapnine, and oxosarcocapnidine], and an isoquinolone [corydaldine (**26**)], against HSV and PI-3 using MDBK and Vero cell lines [44], whose isolation procedures were described elsewhere [45–50] (Figs. 4 and 5). Moreover, the alkaloids were also tested against *E. coli, P. aeruginosa, P. mirabilis, K. pneumoniae, A. baumannii, S. aureus,* and *B. subtilis,* as well as the fungus *C. albicans* by the microdilution method, for their antibacterial and antifungal activities.

According to our findings, all types of the alkaloids appeared to be more active to Gram-negative bacteria than to Gram-positive ones. Most of the alkaloids, including protopine, β-allocryptopine, chelidimerine, fumarophycine, (±)-sibiricine, sibiricine acetate, (±)-dihydrosibiricine, parfumine acetate, α-hydrastine, (+)-bulbocapnine, berberine, (–)-canadine, (–)-ophiocarpine, ophiocarpine-*N*-oxide, corydalmine, oxosarcocapnidine, and corydaldine, showed significant inhibition on *K. pneumoniae* and *A. baumannii,* in particular, better than the rest of the Gram-negatives, at 8 μg/mL concentration as compared to AMP (2 μg/mL). All of the alkaloids, regardless of their structural differences, inhibited *E. coli* and *P. mirabilis* with MIC of 32 μg/mL, while they inhibited *S. aureus* at 64 μg/mL. Interestingly, the alkaloids that were found to inhibit *K. pneumoniae* and *A. baumannii* at 8 μg/mL also had remarkable inhibition on *C. albicans* (4 μg/mL), while a notable occurrence of antifungal activity was observed with the rest at 8 μg/mL concentration.

The tested isoquinolines were observed to display a selective inhibition against PI-3 as seen in Table 4, except for (+)-isoboldine, (–)-stylopine, and (±)-corydalidzine, that were totally ineffective against both viruses. Protopine, β-allocryptopine, chelidimerine, fumarophycine, α-hydrastine, (+)-bulbocapnine, (+)-isoboldine, (–)-sinactine, palmatine, dehydrocoryda-

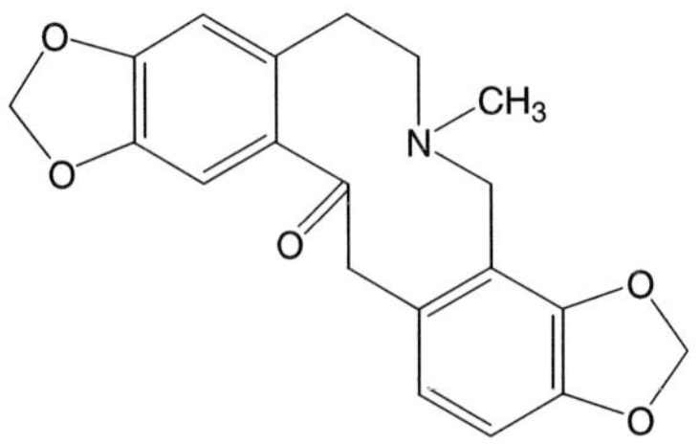

Fig. 4 Chemical structures of the isoquinoline alkaloids with antimicrobial activity

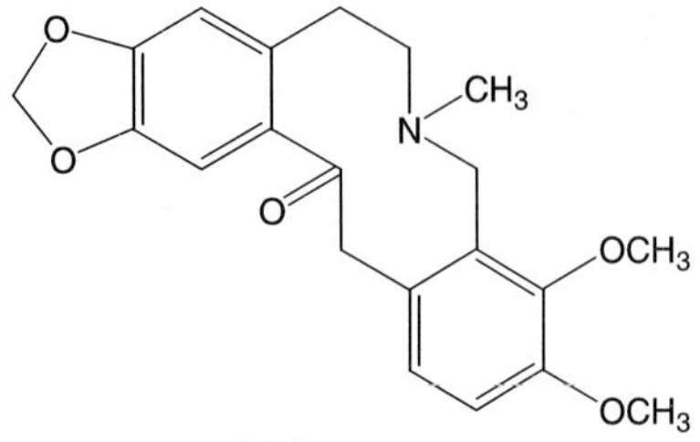

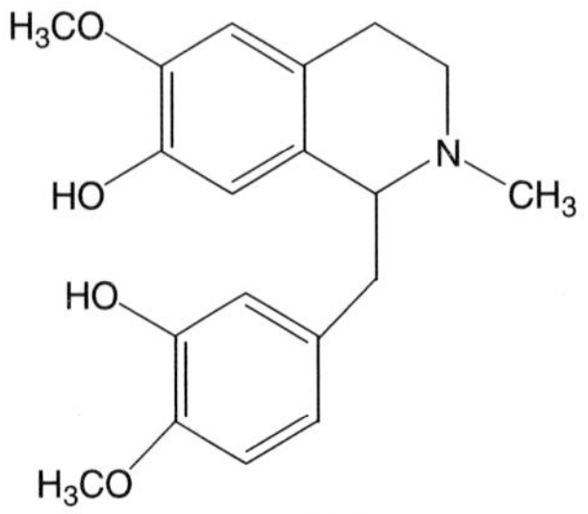

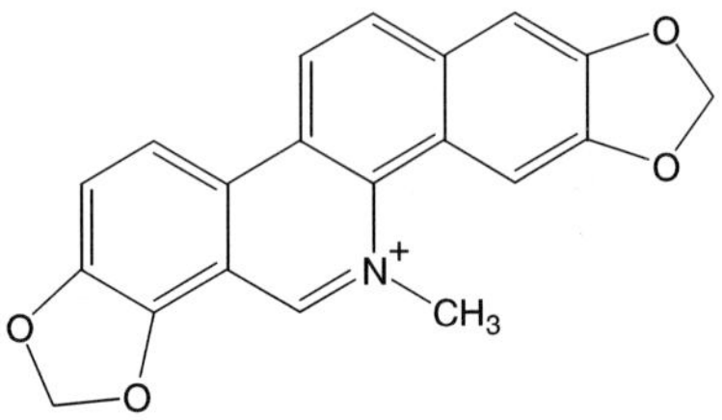

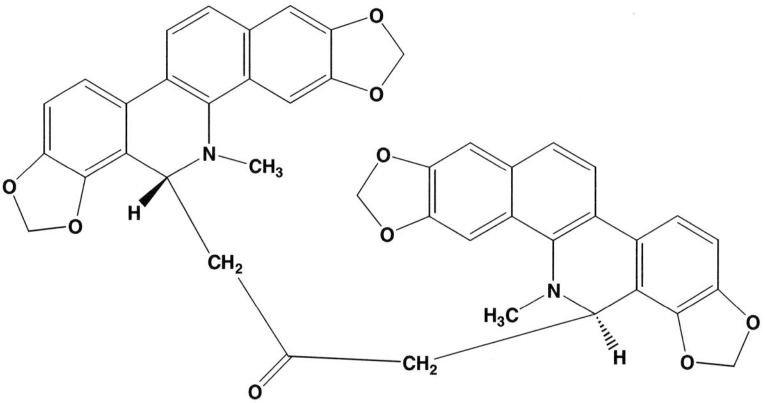

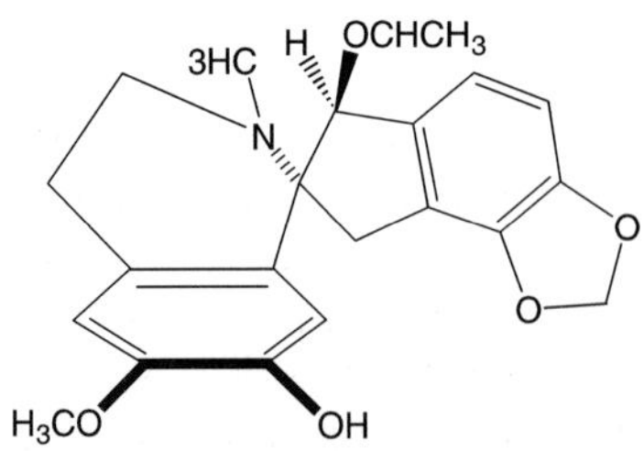

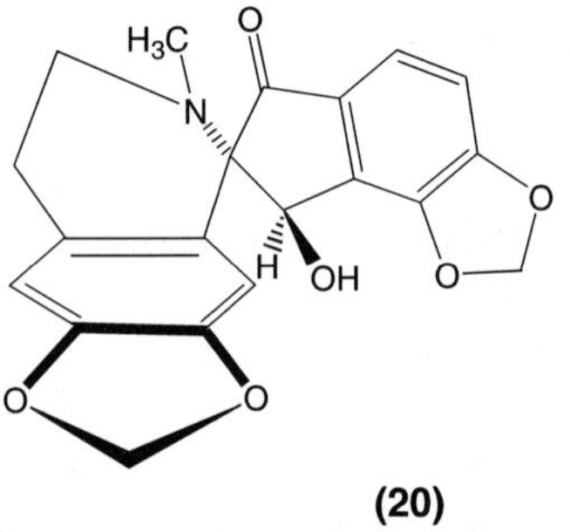

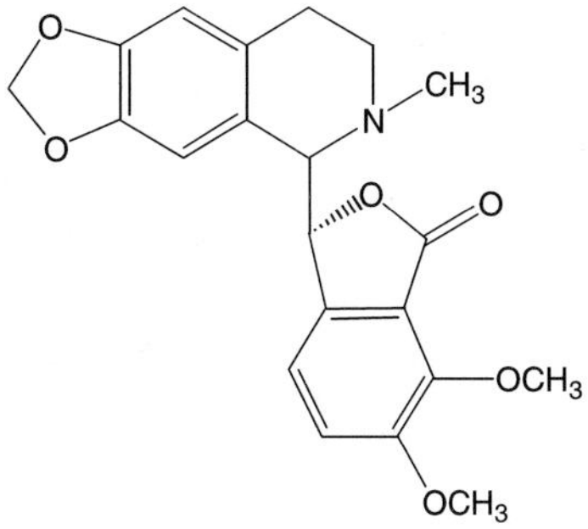

(21)

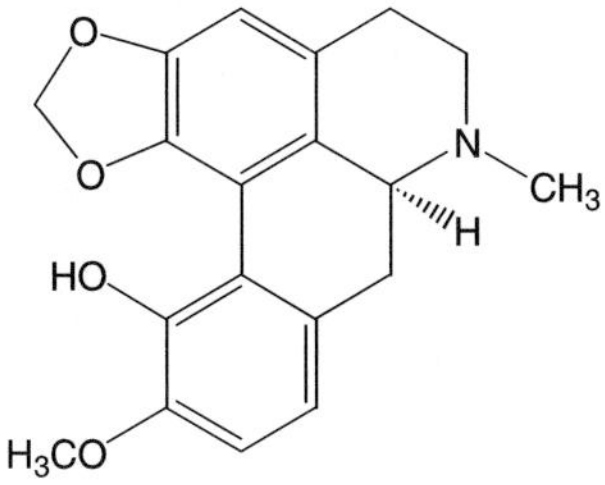

(22)

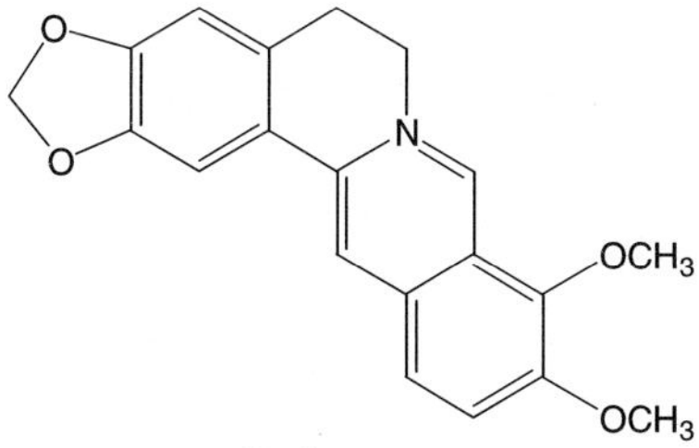

(23)

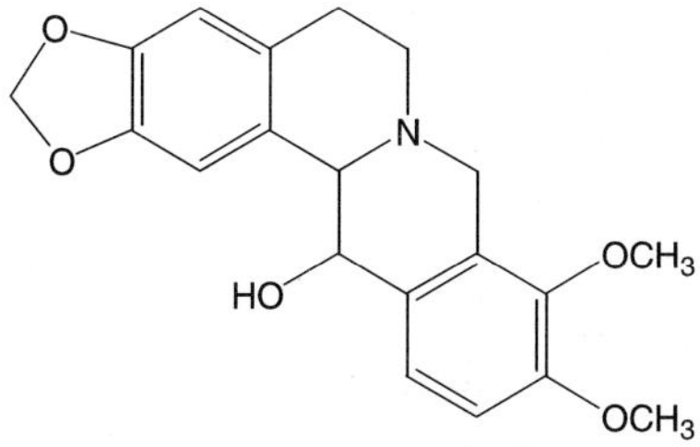

(24)

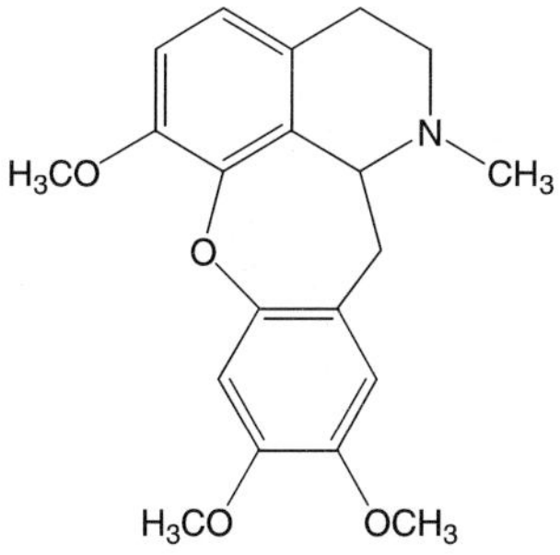

(25)

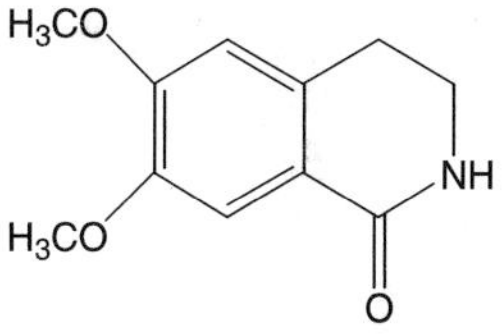

(26)

Fig. 5 continues from Fig. 4

line, dehydrocavidine, (+)-cularicine, oxocularine, and oxosarcocapnine were completely inactive against HSV, whereas maximum CPE concentrations of the rest were the same as acyclovir (16 μg/mL). However, the alkaloids were revealed to be less cytotoxic than acyclovir on MDBK cells, (−)-canadine being the least cytotoxic alkaloid (128 μg/mL). The most active alkaloid with anti-PI-3 effect was shown to be protopine (1–32 μg/mL), followed by fumarophycine (2–32 μg/mL), chelidimerine, (+)-bulbocapnine, and (−)-ophiocarpine (4–32 μg/mL), as well as β-allocriptopine and oxosarcocapni-

Table 4 activity and cytotoxicity of the isoquinoline alkaloids studied

Alkaloids	MDBK cells			Vero cells		
	MNTC	CPE inhibitory conc. vs HSV		MNTC	CPE inhibitory conc. vs PI-3	
		Max.	Min.		Max.	Min.
	(μg/mL)	(μg/mL)	(μg/mL)	(μg/mL)	(μg/mL)	(μg/mL)
Protopine	64	–	–	32	32	1
β-Allocryptopine	64	–	–	32	32	8
(+)-Reticuline	32	16	–	64	32	16
(+)-Norjuziphine	32	16	–	64	32	16
Sanguinarine	32	16	–	32	32	16
Norsanguinarine	32	16	–	32	32	16
Chelidimerine	64	–	–	32	32	4
Fumarophycine	64	–	–	32	32	2
(–)-Fumarophycine acetate	32	16	–	64	32	16
(–)-Corpaine	32	16	–	64	32	16
(±)-Sibiricine	32	16	–	64	32	16
Sibiricine acetate	32	16	–	32	32	16
(±)-Dihydrosibiricine	32	16	–	64	32	16
(+)-Fumariline	32	16	–	64	32	16
(–)-Dihydrofumariline	32	16	–	64	32	16
(+)-Parfumine	32	16	–	64	32	16
Parfumine acetate	32	16	–	64	32	16
(–)-Dihydroparfumine diacetate	32	16	–	64	32	16
α-Hydrastine	64	–	–	64	32	16
(+)-Bicuculline	32	16	–	64	32	16
(–)-Bicuculline	32	16	–	64	32	16
(–)-Adlumidine	32	16	–	64	32	16
(+)-Bulbocapnine	64	–	–	32	32	4
(+)-Isoboldine	–	–	–	–	–	–
Acyclovir	16	16	< 0.25	–	–	–
Oseltamivir	–	–	–	32	32	< 0.25
Berberine	64	–	–	32	–	–
(–)-Stylopine	–	–	–	–	–	16
(–)-Canadine	128	–	–	64	32	16
(–)-Sinactine	32	16	–	64	32	16
(–)-Ophiocarpine	64	–	–	32	32	4
Ophiocarpine-N-oxide	64	–	–	32	32	16
Corydalmine	64	–	–	64	64	32
Palmatine	32	16	–	32	32	16
(±)-Corydalidzine	–	–	–	–	–	–
Dehydrocorydaline	32	16	–	64	–	16
Dehydrocavidine	32	16	–	64	32	16
(+)-Cularicine	32	16	–	32	32	16

Table 4 (continued)

Alkaloids	MDBK cells			Vero cells		
	MNTC	CPE inhibitory conc. vs HSV		MNTC	CPE inhibitory conc. vs PI-3	
		Max.	Min.		Max.	Min.
	(μg/mL)	(μg/mL)	(μg/mL)	(μg/mL)	(μg/mL)	(μg/mL)
Oxocularine	32	16	–	64	32	16
Oxosarcocapnine	32	16	–	64	32	16
Oxosarcocapnidine	64	–	–	32	32	8
Corydaldine	64	–	–	32	32	16
Acyclovir	16	16	< 0.25	–	–	–
Oseltamivir	–	–	–	32	32	< 0.25

dine (8–32 μg/mL). The alkaloids tested exhibited lower or the same degree of cytotoxicity as oseltamivir (32 μg/mL) against Vero cells.

A number of antimicrobial, antiviral, antitumoral, antimalarial, and cytotoxicity studies have been so far reported on various derivatives of natural or synthetic isoquinoline alkaloids [51–57]. In one study, antimicrobial, cytotoxic, and anti-HIV activities of 26 simple isoquinolines and 21 benzylisoquinolines were investigated. The results showed that a quaternary nitrogen atom of isoquinoline or dyhydroisoquinoline type may enhance the potency of antimicrobial activity and cytotoxicity, whereas anti-HIV activity was higher with tetrahydroisoquinolines [58]. In the study of Cui et al., simple isoquinolines, 15 of which were of 1-benzylisoquinoline-type and 19 of which were of protoberberine-derivatives, were screened against Epstein-Barr virus early antigen (EBV-EA) activation induced by 12-O-tetradecanoylphorbol-13-acetate (TPA), which is considered to be an indicator of antitumor-promoting activity. The study was carried out using Raji cells and all 1-benzylisoquinolines and 11 of the protoberberines exerted higher inhibitory activity than β-carotene [59]. Regarding the structure–activity relationship, it was concluded that the inhibitory activity of 1-benzylisoquinolines increased as the number of hydroxyl groups on the aromatic ring increased and, additionally, the size of substituents at the C8 and C13, as well as type and position of the oxygenated substituents on A and D rings, influenced the virus inhibition. Moreover, derivatives of the isoquinoline skeleton attached with carboxamide moiety were declared to be the potent and selective inhibitors of human cytolamegavirus [60].

In another study, the structure–activity relationships of berberine and its derivatives were examined for their antibacterial activity. Among the 13-alkyl-substituted and the 13-unsubstituted protoberberinium salts, an increase in antibacterial activity was observed with the 13-ethyl-9-ethoxyl, the 13-ethyl, and the 13-methyl derivatives against S. aureus by eight-, four-, and twofold, respectively, over berberine, which suggested that steric effects played an noteworthy role in the antibacterial activity [61]. Additionally, replacement of methoxyl groups at the C2 and the C3 of ring A by a methylenedioxy group caused a boost in activity. In this report, it was stated that the quaternary nitrogen atom (such as in protoberberinium salts) an alkylsubstituent at C13, and a methylenedioxy function at C2 and C3 are required for enhanced antibacterial activity. In a study by Nakamoto et al., berberine was revealed to have a strong antifungal effect against *C. albicans*, *C. tropicalis*, and *C. glabrata*, respectively, which is in accordance with our data on berberine [62]. In a recent publication, a high occurrence of antibacterial activity in berberine was shown towards *E. coli*, *K. pneumoniae*, *P. aeruginosa*, *P. fluorescens*, *S. aureus*, *Salmonella typhi*, *Enterococcus* sp., and *Serratia marcescens*, showing a better activity than streptomycin at 50 µg/mL by paper disc diffusion method. Consequently, berberine was concluded to be responsible for the high antibacterial activity of *Coscinium fenestratum* [63]. However, we found that berberine was only active against *K. pneumoniae* and *A. baumannii* by microdilution method, which might result from the application of two different methods. In another former study, berberine obtained from *Berberis heterophylla* was tested against the ATCC strains of *S. aureus*, *Enterecoccus faecalis*, *P. aeruginosa*, *E. coli*, and *C. albicans* by agar diffusion method at 50, 100, and 200 µg/mL concentrations. The alkaloid was highly active against *S. aureus* at 100 and 200 µg/mL, whereas it did not possess any inhibitory effect against *E. faecalis*, *P. aeruginosa*, and *E. coli* [64]. This data was consistent with ours for berberine in the case of *E. coli* and *P. aeruginosa*. It was not active to *S. aureus*, which might again be the result of the use of two dissimilar methods. In the same work, antifungal activity screening was performed with berberine using the clinical strains of several *Candida* sp. such as *C. albicans*, *C. glabrata*, *C. haemulonii*, *C. lusitaniae*, *C. krusei*, and *C. parapsilosis*. Being the most active to *C. krusei*, followed by the rest in decreasing degrees of effectiveness, berberine was expressed as a novel antifungal agent.

In one report, protopine and α-allocryptopine isolated from *Glaucium oxylobum* were tested for their antifungal activity against *Microsporium canis*, *M. gypseum*, *Tricophyton mentagrophytes*, *Epidermophyton floccosum*, *C. albicans*, *Aspergillus niger*, and *Penicillium* sp. [65]. Among these fungi, protopine exerted low activity against *M. canis* and *T. mentagrophytes*, while α-allocryptopine had low activity towards *M. gypseum* and good inhibition of *E. floccosum*. In contrast, protopine was found to be inactive against *C. albicans*, whereas this alkaloid had a high inhibition against the same fungus in our study (4 µg/mL). α-Allocryptopine was also inactive against *C. albicans*, whose β-counterpart exhibited a very good antifungal effect against *C. albicans*. This may reasonably be due to the α- and β-conformation of the compound. In a previous study, the molluscicidal activity of *Argemone mexicana* seeds were tested against the snail *Lymnaea acumi-*

nata, which led to isolation of protopine and sanguinarine as the active components [66].

From the structure–activity point of view, a few features about the isoquinoline alkaloids investigated herein can be pointed out. The quaternary nitrogen atom found on some of the isoquinolines such as dehydrocorydaline, dehydrocavidine, berberine, sanguinarine, and palmatine may have an effect on the decrease of antiviral activity. On the other hand, the synergistic interaction among the isoquinoline alkaloids isolated from *F. vaillantii* may be stated to contribute to the higher antiviral activity of this extract. Protopine-type alkaloids seem to display a higher antiviral effect than the rest.

3
Conclusion

From ancient to modern history, traditional plant-based medicines have played an important role in health care. Many countries in Africa, Asia, and Latin America still rely on traditionally used herbal medicines for primary health care needs. On the other hand, the complex nature of plant extracts and the high probability of competing or synergistic bioactivities within the same extract mean that these results represent the starting point for an activity-guided search for active plant metabolites. It is also evident that a structure–activity relationship exists between the various chemical structures and their antimicrobial activity in most cases. However, antiviral agents, unlike antibacterial drugs which may cover a wide range of pathogens, tend to be narrow in spectrum, and, unfortunately, have a limited efficacy. Historically, the discovery of antiviral drugs has been largely fortuitous. Spurred on by success with antibiotics, drug companies launched huge blind-screening programs with relatively little success. Besides, lead compounds were modified by scientists in an attempt to improve bioactivity. However, there is still a great need to develop more effective antiviral and antimicrobial drug molecules.

As a conclusion, these findings provide additional evidence for the supposition that the assays mentioned above play the part of useful primary screening in the survey of bioactive natural products. For the reasons outlined above, it is very important to focus on plants to discern novel antiviral/antimicrobial compounds. Therefore, we truly hope that our studies, as well as similar reports by different researchers, may help find new antimicrobials/antivirals from herbal sources.

References

1. Alekshun MN, Levy SB (2007) Cell 128:1037
2. Pesavento G, Ducci B, Comodo N, Lo Nostro A (2007) Food Cont 18:196

3. Courvalin P (2006) Digest Liver Dis 38(Suppl2):S261
4. Sabet NS, Subramaniam G, Navaratnam P, Sekaran SD (2007) J Microbiol Meth 68:157
5. Kardas P, Devine S, Golembesky A, Roberts C (2005) Int J Antimicr Agents 26:106
6. Knox K, Lawson W, Dean B, Holmes A (2003) J Hosp Infec 53:85
7. Singer RS, Henrik RF, Wegener C, Bywater R, Walters J, Lipsitch M (2003) Lancet Infect Dis 3:47
8. Sahm DF, Benninger MS, Evangelista AT, Yee YC, Thornsberry C, Brown NP (2007) Otolaryngol-Head and Neck Surg 136:385
9. Lee YL, Cesario T, Gupta G, Flionis L, Tran C, Decker M, Thrupp L (1997) Am J Infec Cont 25:312
10. Donia M, Hamann MT (2003) The Lancet (Inf. Diseases) 3:338
11. Mayer AM, Hamann MT (2004) Mar Biotechnol 6:37
12. Mayer AM, Hamann MT (2005) Comp. Biochem Physiol C Toxicol Pharmacol 140: 265
13. Erdogan I, Şener B (2000) GUEDE-J Gazi Univ Fac Pharm 17:1
14. De Rosa S, De Giulio A, Iodice A (1994) J Nat Prod 57:1711
15. Kubo I, Muroi H, Kubo A (1993) J Agric Food Chem 41:2447
16. Kubo I, Muroi H, Himejima M, Kubo A (1993) Bioorg Med Chem Lett 3:1305
17. Inoue Y, Shiraishi A, Hada T, Hirose K, Hamashima H, Shimada J (2004) FEMS Microbiol Lett 237:325
18. Cushni TPT, Lamb AJ (2005) Int J Antimic Agents 26:343
19. Beschia M, Leonte A, Oencea I (1984) Bull Univ Galati Faso 6:23
20. Panizzi L, Caponi C, Catalano S, Cioni PL, Morelli I (2002) J Ethnopharmacol 79:165
21. Gatto MT, Falcocchio S, Grippa E, Mazzanti G, Battinelli L, Nicolasi G, Lambusta D, Saso L (2002) Bioorg Med Chem 10:269
22. Li YL, Ma SC, Yang YT, Ye SM, But PPH (2002) J Ethnopharmacol 79:365
23. Du J, He ZD, Jiang RW, Ye WC, Xu HX, But PPH (2003) Phytochemistry 62:1235
24. Martini N, Katerere DRP, Eloff JN (2004) J Ethnopharmacol 93:207
25. Kosalec I, Pepeljnjak S, Bakmaz M, Vladimir-Knezevic S (2005) Acta Pharm 55:423
26. Özcelik B, Orhan I, Toker G (2006) Z Naturforsch C 61:632
27. Harborne JB, Williams CA (2000) Phytochemistry 55:481
28. Perez RM (2003) Pharm Biol 41:107
29. Erlund I (2004) Nutr Res 24:851
30. Cushni TPT, Lamb AJ (2005) Int J Antimic Agents 26:343
31. Rios JL, Recio MC (2005) J Ethnopharmacol 100:80
32. Khan MTH, Ather A, Thompson KD, Gambari R (2005) Antiviral Res 67:107
33. Picman AK, Schneider EF, Pieman J (1995) Biochem Syst Ecol 23:683
34. Bylka W, Matlawska I, Pilewski NA (2004) JANA 7:24
35. Veckenstedt A, Puszhai R (1981) Antiviral Res 1:249
36. Veckenstedt A, Güttner J, Beladi I (1987) Antiviral Res 7:169
37. Ohnishi E, Bannai H (1993) Antiviral Res 22:327
38. Wleklik M, Luczak M, Panasiak W, Kobus M, Lammer-Zarawska E (1998) Acta Virol 32:522
39. Şener B, Orhan I, Özcelik B (2007) ARKIVOC 7:265
40. Bello-Ramirez AM, Nava-Ocampo AA (2004) Fund Clin Pharmacol 18:699
41. De Ineìs C, Reina M, Gaviìn JA, González-Coloma A (2006) Z Naturforsch C 61:11
42. Gonzalez-Coloma A, Guadano A, Gutierrez C, Cabrera R, De La Pena E, De La Fuente G, Reina M (1998) J Agric Food Chem 46:286
43. Panter KE, Manners GD, Stegelmeier BL, Lee S, Gardner DR, Ralphs MH, Pfister JA, James LF (2002) Biochem Syst Ecol 30:113

44. Orhan I, Özcelik B, Karaoğlu T, Şener B (2007) Z Naturforsch C 62:19
45. Şener B, Gözler B, Minard RD, Shamma M (1983) Phytochemistry 22:2073
46. Şener B (1983) Int J Crude Drug Res 21:135
47. Şener B (1984) GUEDE-J Fac Pharm Gazi Univ 1:15
48. Şener B (1985) J Nat Prod 48:670
49. Şener B (1986) Int J Crude Drug Res 24:105
50. Şener B (1988) Int J Crude Drug Res 26:61
51. Capilla AS, Romero M, Pujol MD, Caignard DH, Renard P (2001) Tetrahedron 57:8297
52. An TY, Huang RQ, Yang Z, Zhang DK, Li GR, Yao YC, Gao J (2001) Phytochemistry 58:1267
53. Satou T, Koga M, Matsuhashi R, Koike K, Tada I, Nikaido T (2002) Vet Parasitol 104:131
54. Zhang Q, Tu G, Zhao Y, Cheng T (2002) Tetrahedron 58:6795
55. Gomez-Monterrey I, Campiglia P, Grieo P, Diurno MV, Bolognese A, La Colla P, Novellino E (2003) Bioorg Med Chem 11:3769
56. Morrell A, Antony S, Kohlhagen G, Pommier Y, Cushman M (2004) Bioorg Med Chem Lett 14:3659
57. Fischer DCH, Gualdo NCA, Bachiega D, Carvalho CS, Lupo FN, Bonotto SV, Alves MO, Yogi A, Di Santi SM, Avila PE, Kirchgatter K, Moreno PRH (2004) Acta Trop 92:261
58. Iwasa K, Moriyasu M, Tachibana Y, Kim HS, Wataya Y, Wiegrebe W, Bastow KF, Cosentino LM, Kozuka M, Lee KH (2001) Bioorg Med Chem 9:2871
59. Cui W, Iwasa K, Tokuda H, Kashihara A, Mitani Y, Hasegawa T, Nishiyama Y, Moriyasu M, Nishino H, Hanaoka M, Mukai C, Takeda K (2006) Phytochemistry 67:70
60. Chan L, Jin H, Stefanac T, Wang W, Lavallee JF, Bedard J, May S (1999) Bioorg Med Chem Lett 9:2583
61. Iwasa K, Kamiguchi M, Ueki M, Taniguchi M (1996) Eur J Med Chem 31:469
62. Nakamoto K, Sadamori S, Hamada T (1990) J Prost Dent 64:691
63. Nair GM, Narasimhan S, Shiburaj S, Abraham TK (2005) Fitoterapia 76:585
64. Freile ML, Giannini F, Pucci G, Sturniolo A, Rodero L, Pucci O, Balzareti V, Enriz RD (2003) Fitoterapia 74:702
65. Morteza-Semnani K, Amin G, Shidfar MR, Hadizadeh H, Shafiee A (2003) Fitoterapia 74:493
66. Singh S, Singh DK (1999) Chemosphere 38:3319

Author Index Volumes 1–11

The volume numbers are printed in italics

Alamgir M, Black DS C, Kumar N (2007) Synthesis, Reactivity and Biological Activity of Benzimidazoles. *9*: 87–118

Almqvist F, see Pemberton N (2006) *1*: 1–30

Ather A, see Khan MTH (2007) *10*: 99–122

Appukkuttan P, see Kaval N (2006) *1*: 267–304

Ariga M, see Nishiwaki N (2007) *8*: 43–72

Arya DP (2006) Diazo and Diazonium DNA Cleavage Agents: Studies on Model Systems and Natural Product Mechanisms of Action. *2*: 129–152

El Ashry ESH, El Kilany Y, Nahas NM (2007) Manipulation of Carbohydrate Carbon Atoms for the Synthesis of Heterocycles. *7*: 1–30

El Ashry ESH, see El Nemr A (2007) *7*: 249–285

Bagley MC, Lubinu MC (2006) Microwave-Assisted Multicomponent Reactions for the Synthesis of Heterocycles. *1*: 31–58

Bahal R, see Khanna S (2006) *3*: 149–182

Basak SC, Mills D, Gute BD, Natarajan R (2006) Predicting Pharmacological and Toxicological Activity of Heterocyclic Compounds Using QSAR and Molecular Modeling. *3*: 39–80

Benfenati E, see Duchowicz PR (2006) *3*: 1–38

Berlinck RGS (2007) Polycyclic Diamine Alkaloids from Marine Sponges. *10*: 211–238

Besson T, Thiéry V (2006) Microwave-Assisted Synthesis of Sulfur and Nitrogen-Containing Heterocycles. *1*: 59–78

Bharatam PV, see Khanna S (2006) *3*: 149–182

Bhhatarai B, see Garg R (2006) *3*: 181–272

Bianchi N, see Gambari R (2007) *9*: 265–276

Black DS C, see Alamgir M (2007) *9*: 87–118

Branda E, see Buzzini P (2007) *10*: 239–264

Brown T, Holt H Jr, Lee M (2006) Synthesis of Biologically Active Heterocyclic Stilbene and Chalcone Analogs of Combretastatin. *2*: 1–51

Buzzini P, Turchetti B, Ieri F, Goretti M, Branda E, Mulinacci N, Romani A (2007) Catechins and Proanthocyanidins: Naturally Occurring O-Heterocycles with Antimicrobial Activity. *10*: 239–264

Carlucci MJ, see Pujol CA (2007) *11*: 259–281

Castro EA, see Duchowicz PR (2006) *3*: 1–38

Cavaleiro JAS, Tomé Jã PC, Faustino MAF (2007) Synthesis of Glycoporphyrins. *7*: 179–248

Cerecetto H, González M (2007) Benzofuroxan and Furoxan. Chemistry and Biology. *10*: 265–308

Cerecetto H, see González M (2007) *11*: 179–211

Subject Index

Printing: Krips bv, Meppel, The Netherlands
Binding: Stürtz, Würzburg, Germany